Special Topics in Primatology
Volume 5

The Evolution of Hemispheric Specialization in Primates

No contributors to this volume have received remuneration. All profits from the sale of *The Evolution of Hemispheric Specialization in Primates* will go directly to the American Society of Primatologists Conservation Committee for their programs. For information about the Conservation Committee, please contact a Society officer or visit the wet page at: www.asp.org.

Special Topics in Primatology

Janette Wallis, Series Editor

Volume 1 **Primate Conservation: The Role of Zoological Parks**
Edited by Janette Wallis

Volume 2 **The Care and Management of Captive Chimpanzees**
Edited by Linda Brent

Volume 3 **Sexual Selection and Reproductive Competition in Primates: New Perspectives and Directions**
Edited by Clara B. Jones

Volume 4 **Commensalism and Conflict: The Human - Primate Interface**
Edited by James D. Paterson and Janette Wallis

Volume 5 **The Evolution of Hemispheric Specialization in Primates**
Edited by William D. Hopkins

Special Topics in Primatology
Volume 5

The Evolution of Hemispheric Specialization in Primates

Edited by

William D. Hopkins

A Publication of the

American Society of Primatologists

Published on behalf of the **American Society of Primatologists** by Elsevier Ltd.

Academic Press is an imprint of Elsevier
84 Theobald's Road, London WC1X 8RR, UK
Radarweg 29, PO Box 211, 1000 AE Amsterdam, The Netherlands
Linacre House, Jordan Hill, Oxford OX2 8DP, UK
30 Corporate Drive, Suite 400, Burlington, MA 01803, USA
525 B Street, Suite 1900, San Diego, CA 92101-4495, USA

First edition 2007

ISBN: 978-0-12-374197-4
ISSN: 1936-8526 (Series)

For information on all Academic Press publications
visit our website at books.elsevier.com

Printed and bound in the UK

07 08 09 10 11 10 9 8 7 6 5 4 3 2 1

CONTENTS

CONTRIBUTORS

Michael C. Corballis
Department of Psychology
Faculty of Science
University of Auckland
Auckland, New Zealand
[1]

David B. Hanbury
Department of Psychology
University of Southern Mississippi
Hattiesburg, MS, USA
[229]

J. Dee Higley
Laboratory of Clinical Studies
National Institute on Alcohol
Abuse and Alcholism
Bethesda, MD, USA
[253]

Patrick R. Hof
Department of Neuroscience
Mount Sinai School of Medicine
New York, NY, USA *and*
New York Consortium in
Evolutionary Primatology
New York, NY, USA
[93]

William D. Hopkins
Division of Psychobiology
Yerkes National Primate
Research Center
Atlanta, Georgia, USA *and*
Department of Psychology
Agnes Scott College
Decatur, GA, USA
[147]

Sue Howell
The Mannheimer Foundation, Inc.
Clewiston, FL, USA
[253]

Susan Lambeth
Department of Veterinary Sciences
The University of Texas
M.D. Anderson Cancer Center
Bastrop, TX, USA
[147]

Peter F. MacNeilage
Department of Psychology
University of Texas at Austin
Austin, TX, USA
[59]

Adrien Meguerditchian
Department of Psychology
Research Center in Psychology of
Cognition, Language & Emotion
University of Provence
Aix-en-Provence, France *and* Center of
Primatology – CNRS, Rousset-Sur-Arc, France
[183]

Lesley J. Rogers
Centre for Neuroscience and Animal
Behaviour
School of Biological, Biomedical and
Molecular Sciences
University of New England, NSW
Armidale, Australia
[23]

Jamie L. Russell
Division of Psychobiology
Yerkes National Primate Research Center
Atlanta, GA, USA
[147]

Steven J. Schapiro
Department of Veterinary Sciences
The University of Texas
M.D. Anderson Cancer Center
Bastrop, TX, USA *and* Department of Psychology, Southwestern University, Georgetown, TX, USA
[147]

Natalie M. Schenker
Department of Anthropology
University of California at San Diego
La Jolla, CA, USA
[93]

Katerina Semendeferi
Department of Anthropology
Graduate Program in Neurosciences
University of California at San Diego
La Jolla, CA, USA
[93]

Chet C. Sherwood
Department of Anthropology
The George Washington University
Washington, DC, USA
[93]

Giovanna Spinozzi
Unit of Cognitive Primatology
Institute of Cognitive Sciences and Technologies
CNR, Rome, Italy
[205]

Jared P. Taglialatela
Division of Psychobiology
Yerkes National Primate Research Center
Atlanta, GA, USA
[121]

Jacques Vauclair
Department of Psychology
Research Center in Psychology of Cognition, Language & Emotion
University of Provence
Aix-en-Provence, France
[183]

Sheree L. Watson
Department of Psychology
University of Southern Mississippi
Hattiesburg, MS, USA
[229]

Gregory C. Westergaard
Alpha Genesis, Inc.
Yemassee, SC, USA
[253]

Numbers in brackets correspond to the first page of contributor's chapters.

PREFACE

This volume represents a series of theoretical and empirical papers on the evolution of hemispheric specialization in nonhuman primates. Hemispheric specialization refers to the differential processing of perceptual, motor, and cognitive functions to the left and right cerebral hemispheres. In addition, hemispheric specialization can reflect differences in neuroanatomical morphology and neurophysiology. Arguably, the two main manifestations of hemispheric specialization in humans are right-handedness and left hemisphere dominance in language functions. Historically, hemispheric specialization has been considered unique to human evolution and the assumptions of this view were predicated on the absence of population-level asymmetries in nonhuman animals, including primates. Furthermore, this view was reinforced from the assumption that animals did not possess language. Thus, nonhuman animals, including primates, lacked evidence of language and population-level asymmetries leading many to postulate these two traits were specific to *Hominin* evolution and had no antecedents in other primates. Indeed, Warren [1980] in his seminal review article summed up the field at the time by stating "comparative behavioral and neuropsychological study very strongly suggest that the human pattern of handedness and cerebral laterality is species-unique and that no truly homologous traits are to be found in nonhuman primates" (p 357). Warren himself had studied hand preferences for simple reaching in a number of species including macaques and had concluded that they exhibited individual hand preference but not population-level biases. We now know that simple reaching, as an indicator of hemispheric specialization, is probably not the best measure of hand preference in nonhuman primates and that this behavior is strongly influenced by situation and postural factors, some of which are described in a number of chapters (see Hopkins et al., Spinozzi, Vauclair & Meguerditchian, and Watson & Hanbury).

The view of Warren [1980] was challenged in 1987 when MacNeilage and colleagues published a paper in *The Behavioral and Brain Sciences* arguing that

ISSN: 1936-8526 DOI: 10.1016/S1936-8526(07)05011-7

there was some evidence of population-level asymmetries for handedness in nonhuman primates. MacNeilage et al. went on to propose their Postural Origins theory and this paper stimulated an enormous amount of research in nonhuman animals, including primates. Indeed, in the 20 years that have followed since the MacNeilage et al. [1987] paper, there have been more than 70 papers published on handedness in nonhuman primates, whereas in the 20 years preceding there had been far fewer. Based on these studies, we now know that population-level handedness is not unique to humans and is present in many vertebrates (see Rogers's chapter) including nonhuman primates, although there remains some debate over the interpretation of the differences in the distribution of handedness in humans compared to nonhuman primates (see Corballis's chapter). There also remains some debate over whether the pattern of results seen in captive primates generalizes to wild individuals. Several chapters in this volume describe research programs and results of studies aimed at investigating the role of different variables, such as posture and task demands, on the expression of handedness and other behavioral asymmetries (see Hopkins et al., Spinozzi, Vauclair & Meguerditchian, and Watson & Hanbury chapters). In addition to the fundamental question of whether nonhuman primates show population-level asymmetries, in a larger evolutionary framework, one can ask what the adaptive significance of handedness might be for human and nonhuman primates. For example, it has been claimed that left-handed humans may exhibit reduced fitness as reflected in shorter life spans and compromised immune functions [Coren & Halpern, 1991]. Several chapters discuss the specific relevance of nonhuman primate handedness to evolutionary and genetic models of handedness in humans (MacNeilage's and Corballis's chapters) as well as the issue of physiological, immunological, and behavioral phenotypes associated with individual differences in handedness (see Howell et al., and Watson & Hanbury chapters).

In addition to the chapters on handedness, equally important has been the increasing interest in lateralization in communicative functions, as both the behavioral and neuroanatomical level of analysis. In the past 20 years, a number of sophisticated and elaborate studies on asymmetries in the perception and production of species-typical vocalizations and gestures have been conducted in nonhuman primates (and other animals). In light of the fact that much of the theoretical basis for studying hemispheric specialization stems from the claims for uniqueness of language abilities in humans, these studies have been particularly insightful for testing theories on the relationship between asymmetry and language evolutions. Chapters by Taglialatela and Schenker et al., highlight recent studies that have examined lateralization in communication as well as studies on neuroanatomical and miscrostructrual asymmetries in nonhuman primates. The chapter by Rogers also discusses studies of asymmetry in communication in "lower" vertebrates in comparison to results from nonhuman primates. Collectively, this volume captures the range and scope of different approaches and species to the study of hemispheric specialization in primates.

This volume would not have been possible without the assistance of many people, particularly the contributors to whom I am very grateful. I am particularly thankful for the assistance of Janette Wallis, the Series Editor, for inviting me to submit a proposal for this volume, coordinating the reviews of the chapters, and overseeing the development of the book. Of course, we thank those who were willing to review the chapters and the members of the American Society of Primatologists' Publications Committee, who were kind enough to approve this book. I am very appreciative of the many helpful and encouraging primatologists in the field who encouraged me to edit this volume, particularly Jeff French and Steve Schapiro. Grant support for my research has been provided by the National Institutes of Neurological Disorders and Stroke as well as the National Institutes of Child Health and Human Development and I remain indebted to them for continuing to support my research program (and many others) that aim to understand the proximal and distal mechanisms involved on the emergence of hemispheric specialization. In many ways, the investigation of behavioral and brain asymmetries is in its infancy. In fact, much has been written about the evolution of behavioral and brain asymmetries in the absence of any real substantive data in nonhuman primates. I see this volume as part of a larger effort to "re"-educate the scientific community and public on the evidence of behavioral and brain asymmetries in nonhuman primates and to communicate how it relates to human behavior.

William D. Hopkins
Yerkes National Primate Research Center &
Agnes Scott College
Atlanta, GA
July, 2007

REFERENCES

Coren, S.; Halpern, D. F. Left-handedness: A markers of decreased survival fitness. Psychological Bulletin 109: 90–106, 1991.

MacNeilage, P. F.; Studdert-Kennedy, M. G.; Lindblom, B. Primate handedness reconsidered. Behavioral and Brain Sciences 10: 247–303, 1987.

Warren, J. M. Handedness and laterality in humans and other animals. Physiological Psychology 8: 351–359, 1980.

The chimpanzee Leslie living at the Yerkes National Primate Research Center gestures with the right hand to request food from a conspecific. (Photograph by Jared Taglialatela.)

The Evolution of Hemispheric Specialization in Primates
W.D. Hopkins (Editor)

CEREBRAL ASYMMETRY AND HUMAN UNIQUENESS

Michael C. Corballis
Department of Psychology, Faculty of Science, University of Auckland, Auckland, New Zealand

ABSTRACT

What are the qualities of the human brain and behavior that distinguishes us from other primates? This chapter provides a brief historical view of the claim that hemispheric specialization might be one of these traits. Arguments for and against continuity of hemispheric specialization in animals are presented and discussed in the context of the evolution of higher cognitive functions such as language and theory of mind.

Keywords: human handedness, evolution of brain, FOXP2, language, gestural origins

INTRODUCTION

There is something of a tradition to regard cerebral asymmetry as a uniquely human attribute, making humans a distinctively lopsided ape [e.g., Corballis, 1991]. Crow [2002] goes so far as to claim that cerebral asymmetry even defines *Homo sapiens* as a distinct species, and carries with it such distinctively human characteristics as language, theory of mind, and a disposition to psychosis. This tradition carries through to genetic models of handedness, which work on the assumption that some genetic mutation produced a genetic disposition toward right-handedness, such as Annett's [2002] "right shift" [RS+] gene, or McManus's [1999] "dextral" (D) allele. Both Annett and McManus, like Crow, assume that human right-handedness is a consequence of left-cerebral dominance, which also incorporates left-cerebral dominance for speech, and that the genetic mutation occurred at some point in the course of hominid evolution.

The view that handedness and cerebral asymmetry are uniquely human is increasingly challenged by the evidence that cerebral asymmetries are widespread in vertebrates, and probably go far back in evolution [Rogers & Andrew, 2002]. The issue is nevertheless controversial, and was sharply drawn in an exchange between Rogers [2004] and Crow [2004] in a recent issue of the journal *Laterality*. Crow [2004, p 233] vigorously maintains his view that "(D)irectional asymmetry is the key to the origin of modern *Homo sapiens*," and dubs this the "Broca-Annett axiom." His view is that the critical event in the course of hominid evolution was a translocation and subsequent inversion in the Xq21.3/Yp

ISSN: 1936-8526 DOI: 10.1016/S1936-8526(07)05001-4

homologous regions of the X and Y chromosomes. Rogers [2004], on the other hand, draws attention to the extensive literature on cerebral asymmetries in nonhuman species, and writes that "we should be cautious of the current fashion to reduce complex behaviour to unitary genetic explanations (p 230)." In some respects, the debate is reminiscent of the 19th-century debates between T. H. Huxley and Bishop Wilberforce over Darwin's theory of evolution by natural selection. Although the religious element is no longer evident, we again have the Darwinian idea of continuity between humans and other animals pitted against the Cartesian notion of a fundamental discontinuity that somehow placed humans on a different plane.

THE ARGUMENT FOR CONTINUITY

In favor of continuity is the rapidly accumulating evidence for cerebral asymmetries in nonhuman species. Handedness is the most outwardly obvious manifestation of a cerebral asymmetry in humans, and the work of Hopkins and his colleagues has now shown reasonably convincingly that captive chimpanzees display population-level right-handedness on at least some tasks [e.g., Hopkins, 1996; Hopkins & Leavens, 1998; Hopkins et al., 2003]. But other species may also show evidence for right-handedness. Toads, for example, are right-handed (or right-pawed) for some activities [Bisazza et al., 1996], and it has been claimed that 77% of walruses display a preference for the right flipper when feeding. Several of their bones (scapula, humerus, and ulna) are longer in the right than in the left flipper [Levermann et al., 2003]. Although Peterson [1934] found that left-pawed and right-pawed rats persisted in approximately equal numbers even after eight generations of selective breeding for pawedness, two other studies have reported an overall preference for the right paw [Güven et al., 2003; Tsai & Maurer, 1930].

Handedness may not be an especially revealing indicator of asymmetry in nonhuman species, especially if they are quadrupedal and the forelimbs are primarily involved in locomotion. More compelling evidence comes from asymmetries of the brain. A particularly striking example of a human-like asymmetry in the chimpanzee concerns the temporal planum, which in humans overlaps Wernicke's area, an area on the left side of the brain that is involved in the processing of language. In 17 out of 18 chimpanzee brains examined during post-mortem, the temporal planum was larger on the left than on the right [Gannon et al., 1998]. This is a significantly larger proportion than one would predict from the data on humans, which suggest that the left temporal planum is larger than the right in only about two-thirds of the population [Foundas et al., 1995; Geschwind & Levitsky, 1968; Jäncke & Steinmetz, 1993]. Of course the size of a structure may not reveal all, and there is evidence that cell columns in the human temporal planum are more widely spaced on the left than on the right, but that this asymmetry is not present in the chimpanzee, and is weakly reversed in rhesus monkeys [Buxhoeveden & Casanova, 2000]. Even so, the presence of human-like asymmetries in the chimpanzee brain poses a challenge to the Broca–Annett axiom.

Functional brain asymmetries in nonhuman species also often show parallels with human asymmetries, some of them apparently related to the well-established left cerebral dominance for the production and perception of speech and language [Corballis, 1991]. Even in the frog, for example, vocalization seems to be controlled by the left hemisphere [Bauer, 1993], suggesting an ancient origin for the left-hemispheric dominance in humans for speech. Hook-Costigan and Rogers [1998] found that marmosets opened the right side of the mouth wider when making social contact calls, implying left cerebral dominance, but the right side of the mouth wider when expressing fear, implying right cerebral dominance for emotion. This finding mimics that found in humans, with the right side of the mouth dominant for speech [e.g., Graves & Potter, 1988] and the left for emotional expression. These asymmetries are evident in 5–12-month-old human babies, who open the right side of the mouth wider when babbling, and the left side when smiling [Holowka & Petitto, 2002]. Further, there is evidence for left-hemispheric specialization in the perception of species-specific vocalizations in mice [Ehert, 1987], rhesus monkeys [Hauser & Andersson, 1994], and Japanese macaques [Heffner & Heffner, 1984].

Another striking example of human-like cerebral asymmetry comes from the New Caledonian crow, which appears to favor the right eye, and therefore the left hemisphere, when constructing digging tools from Pandanus leaves [Hunt et al., 2001]. This asymmetry may be the crows' equivalent of human right-handedness and left cerebral dominance for tool use and manufacture. In chicks, there is a right-eye advantage in discriminating grain from nonedible grit, and a right hemisphere specialization for more global visual analysis, such as monitoring a hovering predator [Rogers, 2002a], or making use of large-scale geometry of the environment in spatial orientation [Vallortigara et al., 2004]. A left-hemispheric advantage for fine-grained visual analysis and a right-hemisphere advantage for more global vision have also been well-documented in humans [Ivry & Robertson, 1998].

A right-hemisphere bias has been documented for social responses in a number of species of fish [Sovrano et al., 2001], tadpoles [Dadda, 2005], chicks [Vallortigara & Andrew, 1994], sheep [Peirce et al., 2000], and monkeys [Vermeire et al., 1998], and may relate to the right-hemispheric involvement in social understanding in humans [e.g., Sperry et al., 1979]. It may also relate to a more general right-hemisphere dominance in emotional reactions. Tadpoles have been shown to have a bias to turn left when escaping a predator, implying right-hemisphere dominance, but a bias to turn right when turning to take in air at the surface [Rogers, 2002b]. The right hemisphere appears to be dominant for aggression in a number of species, including toads [e.g., Rogers, 2002b], lizards [Deckel, 1995], chicks [Howard et al., 1980], baboons [Casperd & Dunbar, 1996], and humans [Devinsky et al., 1994]. In humans, it has been proposed that the right hemisphere is dominant in avoidance and the left hemisphere in approach [Davidson, 1995], although there is also evidence that the right hemisphere is dominant for emotional reactions in general [e.g., Brockmeier & Ulrich, 1993; Jansari et al., 2000].

Complementary to the right-hemispheric dominance for attack, there is a left-hemisphere dominance for feeding. Chicks [Deng & Rogers, 1997], pigeons [Güntürkun, 1985], zebra finches [Alonso, 1998], and toads [Vallortigara et al., 1998] respond to prey or to feeding matter preferentially with the right eye. Andrew et al. [2000] have suggested that this asymmetry may be related to left-hemispheric control of the mouth structures, an asymmetry that may be widespread in vertebrates and may relate to the left-hemispheric control of vocalization. It is conceivable that the left-hemisphere dominance for feeding may be a precursor of human right-handedness.

This brief summary does not do full justice to the now voluminous literature on cerebral asymmetries in nonhuman species, but should serve to illustrate something of the scope of the evidence, and the parallels with asymmetries observed in our own species.

THE ARGUMENT FOR DISCONTINUITY

There is sometimes a danger in science that one can always find what one is looking for, even if only by capitalizing on chance. Moreover, increasing refinement of measuring techniques may reveal slight asymmetries previously regarded as unimportant – one could probably demonstrate systematic asymmetries in billiard balls given precise enough techniques. Not surprisingly, then, Crow [2004] is at pains to dismiss at least some of the evidence, and to point out examples where population-level asymmetries have been sought and found lacking, as in Collins's [1977] classic studies of handedness (or pawedness) in mice. He notes the meta-analysis of studies of handedness in primates in the wild by McGrew and Marchant [1997], who conclude that, among nonhuman primates, "only chimpanzees show signs of a population bias ... to the right, but only in captivity and only incompletely" (p 201). In Hopkins's studies, only about two-thirds of the animals show the right-hand bias, which is clearly less than the 90% bias in the human population. Further, Hopkins's data are based on captive chimpanzees, and McGrew and Marchant [2001] suggest that right-handedness in these animals may have been a result of contact with right-handed humans, although subsequent research suggests that this is unlikely [Hopkins et al., 2004].

As suggested earlier, though, handedness may not be a good indicator of cerebral asymmetry in nonhuman primates. There is one respect in which humans are different from other primates – they are bipedal. This releases the hands from involvement in locomotion, and perhaps makes it more likely that cerebral asymmetry will be manifest in manual behavior, and especially in skilled activities such as writing, throwing, or the manufacture and use of tools. Handedness is a convenient and salient way of assessing cerebral asymmetry in humans, but may be much less appropriate for quadrupedal species. As we have seen, other more direct measures of functional and anatomical cerebral asymmetry now cast considerable doubt on the proposition that human cerebral asymmetry is unique, or depended on some saltational event.

In my 1991 book *The Lopsided Ape*, I wrote "Although I was fearful that evidence on animal asymmetries would overtake me, it still seems to me to be the case that right-handedness and cerebral asymmetry are unique to humans." To my subsequent relief, I did go on to qualify this statement by adding "—unique not so much in their presence as in their extent, pattern, and population bias" [Corballis, 1991, p vi]. Fifteen years later, I believe the evidence has indeed overtaken us, despite the tenacity of those who still hold to the idea that cerebral asymmetry is uniquely human. On the question of population bias, for example, Ghirlanda and Vallortigara [2004] note that the bias ranges from about 65% to about 90% across a wide variety of species. Indeed this range is apparent even *within* our own species, since about 90% of the human population is right-handed, whereas only about 65% display the characteristic right head turn at birth, along with associated postural asymmetries [Previc, 1991].

If there is a case for discontinuity, it may arise not from the fact of cerebral asymmetry, which is a characteristic of many, if not most, nonhuman species, but rather from the nature of functions that are lateralized in humans. The most obvious of these is language.

THE CASE OF LANGUAGE

Language may indeed provide the strongest evidence for human uniqueness, irrespective of its lateralization in the brain. The idea that language is uniquely human can be attributed to Descartes, but was strongly revived by the Cartesian linguist, Noam Chomsky. The following quotation, for example, also links the uniqueness of language to cerebral asymmetry:

> ... a chimpanzee is very smart and has all kinds of sensorimotor constructions (causality, representational functions, semiotic functions, and so forth), but one thing is missing: that little part of the left hemisphere that is responsible for the very specific functions of human language [quoted in Piattelli-Palmarini, 1980, p 182].

Indeed, that statement seems to carry the essence of the Broca–Annett axiom, albeit without the embellishments added by Crow. Nevertheless, the idea that language is uniquely human has been whittled away over the years since Chomsky first addressed the problem, in part through the moderate success in teaching forms of language to great apes [e.g., Savage-Rumbaugh et al., 1998], and in part through increasing understanding of animal communications in the wild [e.g., Hauser, 1998].

Chomsky himself seems to have retreated somewhat from his earlier stance. He is co-author of a paper by Hauser et al. [2002] arguing that there is a distinction between what they call the faculty of language in the broad sense (FLB) and the faculty of language in the narrow sense (FLN). It is clear that most animals, including primates, are able to make communicative sounds, and perform intentional acts. It is also clear that primates can use sounds or actions in symbolic fashion. For example, vervet monkeys give different warning cries to distinguish between a number of different threats, such as snakes, hawks, eagles, or leopards. When a monkey makes one of these cries, the troop acts appropriately, clambering up trees in response to a leopard call or running into the bushes in response to an eagle call [Cheney & Seyfarth, 1990]. These cries bear no obvious relation to the sounds emitted by the predators they stand for, and are in that sense abstract. More compellingly, perhaps, the bonobo Kanzi is able to use visual symbols on a keyboard to refer to objects and actions; these symbols are again abstract in the sense that they were deliberately chosen by the human keepers so as not to resemble what they stand for [Savage-Rumbaugh et al., 1998].

Based on such arguments, Hauser et al. [2002] argue that FLB is shared by other species, including birds and other mammals, although they also point out that the use of symbols in these examples does not mean that the symbols have all of the properties of *words*. Nevertheless, the critical ingredient that is missing from FLB, and that characterizes FLN, is *recursion*. Recursion is a computational procedure that calls itself, or that calls an equivalent kind of procedure, as exemplified in sentences with embedded clauses, such as *The malt that the rat ate lay in the house that Jack built*. Recursion lies at the heart of grammar, and enables us to create a potential infinity of sentences that convey a potential infinity of meanings. Recursion may also be said to underlie other aspects of human thought, such as theory of mind, which is the understanding that others have beliefs, desires, or intentions.

The question of whether nonhuman primates are capable of theory of mind was first raised by Premack and Woodruff [1978], and remains a source of dispute. For example, Povinelli has tried a number of tests to determine whether a chimpanzee can understand what is going on in the mind of a person [see Povinelli et al., 2000, for a summary], but the results have been largely negative. In one series of studies, chimpanzees readily approached humans to beg for food, but did not seem to be influenced by whether the person could see or not. If given the choice of two individuals to beg from, one with a blindfold over his/her eyes, the animals did not systematically choose the one who could see. The same was true when one of the people had a bucket over his/her head, or covered his/her eyes with his/her hands. These and similar results have been queried, however, partly on the grounds that chimpanzees are competitive, and may not cooperate fully in experiments with humans. Hare et al. [2000] have shown that chimpanzees are aware of what other *chimpanzees* can see, and modify their behavior accordingly. For example, a chimpanzee will approach food when a more dominant chimpanzee cannot see the food, but will be reluctant to do so when they can see that the dominant chimpanzee is watching. Call et al. [2004] have also shown that chimpanzees appear able to tell whether humans are unwilling or unable to perform some action, and modify their own behaviour accordingly, which further suggests at least some capacity for theory of mind.

Whatever the outcome of the debate over theory of mind in chimpanzees or other apes, it is clear that they cannot match the flexible, recursive theory of mind that governs much of human social life. Recursive theory of mind is illustrated in such extendable propositions as *George knows that Alice is suspicious of Henry's liking for Annabel.* Recursion may also underlie the human capacity for generative manufacture, with seemingly no end to the complexity and diversity of mechanical devices that we have wrought upon the physical world [Corballis, 1991]. Again, there seems to be a marked discontinuity between human manufacture and the manufacturing capabilities of other species.

These observations suggest that Crow may very well be correct in insisting on some fundamental discontinuity between humans and other species. It seems unlikely, though, that this discontinuity depends on cerebral asymmetry *per se*. Rather, the unique properties of the human mind may have grown out of functions that were already lateralized well before the emergence of our species. One possibility is that lateralization for language has its origins in left-hemispheric specialization for serial processing and manipulatory control observed in nonhuman species [Andrew & Rogers, 2002]. A related idea is that both handedness and the left-cerebral dominance for language, but not language itself, may relate to an ancient left-hemispheric dominance for vocalization [Corballis, 2003]. In this scenario, language may have originated in the so-called "mirror-system" in the primate brain, which is a system for mapping perceived manual gestures onto the production of those gestures [Rizzolatti et al., 2001]. At first, then, the communication system that led eventually to language may have been primarily manual, but gradually incorporated vocalizations. Since vocalization was predominantly left-hemispheric, its incorporation into the mirror system may have lateralized both language and the manual system. There are reasons to suppose that the final shift to autonomous speech occurred very late in hominid evolution, and possibly even after *Homo sapiens* evolved [Corballis, 2004b].

INDIVIDUAL DIFFERENCES AND THE GENETICS OF ASYMMETRY

Another reason to doubt that cerebral asymmetry was critical to human speciation is that not all people are left-cerebrally dominant. What are we to make of people who are left-handed, have language represented in the right cerebral hemisphere, or who simply lack consistent handedness or cerebral asymmetry? Must we somehow deny them human status? I think not.

Palmer [2004] distinguishes between two types of asymmetry in nature. One is antisymmetry, in which there are equal numbers of sinistral and dextral forms, as in the claws of male fiddler crabs, which are sometimes larger on the left and sometimes larger on the right. The other is directional asymmetry, in which most of the members of a species are asymmetrical in the same direction, as in the case of the vertebrate heart. In the case of antisymmetry, the direction of asymmetry is almost never inherited, while in the case of directional asymmetry, it typically is. Directional asymmetry might arise directly, through some genetic mutation, from symmetry, or it might arise through genetic assimilation from antisymmetry – that is, environmental conditions may have favored one or the other symmetrical form, leading to a preponderance of that form.

It seems likely that handedness, at least, might have arisen through genetic assimilation of one form of asymmetry from a character that was previously antisymmetric. Mice are equally divided into left- and right-handers, and selection for left-handedness fails to influence the relative proportions in succeeding generations [Collins, 1969]. There does appear to be a genetic component underlying the direction of handedness in chimpanzees, however, where the evidence suggests a 65:35 split in favor of right-handedness [Hopkins et al., 2001]. In humans, the split is about 90:10 in favor of right-handedness, and there is strong evidence for a genetic component to human handedness [Annett, 2002; McManus, 1999]. This progression suggests canalization (resistance to environmental perturbation) and genetic assimilation, and may apply more generally to cerebral asymmetry.

The asymmetry of the heart, which in all vertebrates is asymmetrical and displaced to the left, may also have arisen from ancestral antisymmetry, since there seems to have been a general trend for spontaneous reversals to decline throughout vertebrate evolution. The rate of reversal is about 5% in fish, 1–2% in amphibians and birds, less than 0.1% in mammals, and only 0.01% in humans [Palmer, 2004]. Moreover, there is an inbred strain of mice in which the heart and other visceral organs are antisymmetric, with the direction determined by chance [Layton, 1976], which may perhaps represent an ancestral state of affairs. This again suggests increasing canalization over evolutionary time.

This raises the question of where laterality is bred – in the heart or in the head? It has been suggested, in fact, that the asymmetry of the brain may have been ancestral to the asymmetry of the heart [Palmer, 2004] – although reversal of the viscera apparently does not result in reversal of language dominance [Kennedy et al., 1999]. Genetic regulation of brain asymmetry itself has been most studied in zebrafish, which show asymmetries of the habenula and pineal complex on the left side. Genes involved in the Nodal pathway, notably *lefty* and *pitx2*, have been shown to influence the lateralization of these areas [Halpern et al., 2003]. Similar asymmetries exist in other species and, oddly enough, may be biased in different directions in different species; for example, in the albino rat, they favor the left side while in the albino mouse they favor the right [Conchla & Wilson, 2001]. If the asymmetries of the organs are indeed part of a complex that includes the brain, then it may indeed be naïve to suppose that the cerebral asymmetries underlying handedness and dominance for language are concocted from a *tabula rasa*. Moreover, a cascade of events involving a large number of genes orchestrates the asymmetries of the internal organs

[Levin, 2005]. The majority of these are nonconserved, which explains some of the differences between species [Palmer, 2004]. It may make better sense, then, to see asymmetry in the human cerebral cortex as part of this more general genetic cascade, perhaps with some uniquely human elements, rather than the outcome of some unique genetic event.

One might ask whether human handedness and cerebral asymmetry will follow the heart, and stabilize at near-universal right-handedness and left-cerebral dominance for language. There are some reasons to suppose that the present proportions may be retained. First, they have probably been stable for something like 5,000 years [Coren & Porac, 1977], although this is probably too short a time to demonstrate an evolutionary trend. Second, there are *a priori* arguments in favor of a balance of asymmetries, perhaps maintained through frequency-dependent selection. Suppose, for example, that members of a group tend to stick together to avoid predation, and run off to the left when a predator threatens. By being one of many, each individual is less likely to be singled out by the predator. The predator may nevertheless choose to attack the mob rather than the strays, since the chances of catching at least one victim is maximized. Some individuals may therefore benefit from joining a minority that veers off to the right – a strategy that works only if this group remains a minority. This may have resulted in a subtle selection dynamic that held left- and right-turning in balance [Ghirlanda & Vallortigara, 2004], but with left-turning, implying a right-hemisphere dominance for this behavior, maintained for the majority. It has been argued similarly that left-handers hold an advantage in fighting, but only so long as they are in the minority [Raymond et al., 1996].

An alternative approach is to suppose that a balanced polymorphism is achieved though a heterozygotic advantage. The simplest models that incorporate this idea are those of Annett [2002] and McManus [1999]. Both can be described as two-allele models, and share the common theme that one allele codes for right-handedness and left-cerebral dominance for speech, while the other essentially leaves the direction of asymmetry to chance. If we suppose, following McManus [1999] that a dextral allele, D, codes for right-handedness [and left-cerebral dominance for language], while a chance allele, C, leaves the direction of asymmetry to chance, then balanced polymorphism is sustained by an advantage in reproductive fitness of DC heterozygotes over DD and CC homozygotes. This model allows for possible cultural or species differences, since the relative proportions of the two alleles at fixation in the population will depend on the relative *disadvantages* of the DD and CC genotypes, and these may vary between cultures. Specifically, the ratio of p(D):p(C) is given by $(1-f_{CC}):(1-f_{DD})$, where f_{DD} and f_{CC} are the fitnesses of DD and CC genotypes relative to that of the DC genotype. Thus if the probabilities of survival of DD and CC genotypes are 95% and 90% of that of DC genotypes, respectively, the proportion of D alleles will outnumber that of C alleles by 2:1.

Why might heterozygotes be at an advantage? In a general sense, there might be a tradeoff between symmetry and asymmetry. Bilateral symmetry is itself an adaptation to the equivalence of left and right in the natural world [Corballis & Beale, 1976], and goes back at least to the origins of the Bilateria some 600 million years ago, and perhaps even earlier [e.g., Finnerty et al., 2004]. But bilateral symmetry is also a constraint on function, and is abandoned, or at least relaxed, when lateral specialization is more adaptive. This suggests a tradeoff between pressures to symmetry and pressures to lateralization, and heterozygotes might preserve the most adaptive balance.

More specifically in relation to human capacities, Annett [2002] has summarized evidence that DD homozygotes may be deficient in spatial processing, while CC individuals may be at risk for verbal impediments. CD individuals, like Baby Bear's porridge in the Goldilocks story, are just right – neither too hot nor too cold. Another perspective is provided by evidence that bilaterality (CC) may be associated with more general deficits in academic ability [Crow et al., 1998], and also with a tendency to magical ideation [Barnett & Corballis, 2002; Nicholls et al., 2005]. Perhaps, then, the two alleles may express, not only the tension between symmetry and lateralization, but also the age-old conflict between reason and superstition. The heterozygotes are there to keep the peace.

The two-allele models of Annett and McManus are at odds with Crow's idea that the laterality gene is located in homologous regions of the X and Y chromosomes, since polymorphisms on the Y chromosome are unstable under any selective regime; the proportion of D alleles on that chromosome will stabilize at 0 or 1 [Corballis, 1997]. It is conceivable, however, that the gene may be on the X chromosome alone [Corballis, 2001; McKeever, 2000]. Nevertheless a genome-wide search for the handedness gene has since offered little support for X-linkage, and suggests that the region 2p11.2-12 on chromosome 2 may be a better bet [Francks et al., 2002]. Although the authors report that this failed to replicate in an independent sample, a further analysis has revealed significant paternal linkage within this site [Francks et al., 2003], suggesting that genetic imprinting may play a role. (Imprinting implies that gene expression depends on the parent of origin, in this case the father.) In view of the complexity of genetic influences over visceral asymmetries, it is perhaps unlikely that cerebral asymmetry depends on a single gene, although the models proposed by Annett and McManus provide a useful start.

COULD LANGUAGE ITSELF BE THE RESULT OF A SINGLE MUTATION?

The Broca–Annett axiom proposed by Crow is unlikely to be true, at least insofar as it pertains to cerebral asymmetry. Cerebral asymmetry is probably polymorphic, and handedness almost certainly is, and the nature of that polymorphism suggests a tradeoff between bilateral symmetry and asymmetry. Nevertheless, it may well be true that characteristics of the brain that *are* uniquely human are also lateralized. That is, our uniqueness may lie not so much in lateralization itself as in the functions that are lateralized.

These functions may include language, theory of mind, and perhaps recursive manufacture. One function that may be uniquely lateralized in humans but not in other primates is spatial attention [Hillis, 2006]. In humans, this function is largely under the control of the right cerebral hemisphere, whereas there is no evidence for lateralization in monkeys [Driver & Vuilleumier, 2001]. This asymmetry, however, may be a secondary consequence of the left-hemispheric dominance for language [Karnath et al., 2001].

Although there is fairly general agreement that language, at least, is uniquely human, can it be attributed to a single mutation, a "big bang" in hominid evolution? This idea has been advanced by a number of authors besides Crow, including Bickerton [1995], who wrote that "... true language, via the emergence of syntax, was a catastrophic event, occurring within the first few generations of *Homo sapiens sapiens*" (p 69). Further, evolutionary psychologists have promoted the notion of a "monomorphic mind" [Tooby & Cosmides, 1992, p 79], arguing that individual differences in human cognition are not due to genetic differences. The language faculty, then, is said to be common to all humans, and variations are due to environmental influences. This is at least consistent with the notion that language may be the result of a single mutation.

It seems unlikely, though, that a faculty as complex as language, even in the narrow sense of FLN, could have arisen in a single step. Jackendoff [2002] has proposed that the evolution of syntax probably involved a sequence of steps, including the formation of hierarchical phrase structure, the formation of grammatical categories, and the emergence of inflections and grammatical functions to convey semantic relations. These steps, however, probably unfolded in the course of late hominid evolution, beginning perhaps with the emergence of the genus *Homo* from around 2 million years ago. One of the main distinguishing characteristics of *Homo* was the increase in brain size [Deacon, 1997], perhaps heralding the emergence of more complex cognition, culminating in recursive thought.

Indeed Chomsky [1975] once argued that language may have arisen simply as a consequence of possessing an enlarged brain, without the assistance of natural selection:

> We know very little about what happens when 10^{10} neurons are crammed into something the size of a basketball, with further conditions imposed by the specific manner in which this system developed over time. It would be a serious error to suppose that all properties, or the interesting structures that evolved, can be 'explained' in terms of natural selection (p 59).

Nevertheless, the increase in brain size itself may have depended on natural selection, perhaps driven by adaptive pressure for more advanced communication and social understanding. It is possible that this may have involved changes in the Xq21.3/Yp homologous regions of the X and Y chromosomes, as claimed by Crow [2002]. However, recent research has brought to light other genetic mutations that may be critical. It is of some interest that two of these mutations resulted in the inactivation of genes, suggesting that we may owe our humanity at least in part to the *loss* of genes, rather than the incorporation of new ones.

One of these mutations was on a gene on chromosome 7 that encodes the enzyme CMP-*N*-acetylneuraminic acid (CMP-Neu5Ac) hydroxylase (CMAH). An inactivating mutation of this gene has resulted in the absence in humans of the mammalian sialic acid *N*-glycolylneuraminic acid (Neu5Gc). This acid appears to be absent in Neanderthal fossils as well as in humans, but is present in modern-day nonhuman primates. It has been down-regulated in the brain in the chimpanzee, and through mammalian evolution, leading to speculation that inactivation of the CMAH gene may have progressively removed a constraint on brain growth in human ancestry [Chou, et al., 2002]. Chou et al. applied molecular-clock analysis to the CMAH genes in chimpanzees and other great apes, as well as to the pseudogene in humans, which indicated that the mutation occurred some 2.8 million years ago, leading up to the expansion in brain size.

The other inactivating mutation that may have contributed to the increase in brain size has to do with a gene on chromosome 7 that encodes myosene heavy chain (MYH16), responsible for the heavy masticatory muscles in most primates, including chimpanzees and gorillas, as well as the early hominids. Molecular-clock analysis shows that this gene was inactivated an estimated 2.4 million years ago, leading to speculation that the diminution of jaw muscles and their supporting bone structure removed a further constraint on brain growth [Stedman et al., 2004]. It is a matter of further speculation as to why this seemingly deleterious mutation became fixed in the ancestral human population. It may have had to do with the change from a predominantly vegetable diet to a meat-eating one, or it may have had to do with the increasing use of the hands rather than the jaws to prepare food [Currie, 2004]. After an initial flurry of interest in this mutation, though, there has recently been doubt expressed as to whether it was really involved in the increase in brain size [McCollom et al., 2006].

It is likely, though, that there are a host of genes involved in the development of brain size, and the search for some single mutation that explains human uniqueness is a lingering manifestation of Cartesian dualism – it is as though a miraculous mutation has somehow replaced God as the instigator of all things human. One gene known to be a specific regulator of brain size is the *abnormal spindle-like microcephaly associated* (ASPM) gene, and the evidence suggests strong positive selection of this gene in the lineage leading to *Homo sapiens* [Evans et al., 2004, 2005]. Indeed, a selection sweep appears to have occurred as recently as 5,800 years ago, suggesting that the human brain is still undergoing rapid evolution [Mekel-Bobrov et al., 2005]. Another gene known as microcephalin (*MCPH6*) has also been shown to regulate brain size, and one variant in modern humans arose an estimated 37,000 years ago. Other genes involved in the control of brain size that have undergone accelerated rates of protein evolution at points in human lineage have been noted by Dorus et al. [2004].

One further gene that may be involved specifically in the evolution of language and speech, and perhaps of cerebral asymmetry, is the *FOXP2* (forkhead box P2) gene. A mutation of this gene appears to be responsible for a disorder of speech and language in half of the members of an extended family, known as the KE family, in England [Fisher et al., 1998]. The nature of the deficit in the affected members of the KE family, and therefore the role of the *FOXP2* gene, has been debated. Some have argued that *FOXP2* gene is involved in the development of morphosyntax [Gopnik, 1990], and it has even been identified more broadly as the "grammar gene" [Pinker, 1994]. Subsequent investigation suggests, however, that the core deficit is one of articulation, with grammatical impairment a secondary outcome [Watkins et al., 2002].

The *FOXP2* gene is highly conserved, differing in only three places from that in the mouse, but two of the three mutations occurred in the hominid line, after the split from the apes. According to one estimate, the more recent of these mutations occurred within the past 100,000 years [Enard et al., 2002]. This raises the possibility that this last mutation of the *FOXP2* gene was the final step in the emergence of autonomous speech from a language that earlier depended, at least in part, on manual gestures [Corballis, 2004a]. Indeed, a recent study based on functional magnetic resonance imaging (fMRI) suggests a link between the *FOXP2* mutation and cerebral asymmetry. Unaffected members of the KE family showed the expected activation of Broca's area, the primary speech area on the left side of the brain, while they covertly generated verbs, but affected members of the family showed *under*activation of Broca's area and diffuse bilateral activation [Liégeois et al., 2003]. Although the lateralization of vocal production may have a long evolutionary history, as suggested earlier, the lateralization of speech circuits is almost certainly unique to humans – if only because speech itself is a uniquely human accomplishment.

CONCLUSIONS

Given the wealth of evidence for cerebral and manual asymmetries in nonhuman species, the parallels between these asymmetries and those in humans, and the complexity of genetic influences on asymmetries, it is very unlikely that the Broca-Annett axiom enunciated by Crow [2002] is correct. Cerebral asymmetry may even be ancestral to that of the heart, and numerous genes have shaped the asymmetries in present-day organisms, including, no doubt, humans.

Nevertheless, Crow may well be correct in insisting that humans are endowed with some special properties, including language, theory of mind, and a disposition to psychosis – and perhaps one might add a voracious capacity for environmental destruction. The critical ingredient in all of these may be recursive thought. These properties do appear to be lateralized in the human brain, but their asymmetrical representation probably arose because their precursors were lateralized. For example, the lateralization of speech may be simply a manifestation of a widespread lateralization for vocalization that may even go back to the origins of the vocal cords [Bauer, 1993].

It is also unlikely that human language or recursive abilities emerged as a result of a single mutation. Big bang theories provide convenient solutions, even if smacking of creationism, but evolution is incremental. Crow [2004] refers to Gould's [2002] idea that speciation occurs through saltations ("punctuated evolution"), but the idea that human speciation, along with a suite of special characteristics, was the result of a single mutation is extreme even in terms of Gould's proposal. Gould's theory of punctuated evolution is couched in geological time, and the shaping of *Homo sapiens* over the 2 million or so years since the emergence of the genus *Homo* would surely be considered saltational in a geological time-scale. We arrived quickly, but not *that* quickly.

Nevertheless, Crow has performed a valuable service by articulating a theory that is admirably explicit, and that serves to remind biologists that there are still real problems in explaining how our extraordinary species came about. I think Crow is almost certainly wrong, but those with a more gradualist approach do need to recognize that there are functions, such as language, that far outstrip any communication system that has been recorded in any other species. It will be a very long time, I think, before any other species will engage in this debate, and it may need a little balanced polymorphism to see it through equably.

REFERENCES

Alonso, V. Lateralization of visual guided behaviour during feeding in zebra finches (*Taeniopygia guttata*). Behavioral Processes 43: 257–263, 1998.

Andrew, R. J.; Rogers, L. J. The nature of lateralization in tetrapods. Pp. 94–125 in L. J. Rogers; R. J. Andrew, eds. Comparative Vertebrate Lateralization. Cambridge, Cambridge University Press, 2002.

Andrew, R. J.; Tommasi, L.; Ford, N. Motor control by vision and the evolution of cerebral lateralization. Brain and Language 73: 220–235, 2000.

Annett, M. Handedness and Brain Asymmetry: The Right Shift Theory. Hove, United Kingdom, Psychology Press, 2002.

Barnett, K. J.; Corballis, M. C. Ambidexterity and magical ideation. Laterality 7: 75–84, 2002.

Bauer, R. H. Lateralization of neural control for vocalization by the frog (*Rana pipiens*). Psychobiology 21: 243–248, 1993.

Bickerton, D. Language and Human Behavior. Seattle, WA, University of Washington Press, 1995.

Bisazza, A.; Cantalupo, C.; Robins, A.; Rogers, L. J.; Vallortigara, G. Right pawedness in toads. Nature 379: 408, 1996.

Brockmeier, B.; Ulrich, G. Asymmetries of expressive facial movements during experimentally induced mood states – a video-analytical study. Cognition and Emotion 7: 393–405, 1993.

Buxhoeveden, D.; Casanova, M. Comparative lateralization patterns in the language areas of human, chimpanzee, and rhesus monkey brains. Laterality 5: 315–330, 2000.

Call, J.; Hare, B.; Carpenter, M.; Tomasello, M. 'Unwilling' versus 'unable': Chimpanzees' understanding of human intentional action. Developmental Science 7: 488–498, 2004.

Casperd, J. M.; Dunbar, R. I. M. Asymmetries in the visual processing emotional cues during agonistic interactions in gelada baboons. Behavioral Process 37: 57–65, 1996.

Cheney, D. L.; Seyfarth, R. M. How Monkeys See The World. Chicago, University of Chicago Press, 1990.

Chomsky, N. Reflections on Language. New York, Pantheon, 1975.

Chou, H. H.; Hakayama, T.; Diaz, S.; Krings, M.; Indriati, E.; Leakey, M.; *et al.* Inactivation of CMP-*N*-acetylneuraminic acid hydroxylase occurred prior to brain expansion during human evolution. Proceedings of The National Academy of Sciences, USA 99: 11736–11741, 2002.

Collins, R. L. On the inheritance of handedness: II. Selection for sinistrality in mice. Journal of Heredity 60: 117–119, 1969.

Collins, R. L. Origins of the sense of asymmetry: Mendelian and non-Mendelian models of inheritance. Annals of the New York Academy of Sciences 299: 283–305, 1977.

Conchla, M. L.; Wilson, S. W. Asymmetry in the epithalamus of vertebrates. Journal of Anatomy 199: 63–84, 2001.

Corballis, M. C. The Lopsided Ape. New York, Oxford University Press, 1991.

Corballis, M. C. The genetics and evolution of handedness. Psychological Review 104: 714–727, 1997.

Corballis, M. C. Is the handedness gene on the X chromosome? Psychological Review 108: 805–810, 2001.

Corballis, M. C. From mouth to hand: Gesture, speech, and the evolution of right-handedness. Behavioral and Brain Sciences 26: 199–260, 2003.

Corballis, M. C. FOXP2 and the mirror system. Trends in Cognitive Sciences 8: 95–96, 2004a.

Corballis, M. C. The origins of modernity: Was autonomous speech the critical factor? Psychological Review 111: 543–552, 2004b.

Corballis, M. C.; Beale, I. L. The Psychology of Left and Right. Hillsdale, NJ, Lawrence Erlbaum Associates, 1976.

Coren, S.; Porac, C. Fifty centuries of right-handedness: The historical record. Science 198: 631–632, 1977.

Crow, T. J. Sexual selection, timing, and an X–Y homologous gene: Did *Homo sapiens* speciate on the Y chromosome? Pp. 197–216 in T. J. Crow, ed. The Speciation of Modern *Homo Sapiens*. Oxford, Oxford University Press, 2002.

Crow, T. Directional asymmetry is the key to the origin of modern *Homo sapiens* (the Broca–Annett axiom): A reply to Rogers' review of The Speciation of Modern Homo Sapiens. Laterality 9: 233–242, 2004.

Crow, T. J.; Crow, L. R.; Done, D. J.; Leask, S. Relative hand skill predicts academic ability: Global deficits at the point of hemispheric indecision. Neuropsychologia 36: 1275–1282, 1998.

Currie, P. Muscling in on hominid evolution. Nature 428: 373–374, 2004.

Dadda, M. Social aggregation and lateralized response to social stimuli in tadpoles (*Bufo bufo*): Influence of developmental stage. Laterality 10: 345–352, 2005.

Davidson, R. J. Cerebral asymmetry, emotion, and affective style. Pp. 361–389 in R. J. Davidson; K. Hugdahl, eds. Brain Asymmetry. Cambridge, MA, MIT Press, 1995.

Deacon, T. The Symbolic Species: The Co-Evolution of Language and the Brain. New York, W.W. Norton, 1997.

Deckel, A. W. Laterality of aggressive responses in *Anolis*. Journal of Experimental Zoology 272: 194–200, 1995.

Deng, C.; Rogers, L. J. Differential contributions of the two visual pathways to functional lateralization in chicks. Behavioural Brain Research 87: 173–182, 1997.

Devinsky, O.; Ronsaville, D.; Cox, C.; Witt, E.; Fedio, P.; Theodore, W. H. Interictal aggression in epilepsy: The Buss–Durkee Hostility Inventory. Epilepsia 35: 585–590, 1994.

Dorus, S.; Vallender, E. J.; Evans, P. D.; Anderson, J. R.; Gilbert, S. L.; Mahowald, M.; *et al.* Accelerated evolution of nervous system genes in the origin of *Homo sapiens*. Cell 119: 1027–1040, 2004.

Driver, J.; Vuilleumier, P. Perceptual awareness and its loss in unilateral neglect and extinction. Pp. 39–88 in S. Dehaene, ed. The Cognitive Neuroscience of Consciousness. Amsterdam, Elsevier, 2001.

Ehert, G. Left hemisphere advantage in the mouse brain for recognizing ultrasonic communication calls. Nature 325: 249–251, 1987.

Enard, W.; Przeworski, M.; Fisher, S. E.; Lai, C. S. L.; Wiebe, V.; Kitano, T.; *et al.* Molecular evolution of FOXP2, a gene involved in speech and language. Nature 418: 869–871, 2002.

Evans, P. D.; Anderson, J. R.; Vallender, E. J.; Gilbert, S. L.; Malcom, C. M.; Dorus, S.; *et al.* Adaptive evolution of ASPM, a major determinant of cerebral cortical size in humans. Human Molecular Genetics 13: 489–494, 2004.

Evans, P. D.; Gilbert, S. L.; Mekel-Bobrov, N.; Vallender, E. J.; Anderson, J. R.; Vaez-Azizi, L. M.; Tishkoff, S. A.; Hudson, R. R; Lahn, B. T. Microcephalin, a gene regulating brain size, continues to evolve adaptively in humans. Science 309: 1717–1720, 2005.

Finnerty, J. R.; Pang, K.; Burton, P.; Paulson, D.; Martindale, M. Q. Origins of bilateral symmetry: Hox and Dpp expression in a sea anenome. Science 304: 1335–1337, 2004.

Fisher, S. E.; Vargha-Khadem, F.; Watkins, K. E.; Monaco, A. P.; Pembrey, M. E. Localisation of a gene implicated in a severe speech and language disorder. Nature Genetics 18: 168–170, 1998.

Foundas, A. L.; Leonard, C. M.; Heilman, K. M. Morphological cerebral asymmetries and handedness – the pars triangularis and planum temporale. Archives of Neurology 52: 501–508, 1995.

Francks, C.; Fisher, S. E.; McPhie, I. L.; Richardson, A. J.; Marlow, A. J.; Stein, J. F.; Monaco, A. P. A genome-wide linkage screen for relative hand skill in sibling pairs. American Journal of Human Genetics 70: 800–805, 2002.

Francks, C.; DeLisi, L. E.; Shaw, S. H.; Fisher, S. E.; Richardson, A. J.; Stein, J. F.; Monaco, A. P. Parent of origin effects on handedness and schizophrenia susceptibility on chromosome 2p12-q11. Human Molecular Genetics 12: 3225–3230, 2003.

Gannon, P. J.; Holloway, R. L.; Broadfield, D. C.; Braun, A. R. Asymmetry of chimpanzee planum temporale: Human-like brain pattern of Wernicke's area homolog. Science 279: 220–221, 1998.

Geschwind, N.; Levitsky, W. Human brain: Left-right asymmetries in temporal speech region. Science 161: 186–187, 1968.

Ghirlanda, S.; Vallortigara, G. The evolution of brain lateralization: A game-theoretical analysis of population structure. Proceedings of the Royal Society of London 271: 853–857, 2004.

Gopnik, M. Feature-blind grammar and dysphasia. Nature 344: 715, 1990.

Gould, S. J. The Structure of Evolutionary Theory. Harvard, MA, Belknap Press, 2002.

Graves, R. E.; Potter, S. M. Speaking from two sides of the mouth. Visible Language 22: 129–137, 1988.

Güntürkun, O. Lateralization of visually controlled behavior in pigeons. Physiology and Behavior 34: 575–577, 1985.

Güven, M.; Elalmis, D. D.; Binokay, S.; Tan, U. Population right-paw preference in rats assessed by a new computerized food-reaching test. International Journal of Neuroscience 113: 1675–1689, 2003.

Halpern, M. E.; Liang, J. O.; Gamse, J. T. Leaning to the left: Laterality in the zebrafish forebrain. Trends in Neuroscience 26: 308–315, 2003.

Hare, B.; Call, J.; Agnetta, B.; Tomasello, M. Chimpanzees know what conspecifics do and do not see. Animal Behaviour 59: 771–785, 2000.

Hauser, M. D. The Evolution of Communication. Cambridge, MA, The MIT Press, 1998.

Hauser, M. D.; Andersson, K. Functional lateralization for auditory temporal processing in adult, but not infant rhesus monkeys: Field experiments. Proceedings of the National Academy of Sciences, USA 91: 3946–3948, 1994.

Hauser, M. D.; Fitch, W. T.; Chomsky, N. The faculty of language: What is it, who has it, and how did it evolve? Science 298: 1569–1579, 2002.

Heffner, H. E.; Heffner, R. S. Temporal lobe lesions and perception of species-specific vocalizations by Japanese macaques. Science 226: 75–76, 1984.

Hillis, A. E. Neurobiology of unilateral spatial neglect. The Neuroscientist 12: 153–174, 2006.

Holowka, S.; Petitto, L. A. Left hemisphere cerebral specialization for babies while babbling. Science 297: 1515, 2002.

Howard, K. J.; Rogers, L. J.; Boura, A. L. A. Functional lateralisation of the chicken forebrain revealed by use of intracranial glutamate. Brain Research 188: 369–382, 1980.

Hook-Costigan, M. A.; Rogers, L. J. Lateralized use of the mouth in production of vocalizations by marmosets. Neuropsychologia 36: 1265–1273, 1998.

Hopkins, W. D. Chimpanzee handedness revisited: 55 years since Finch (1941). Psychonomic Bulletin and Review 3: 449–457, 1996.

Hopkins, W. D.; Dahl, J. F.; Pilcher, D. Evidence in support of the right-shift theory and developmental instability. Psychological Science 12: 299–303, 2001.

Hopkins, W. D.; Hook, M.; Braccini, S.; Schapiro, S. J. Population-level right handedness for a coordinated bimanual task in chimpanzees: Replication and extension in a second colony of apes. International Journal of Primatology 24: 677–689, 2003.

Hopkins, W. D.; Leavens, D. A. Hand use and gestural communication in chimpanzees (*Pan troglodytes*). Journal of Comparative Psychology 112: 95–99, 1998.

Hopkins, W. D.; Wesley, M. J.; Izard, M. K.; Hook, M.; Schapiro, S. J. Chimpanzees (*Pan troglodytes*) are predominantly right-handed: Replication in three populations of apes. Behavioural Neuroscience 118: 659–663, 2004.

Hunt, G. R.; Corballis, M. C.; Gray, R. D. Laterality in tool manufacture by crows. Nature 414: 707, 2001.

Ivry, R. B.; Robertson, L. C. The Two Sides of Perception. Cambridge, MA, MIT Press, 1998.

Jackendoff, R. Foundations Of Language: Brain, Meaning, Grammar, Evolution. Oxford, UK, Oxford University Press, 2002.

Jäncke, L.; Steinmetz, H. Auditory lateralization and planum temporale asymmetry. Neuroreport 5: 69–72, 1993.

Jansari, A.; Tranel, D.; Adolphs, R. A valence-specific lateral bias for discriminating emotional facial expressions in free field. Cognition and Emotion 14: 341–353, 2000.

Karnath, H. O.; Ferber, S.; Himmelbach, M. Spatial awareness is a function of the temporal not the posterior parietal lobe. Nature 411: 950–953, 2001.

Kennedy, D. N.; O'Craven, K. M.; Ticho, B. S.; Goldstein, A. M.; Makris, N.; Henson, J. W. Structural and functional brain asymmetries in human *situs inversus totalis*. Neurology 55: 1260–1265, 1999.

Levermann, N.; Galatius, A.; Ehlme, G.; Rysgaard, S.; Born, E. W. Feeding behaviour of free-ranging walruses with notes on apparent dextrality of flipper use. BMC Biology 3: 9, 2003.

Layton, W. M. Jr. Random determination of a developmental process. Journal of Heredity 67: 336–338, 1976.

Levin, M., Left-right asymmetry in embryonic development: A comprehensive review. Mechanisms of Development 122: 621–621, 2005 (Abstract only, but text available from author).

Liégeois, F.; Baldeweg, T.; Connelly, A.; Gadian, D. G.; Mishkin, M.; Vargha-Khadem, F. Language fMRI abnormalities associated with FOXP2 gene mutation. Nature Neuroscience 6: 1230–1237, 2003.

McCollom, M. A.; Sherwood, C. C.; Vinyard, C. J.; Lovejoy, C. O.; Schachat, F. Of muscle-bound crania and human brain evolution: The story behind the *MYH16* headlines. Journal of Human Evolution 50: 232–236, 2006.

McGrew, W. C.; Marchant, L. F. On the other hand: Current issues in a meta-analysis of the behavioural laterality of hand function in nonhuman primates. Yearbook of Physical Anthropology 40: 201–232, 1997.

McGrew, W. C.; Marchant, L. F. Ethological study of manual laterality in the chimpanzees of the Mahale Mountains, Tanzania. Behaviour 138: 329–358, 2001.

McKeever, W. F. A new family handedness sample with findings consistent with X-linked transmission. British Journal of Psychology 91: 21–39, 2000.

McManus, C. Handedness, cerebral lateralization, and the evolution of handedness. Pp. 194–217 in M. C. Corballis; S. E. G. Lea, eds. The Descent of Mind. Oxford, Oxford University Press, 1999.

Mekel-Bobrov, N.; Gilbert, S. L.; Evans, P. D.; Vallender, E. J.; Anderson, J. R.; Hudson, R. R.; Tishkoff, S. A.; Lahn, B. T. On-going adaptive evolution of *ASPM*, a brain size determinant in *Homo sapiens*. Science 309: 1720–1722, 2005.

Nicholls, M. E. R.; Orr, C. A.; Lindell, A. K. Magical ideation and its relation to lateral preference. Laterality 10: 503–515, 2005.

Palmer, A. R. Symmetry breaking and the evolution of development. Science 306: 828–833, 2004.

Peirce, J. W.; Leigh, A. E.; Kendrick, K. M. Configurational coding, familiarity and the right hemisphere advantage for facial recognition in sheep. Neuropsychologia 38: 475–483, 2000.

Peterson, G. M. Mechanism of handedness in the rat. Comparative Psychology Monograph 9: 46, 1934.

Piattelli-Palmarini, M. Language and Learning: The Debate between Jean Piaget and Noam Chomsky. Cambridge, MA, Harvard University Press, 1980.

Pinker, S. The Language Instinct. New York, Morrow, 1994.

Povinelli, D. J.; Bering, J. M.; Giambrone, S. Toward a science of other minds: Escaping the argument by analogy. Cognitive Science 24: 509–541, 2000.

Premack, D.; Woodruff, G. Does the chimpanzee have a theory of mind? Behavioral and Brain Sciences 4: 515–526, 1978.

Previc, F. A general theory concerning the prenatal origins of cerebral lateralization in humans. Psychological Review 98: 299–334, 1991.

Raymond, M.; Pontier, D.; Dufour, A.; Moller, A. P. Frequency-dependent maintenance of left-handedness in humans. Proceedings of the Royal Society of London B 263: 1627–1633, 1996.

Rizzolatti, G.; Fogassi, L.; Gallese, V. Neurophysiological mechanisms underlying the understanding and imitation of action. Nature Reviews 2: 661–670, 2001.

Rogers, L. J. Evolution of hemispheric specialization: Advantages and disadvantages. Brain and Language 73: 236–253, 2002a.

Rogers, L. J. Lateralized brain function in anurans: Comparison to lateralization in other vertebrates. Laterality 7: 219–240, 2002b.

Rogers, L. J. Leaping to laterality and language: The case against. Laterality 9: 225–232, 2004.

Rogers, L. J.; Andrew, R. J. Comparative Vertebrate Lateralization. Cambridge, Cambridge University Press, 2002.

Savage-Rumbaugh, S.; Shanker, S. G.; Taylor, T. J. Apes, Language, and the Human Mind. New York, Oxford, University Press, 1998.

Sovrano, V. A.; Bisazza, A.; Vallortigara, G. Lateralization of responses to social stimuli in fishes: A comparison between different methods and species. Physiology And Behavior 74: 237–244, 2001.

Sperry, R. W.; Zaidel, E.; Zaidel, D. Self-recognition and social awareness in the disconnected right hemisphere. Neuropsychologia 17: 153–166, 1979.

Stedman, H. H.; Kozyak, B. W.; Nelson, A.; Thesier, D. M.; Su, L. T.; Low, D. W.; *et al.* Myosin gene mutation correlates with anatomical changes in the human lineage. Nature 428: 415–418, 2004.

Tooby, J.; Cosmides, L. The psychological foundations of culture. Pp. 19–136 in J. H. Barkow; L. Cosmides; J. Tooby, eds. The Adapted Mind: Evolutionary Psychology and the Generation of Culture. New York, Oxford University Press, 1992.

Tsai, L. S.; Maurer, S. Right-handedness in white rats. Science 72: 436–438, 1930.

Vallortigara, G.; Andrew, R. J. Differential involvement of right and left hemisphere in individual recognition in the domestic chick. Behavioural Processes 33: 41–58, 1994.

Vallortigara, G.; Pagni, P.; Sovrano, V. A. Separate geometric and nongeomatric modules for spatial reorientation: Evidence from a lopsided animal brain. Journal of Cognitive Neuroscience 16: 390–400, 2004.

Vallortigara, G.; Rogers, L. J.; Bisazza, A.; Lippolis, G.; Robins, A. Complementary right and left hemifield use for predatory and agonistic behavior. Neuroreport 9: 3341–3344, 1998.

Vermeire, B. A.; Hamilton, C. R.; Erdmann, A. L. Right-hemispheric superiority in split-brain monkeys for learning and remembering facial discriminations. Behavioral Neuroscience 112: 1048–1061, 1998.

Watkins, K. E.; Dronkers, N. F.; Vargha-Khadem, F. Behavioural analysis of an inherited speech and language disorder: Comparison with acquired aphasia. Brain 125: 452–464, 2002.

An adult female Bornean Orangutan holding a piece of food in her right hand. This Orangutan has been rehabilitated and has been free-ranging at Sepilok, Sabah, for several years. Her offspring can be seen partly hidden behind her. (Photograph by Gisela Kaplan.)

The Evolution of Hemispheric Specialization in Primates
W.D. Hopkins (Editor)

LATERALIZATION IN ITS MANY FORMS, AND ITS EVOLUTION AND DEVELOPMENT

Lesley J. Rogers
Centre for Neuroscience and Animal Behaviour, School of Biological, Biomedical and Molecular Sciences, University of New England, NSW, Armidale, Australia

ABSTRACT

Lateralization of the central nervous system and side biases in behavior are much more common than once thought and it now seems that they are so widespread that it may be symmetry, rather than asymmetry, that demands an explanation. This chapter discusses examples of lateralization in vertebrates and invertebrates. It summarizes research on the development of lateralization in two model species and presents a gene-environment perspective, including the influence of hormones on the development of lateralization. It then considers evidence indicating why lateralization evolved, and does so from the perspective of both the lateralized individual and the lateralized population. After discussion about the likelihood that preferred use of one limb (hand) over the other is an unreliable indicator of the presence/absence of hemispheric specialization (viz., absence of handedness does not necessarily mean absence of hemispheric lateralization), association between brain lateralization, limb preference and temperament is given some special attention.

Keywords: vertebrate asymmetry, marmoset handedness, evolution of laterality, cognition

LATERALIZATION IN VERTEBRATES

Examples

Lateralization refers to frequencies skewed away from a normal distribution in the population, one example of which is directional bias, such that the majority of individuals are lateralized in the same direction, and another example is antisymmetry, in which the majority of individuals are lateralized but the number of left- and right-biased individuals is approximately equal [for examples of these distributions, see Rogers, 2002a].

A large body of evidence shows directionally lateralized motor and perceptual functions in vertebrate species, from fish to mammals. Since this has been reviewed in detail previously [Bradshaw & Rogers, 1993; Rogers &

ISSN: 1936-8526 DOI: 10.1016/S1936 8526(07)05002-6

Andrew, 2002; Rogers, 2002a; Bisazza et al., 1998; Halpern et al., 2005], here I will give just a few examples that illustrate the possibility of a quite general pattern of hemispheric specialization across vertebrate species for some basic functions important for survival.

Lateralization in the domestic chick, across a range of functions, has been studied the most extensively so far. This species shows a clear bias to use the right eye (and left hemisphere, since the visual input is mainly processed by the hemisphere contralateral to the eye) to process inputs required to direct pecks at food grains scattered amongst a distracting background of small pebbles stuck down to the floor: chicks tested with a patch over the left eye can perform the task, whereas chicks tested with a patch over the right eye are unable to avoid pecking at the pebbles [Rogers & Anson, 1979; Rogers, 1997]. A recent study has shown that the main reason for the superior ability of the right eye/left hemisphere system is its ability to inhibit responding to distracting targets (i.e., pecking at the pebbles) until a decision about the appropriate target is made, whereas the left eye/right hemisphere is less able do so [Rogers et al., 2007].

The chick also shows this lateral bias when it has to manipulate an object with its beak: Andrew et al. [2000] have shown that chicks approach a bowl with a lid that they must remove to obtain a food reward by taking a path veering to the left so that their right eye is fixating the target. Approach to a bowl simply containing food and without a lid is made on a path veering rightward.

Young chicks use the left eye and right hemisphere when they need to examine an advancing, overhead predator [Rogers, 2000; Rogers et al., 2004] and so do adult chickens [Evans et al., 1993]. Chicks also use the right hemisphere when they have to find food using distal spatial cues as a cognitive map [Tommasi & Vallortigara, 2004; Tommasi et al., 2003], when they choose between a model social partner and a slightly unfamiliar stimulus [Vallortigara & Andrew, 1994; Deng & Rogers, 2002a] and when they engage in agonistic interactions with conspecifics [Rogers et al., 1985; Vallortigara et al., 2001].

All of these examples of lateralization have been shown by testing groups of chicks, usually around 10–12 chicks per group but repeated many times on different occasions and also in different laboratories, meaning that the sample sizes are very respectable. Usually, the chicks are tested monocularly with a patch over the left or right eye or after pharmacological treatment of the left or right hemisphere. The types of lateralization revealed in this way are, therefore, directional biases present at group level. They would not be revealed by these techniques if they were merely present as lateralization at the individual level but with no bias for the direction of lateralization to be the same for the majority of individuals in the group. Since, in most cases, the chicks have been tested using only the left or the right eye, it is not possible to determine the strength of lateralization for each individual. Nevertheless, comparison of the scores for the left and right groups shows very little overlap, suggesting that the lateralization is strong. One experiment [Rogers et al., 1985] did test the chicks first with one eye and then the other, in random order, measuring their attack and copulation responses to a simulated conspecific, and found marked differences in scores for left- and right-eyed performance per individual. In other words, the lateralizations at individual and group levels appear to be as strong as the lateralization present in humans.

Remarkably similar patterns of lateralization have been found in several other species. Toads, for example, are more responsive to a predator approaching on their left side compared to their right side [Lippolis et al., 2002], and both frogs and toads lodge agonistic strikes more frequently at conspecifics on their left side [Robins et al., 1998; Robins & Rogers, 2006]. The same leftward bias for displaying agonistically to conspecifics is seen in the males of two species of lizards [Deckel, 1995; Hews & Worthington, 2001] and in females of another species of lizard [Hews et al., 2004]. Also similar to the chick, toads strike preferentially at prey seen in their right visual hemifield, whereas they ignore prey in their left visual hemifield [Robins & Rogers, 2004]. Furthermore, some similar lateralizations have been shown in mammalian species: dunnarts respond more strongly to a predator advancing on their left side [Lippolis et al., 2005] and gelada baboons are more likely to express agonistic behavior to a conspecific approaching on their left side than on their right side [Casperd & Dunbar, 1996].

The specialization of the right hemisphere for recognizing faces, shown in primates [Hauser & Akre, 2001; Weiss et al., 2002] and sheep [Broad et al., 2000; Costa et al., 2004], might also have something in common with the chick's use of the left eye and right hemisphere to recognize model partners. Indeed, the pattern emerging for a range of vertebrates is not fundamentally different from that known for humans. Of particular note is the specialization of the right hemisphere for face recognition, spatial cognition and the expression of intense emotion in humans [Davidson & Hugdahl, 1995] and the corresponding function of the right hemisphere in a range of vertebrate species.

The fundamental and complementary specializations of the hemispheres in vertebrates may be summarized as the following pattern: use of the right hemisphere for controlling rapid responses, often relying on a use of an established template or cognitive map, and use of the left hemisphere for processing information and controlling motor responses that require some degree of inhibition initially before a decision is made. This may explain the specialized role of the left hemisphere in control of motor responses that involve manipulation and it may be a precursor to use the right hand for fine manipulation in higher primates. In fact, it may also explain the preferred use of the right forelimb in other vertebrates when tasks requiring fine manipulation are performed. It is tempting to suggest that the preferred use of the right forelimb and paw displayed by toads (three species) when they attempt to remove an offending object from the head or snout reflects this specialization of the left side of the brain (discussed more later). In addition, the right-paw directional bias, now found in rats using improved techniques to score reaching for pieces of food

[Tang & Verstynen, 2002; Güven et al., 2003], may also reflect specialized control by the left hemisphere for manipulating objects.

These examples show that similar hemispheric specializations occur in species with laterally placed eyes, and hence with only a small binocular overlap, as well as complete crossing over of the optic nerve fibres (e.g., the species of fish and birds studied so far, as well as rodents), and in species with frontally placed eyes, a large binocular overlap and both crossing and non-crossing fibres in the optic nerve (e.g., primates). Somewhat different methods must be used to show the lateralities in these two conditions. Whereas monocular testing reveals lateralization of function in species with laterally placed eyes, in species with large binocular overlap the tests must involve scoring the differential responsiveness to stimuli in the extreme lateral visual fields. The frogs and toads discussed above fall somewhere between these two conditions since their visual fibers decussate completely (as in birds) but they have a large binocular field of vision [Lippolis et al., 2002]. In their case too, lateralization has been revealed by stimuli presented, at least initially, in the lateral, monocular field (i.e., by advancing a stimulus into the left or right monocular field [Lippolis et al., 2002], by rotating stimuli around the toad clockwise and counter clockwise [Robins & Rogers, 2004] or, in more naturalistic settings, by simply noting the location in the monocular fields of targets to which the animals respond [Robins & Rogers, 2006].

Motor asymmetries for limb use are also present in birds. Csermely [2004] discovered that a number of birds of prey show right-foot preferences for grasping their prey. Given that grasping at live prey may require some degree of fine manipulation, this preference for the right foot might be expected. Chicks also have a right-foot preference for scraping the ground while feeding [Dharmaretnam et al., 2002; Rogers & Workman, 1993] and for the first foot to move when stepping off from standing on two feet, the latter with a strength of almost 90% [Casey & Martino, 2000]. However, at least for scratching while feeding, the preferred foot seems to be secondary to preferred use of the right eye in guiding pecking at food targets, as shown by monocular testing [Tommasi & Vallortigara, 1999].

If use of the left hemisphere and right foot for manipulation of objects/prey is a characteristic of avian species, we need to explain the preferred use of the left foot to hold food objects, as shown in a number of species of parrots [Harris, 1989; Rogers, 1980]: in some cases, such as the sulfur-crested cockatoo, the foot preference is greater than 90% [Rogers, 1980]. The answer to the question of why the left foot is preferred may lie in the use of the beak for fine manipulation in these species, the left foot being used merely to anchor or hold fast the food object. Observation of the motor movements of cockatoos feeding on Banksia seed heads confirms that it is indeed the beak that is used for manipulation (personal observations of yellow-tailed black cockatoos). Moreover, use of the left foot for holding may enable use of the bird's right eye (left hemisphere) to guide manipulation of the beak and coordinate it with respect to

Figure 1. Photograph of a wild yellow-tailed black-cockatoo, *Calyptorhynchus funereus*, feeding on a *Banksia* seedpod. Note that the bird is holding the pod in its left foot, manipulating the seeds to open them with its beak and viewing with its right eye. This is a typical posture adopted when feeding and it suggests that the monocular field of the right eye may guide the beak position during this fine manipulation task. It must be said, however, that monocular field of the left eye could also be used with other head postures although my own observations suggest that the posture shown is more common. This species has a directional bias of preferred use of the left foot for holding the food [Rogers, 1980, and subsequent unpublished observations]. (Photograph by G. Kaplan.)

the grasping foot, as Figure 1 suggests. Further, detailed investigations of coordinated foot and beak use in parrots are needed to test this hypothesis.

Homoplasy or Homology?

Some previously strong advocates of the uniqueness of lateralization to the human species now recognize the mounting evidence for lateralization in other vertebrate species. Corballis [1991] is one of them: recently he has written that belief in the notion that cerebral asymmetry defines the human condition has been one of the myths of our time [Corballis, 2005]. His willingness to embrace the available evidence contrasts to Crow's denial of this evidence and insistence that lateralization is unique to humans [Crow, 2002, 2005]. A recent review by

Sun and Walsh [2006] also singles humans out as being unique in having a more than 90% population bias to use a preferred limb and the authors do so by underplaying evidence for handedness in primates and ignoring the evidence for limb preferences and other lateralized behavior known to be present in other species. As noted above, the latter preferences are often as strong as 90%.

Corballis writes, "Some of our lateralized activities may well be distinctive to our own species, but cerebral lateralization itself is not" [Corballis, 2005, p 595]. I agree with this statement and see that now it might be a matter of deciding to what extent humans have the same types of lateralization as do other vertebrates. This brings us to consider whether lateralization of the brain has followed the trajectory of convergent evolution (homoplasy) or whether it has a common ancestry (homology). McManus [2005], who also embraces the evidence for lateralization in vertebrate species, argues in favor of homoplasy. To my mind, the general pattern of separate specializations of the hemispheres suggests its early evolution and likely homology, as others and I have argued previously [Andrew & Rogers, 2002; Rogers, 2002a; Vallortigara & Rogers, 2005]. Based on clear evidence that teleost fish have lateralization [Miklósi & Andrew, 1999; Bisazza et al., 2000; Brown et al., 2004], Andrew [2002] has reasoned that lateralization appeared as early as the first vertebrates and as a consequence of their asymmetrical eyes and mouth.

Once the first step had been taken to differentiate one hemisphere to perform certain types of processing and the other hemisphere to perform other types of processing, other lateralities might have built on this foundation. In that sense, we may see the general pattern of hemispheric specialization as a fundamental template conserved throughout vertebrate evolution. Of course, the degree to which this fundamental lateralization is expressed, and elaborated on, could be modified by many factors, some of them the genetic modifications that lead to the evolution of new species and others the influences of experience, as discussed below. Each species may, therefore, express some unique forms of lateralization but these are likely to be resonant with the fundamental pattern of lateralization. Owing to experiential effects on the ontogenesis of lateralization, individuals may express varying degrees of lateralization, or indeed no lateralization at all, and so too might subpopulations of species in particular ecological contexts.

ONTOGENY OF LATERALIZED FUNCTIONS

The emphasis of research on lateralization so far has been on its evolution and inheritance. Much less attention has been paid to its development and the influences of experience on the strength and direction of lateralization. Nevertheless, there are some known examples of lateralization that are markedly influenced by early experience and they should convince us that genetic influences on lateralization are not paramount. The existence of directional lateralization in a species may well be an evolutionary stable strategy

[Vallotigara & Rogers, 2005], but this does not mean that genes have a deterministic role in lateralization. On the contrary, the ontogeny of lateralization depends on the interaction of genes and experience.

Two model systems have been instrumental in demonstrating the interaction between genetic, hormonal and early experiential effects on the development of lateralization. These are the developing rat and domestic chick. Denenberg et al. [1978] [see also Denenberg, 1981] have shown most convincingly that many of the lateralized functions of the adult rat depend on the neonatal pup experiencing a treatment known as handling. Handling involves taking the pups away from their mother and placing them alone in small containers for 3 min each day for the first 20 days of life. The effective aspects of this treatment are likely to be the brief exposure of the pups to novelty and, when the pups are returned to their mother, the stimulated change in maternal–pup and pup–pup behavior [Denenberg, 1999]. This experience, in interaction with the hormonal condition of the pups, appears to alter cerebral lateralization by affecting the growth of the corpus callosum. Handled pups with high levels of testosterone, either naturally in males or by administration of the hormone to females, develop a larger corpus callosum and display lateralized behavior [Denenberg et al., 1991]. Estrogen has the opposite effect [Fitch et al., 1991] and there are sex differences in the response to handling [Sherman et al., 1983]. These findings on the interaction between steroid hormone levels (likely involving the stress steroid hormones also) have direct relevance to interpreting the body of work on the size of the corpus callosum and brain lateralization in humans. As Cowell and Denenberg [2002] have explained, variation between the sexes, between individuals and across the life-span may result from the interactive effects of a number of hormones and multiple environmental factors.

The lateralized behavior that depends on the interactive influences of handling and steroid hormone levels includes activity levels in the open field, direction of moving off in the open field, taste aversion and muricide [Denenberg, 1981; Sherman et al., 1980]. Handling also establishes dominant use of the right hemisphere in control of spatial learning in the Morris maze, as shown by monocular testing [Cowell et al., 1997]. Furthermore, handling also has a long-term effect on hand preference in rats, but in this case the handling removed the lateralization: Tang and Verstynen [2002] found that handling shifted the group distribution of paw preference, measured as reaching for food, from a right directional bias to no significant bias. As in the case of Denenberg's experiments, this change in paw preference as a result of early experience was very long lasting (i.e., stable well into adulthood).

Such early and perhaps rather subtle effects of experience in early life on the development of hand preference could apply also to primates, and I believe that it would be timely to investigate this in some detail. Lyons et al. [2002] have demonstrated an effect of early experience on the prefrontal cortex of *Saimiri* monkeys: availability of the mother affects the development of the right prefrontal cortex. In fact, association between hand preference and general expression of behavior (discussed later) may well be the outcome of early experience. For example, an individual suffering stress in early life might come to rely on predominate use of the right hemisphere, at least in most contexts although not exclusively, and so express left-paw preference in tasks of a general type not requiring fine manipulation. In fact, a stress-induced shift towards preferential reliance on use of the right hemisphere would suggest that the results of Denenberg and Tang and Verstynen are similar since the former showed that handling largely increased the expression of behavior controlled by the right hemisphere and the latter showed a shift away from use of the left hemisphere (right paw) towards use of the right hemisphere (there was a trend towards a left-paw preference).

The second model species showing the influences of early life experience on the development of lateralization is the (domestic) chick. The chick embryo, as true for the majority of avian embryos, is turned in the egg during its final stages of development before hatching with the left side of its head against its own body [illustrated in Rogers, 1990, 1995]. This means that the embryo's body occludes the left eye, whereas the right eye can be stimulated by light passing through the shell and membranes. This asymmetrical stimulation of the eyes by light generates directional lateralization for a number of visual functions, as mentioned above [Rogers, 1982, 1990, 1997], and it causes asymmetry to develop in the thalamofugal visual projections to the Wulst region of the forebrain [Rogers & Sink, 1988; Rogers & Deng, 1999]. As short a period as 2 h of light exposure during a sensitive period when the visual projections to the forebrain are growing is sufficient to lead to the development of these visual lateralizations [Rogers, 1982, 1990]. Following incubation of the eggs in the dark, no lateralization for these visual functions develops and the thalamofugal visual projections develop symmetrically. If the embryo's head is gently eased out of the eggshell on day 19 or 20 of incubation (hatching occurs on day 21) and an eye patch is applied to the right eye while the left eye is exposed to light, the direction of lateralization is reversed: this applies to lateralization of performance on the pebble-floor task, attack and copulation [Rogers, 1990], as well as to lateralization of the thalamofugal visual projections [Rogers & Sink, 1988]. Exposure of both the eyes of the embryo to light from day 19 or 20 until hatching, and thereafter, prevents the development of these structural and functional lateralizations [Rogers & Bolden, 1991].

Visual experience prior to hatching also influences the development of turning preferences in the domestic chick [Casey & Karpinski, 1999] and bobwhite quail [Casey & Lickliter, 1998]. Furthermore, motor behavior during hatching is important in the development of certain forms of lateralization [Casey & Sleigh, 2001]. After pipping the eggshell, the chick proceeds to crack apart the eggshell in an anticlockwise direction, driven by the right side of the chick's body as it uses its limbs, head and neck to rotate within the egg. Disruption of this hatching process prevents the development of motor lateralization [Casey, 2005; Casey & Martino, 2000].

The effect of visual experience pre-hatching on the ontogeny of turning biases in domestic chicks, measured as left-right choice in a T-maze, has an overall similarity to the effect of light exposure on the development of asymmetry in the visual projections and of lateralization of performance on the pebble-floor task, attack and copulation, but there is an interesting difference. In the latter cases, artificial exposure of the embryo's left eye to light reversed the normal asymmetries (i.e., the directional bias in the group is reversed). By contrast, the effect of left eye exposure to light (visual experience) on turning preference is to prevent the development of any directional bias: about half of the chicks receiving this treatment had a left turning preference and the rest a right turning preference, and there were no chicks without a turning preference [Casey & Karpinski, 1999].

The same antisymmetric distribution of turning preferences was produced by exposing both of the embryo's eyes to visual experience. In other words, only when the right eye was exposed (normal condition) did a directional bias in turning develop, and that was a strong 90% leftwards. These differences in the effect of exposing the left, right or both eyes to light on the distributions of the various forms of lateralization would be interesting to examine further since turning seems to have a fall-back position of antisymmetry when any condition other than the normal right eye stimulation occurs, whereas the lateralization of the visual pathways, performance on the pebble-floor, attack and copulation follows the direction of the exposed eye.

In fact, a more direct association between asymmetrical stimulation of the eyes would be expected for development of the visual pathways and visual tasks than for the turning task, especially since the latter task used playback of the maternal call to attract the chick to move in the T-maze. The auditory input may well have primed the attention of one hemisphere, possibly the left hemisphere since there is evidence that this hemisphere of the chick is specialized for habituating to an auditory stimulus [Rogers & Anson, 1979; Howard et al., 1980]. Hence, auditory and visual processing, as well as purely motor biases, might all be involved in the task used to measure turning preference. It should be noted, however, that a similar experiment measuring the effect of pre-hatching visual experience on turning bias in the bobwhite quail did show that exposure of the left eye to light reversed the directional bias for turning [Casey & Lickliter, 1998], suggesting species difference in response to the testing paradigm used, possibly as a result of species differences in attention to the auditory cues in testing.

Without considering further details of the differences between these procedures, the point of relevance here is that the development of different types of lateralization may be influenced in different ways by one type of experience and, of course, different types of sensory stimulation would be expected to influence different aspects of lateralization. For example, lateralization of olfactory processing in the chick remains unaffected by incubating the embryos in the dark [Rogers et al., 1998] and so too does choice to approach a social companion [Andrew et al., 2004]. Also, auditory stimulation of embryos does not affect the development of visual lateralization [Casey & Lickliter, 1998; Zappia & Rogers, 1983].

As is the case in the rat, steroid hormones interact with the experience of light stimulation to affect the development of visual lateralization in the chick [summarized by Deng & Rogers, 2002b]. Injection of testosterone, estrogen or corticosterone into the egg prior to the sensitive period for light stimulation causes the visual projections to develop symmetrically despite exposure of the embryo to light [testosterone, Schwarz & Rogers, 1992; estrogen, Rogers & Rajendra, 1993; corticosterone, Rogers & Deng, 2005].

The chick, therefore, develops lateralization for certain visual functions (food searching, attack, copulation, predator detection and turning bias) as a result of an interactive process between genes (determining orientation of the embryo in the egg), hormones and experience (light stimulation and the motor experiences of hatching). Light experience also leads to the development of visual lateralization in the pigeon [Güntürkün, 1993], bantam chick and bob-white quail [Casey & Sleigh, 2001]. These avian species provide us with model systems for research investigating the relevant ontogenetic events that might apply more broadly to other species, including primates. In my opinion, they point out a need to investigate the influence of experience on the development of both motor and sensory lateralization in higher vertebrates and should not be written off as something different that sets avian lateralization apart from that of mammals [e.g., as suggested by McManus, 2005].

Recently, Denenberg [2005] has hypothesized that a threshold amount of environmental stimulation may be necessary if the brain is to develop lateralization. He encompasses both the rat and chick experiments in this general explanation since, in both examples, it is the addition of stimulation that leads to the development of a lateralized brain. We may add the study of *Saimiri* monkeys to this [Lyons et al., 2002] since here too more stimulation, caused by intermittent separations from their mother, which is known to be stressful, leads to the development of a more lateralized brain. The idea is that higher levels of stimulation in early life prepare the individual for life in a world that will be demanding of the animal's cognitive capacity, the latter being enhanced by lateralization of the brain (discussed later). Reduced stimulation (light deprivation in the case of the chick and no handling/separation from the mother in the case of the rat and monkey) leads the individual to prepare for a world in which there will be little variability and hence few demands on the brain. As Denenberg proposes, brain redundancy (symmetry) results, and this is the animal's back-up, fail-safe system.

How might the absence of light stimulation of the developing embryo during the final stages of incubation be a predictor of a stable, relatively unvarying environment? Let us consider the hen of *Gallus gallus* [jungle fowl from which domestic chickens are derived; see Rogers, 1995]. The hen prevents light exposure of her eggs when she sits on them. Hence, the embryos are exposed to light when she leaves the nest to forage or she is disturbed by social interactions with conspecifics or to the threat of predation. Therefore, light exposure of the embryos, leading to brain lateralization, would predict a more varying world of social interaction or heightened levels of predation. Completely consistent with this line of reasoning, Brown et al. [2004] have discovered that a species of poeciliid fish shows directional lateralization, measured in terms of the eye used to view a predator, when it is present in parts of a stream with high predation but not in those parts of the stream with low predation.

Apart from the potential of light stimulation of the embryo to vary according to the amount of social stress and predation pressure, stressed hens deposit higher levels of corticosterone in their eggs and the late-stage embryo itself secretes corticosterone in response to being stressed. As we know that corticosterone influences the development of lateralization of the thalamofugal visual pathway [Rogers & Deng, 2005], this may be another possible means by which stress in early life modulates the development of lateralization. However, experimental elevation of corticosterone levels during the final stages of incubation actually prevents (or inhibits) the development of lateralization of the visual pathways, which is opposite to the prediction based on Denenberg's hypothesis. Nevertheless, only one dose of corticosterone has been administered so far and, as well known in endocrinology research, dose of hormone and rate of change in the level of a hormone is critical, often leading to opposite effects. Hence, the role of corticosterone in the ontogeny of the lateralized visual pathway is established but the ecological significance of this requires further investigation.

These investigations of the ontogeny of lateralization demonstrate that multiple factors interact to produce the asymmetry or symmetry of a single function or a collection of linked functions. My aim has been to show that, while genetic expression is involved and evolutionary selection for the characteristic of lateralization occurs, both the degree and the direction of lateralization depends on experience and stimulation from the environment. Hence, it is not simply a matter of whether or not genes determining left or right biased lateralization (e.g., handedness) are present but how the expression of those genes might interact with stimulation and experience. Realizing this, leads us to be interested in and excited by the variability in lateralization from species to species and environment to environment, rather than to continue with the focussed search for a linear evolution of lateralization leading to the claimed fully lateralized human brain at the pinnacle.

ADVANTAGE OF BEING LATERALIZED

The existence of lateralization in a wide range of vertebrate species suggests strongly that benefits are bestowed by having a lateralized brain. It has long ago been suggested that the human brain gains cognitive capacity by being lateralized [e.g., Levy, 1977; Dunaif-Hattis, 1984] and the same argument may be applied to other species. Only now, after three decades of research focussed on documenting the presence of lateralization in nonhuman vertebrates, are we beginning to devise ways of testing the hypothesis about lateralization enhancing cognitive capacity.

Güntürkün [2000] provided one of the first pieces of empirical evidence that lateralization enhances cognitive ability in animals. He determined the degree of visual lateralization in pigeons by testing them monocularly (each bird having the left and the right eye tested) on a task requiring them to discriminate food grains from inedible objects and then compared the strength of lateralization with accuracy of discrimination. Strength of lateralization correlated positively with accuracy of pecking at food grains. A similar result has been reported for chimpanzees foraging for termites by McGrew and Marchant [1999]: fishing for termites by inserting sticks into the mound is more efficient in chimpanzees that adopt consistent use of one hand to hold the probe and the other as a surface on which to wipe the probe before scooping up the termites with the lips, compared to chimpanzees using a less lateralized strategy.

Domestic chicks are an excellent model species in which lateralized and non-lateralized individuals can be compared since several of their visual lateralizations are generated by exposing the late-stage embryo to light [Rogers, 1982, 1990], as explained in some detail in the previous section, due to the orientation of the embryo in the egg such that it occludes its left eye. Hence, the cognitive abilities of chicks hatched from eggs exposed to light during the final stages of incubation can be compared to those of chicks hatched from eggs incubated in the dark.

Rogers et al. [2004] designed a task in which chicks, if they have been exposed to light before hatching, would have to use each hemisphere for a different specialized function simultaneously. The task required the chick to search for food in the pebble-floor task [for which it uses the right eye and left hemisphere; Rogers, 1990, 1997] while at the same time monitoring overhead for a model predator, a silhouette of a hawk [for which it uses the left eye and right hemisphere; Rogers, 2000]. Light-exposed and dark-incubated chicks were compared. As predicted the light-exposed, and hence lateralized, chicks could perform both aspects of the task better than the dark-incubated, non-lateralized chicks. The latter were less likely to notice the predator overhead when they were concentrating on pecking for food. They responded to the predator after a longer latency than did the light-exposed chicks. The dark-incubated chicks also performed less well than the light-exposed ones in finding grain. In fact, they had so much difficulty in performing the two tasks at once that they pecked more and more pebbles as the task progressed. They were unable to remember the pebble-floor task the next day. When the dark-incubated chicks were tested on the pebble-floor task without the presence of the predator, they were able to perform as well as the light-exposed chicks and remembered the task well on the next day. Hence, their inability was revealed only when they had to perform two tasks simultaneously, something that is commonly demanded of animals in the natural environment.

A similar inability has been found in non-lateralized fish tested on a dual task requiring foraging for shrimp in the presence of a predator. Dadda and Bisazza [2006] compared the performance of two selected lines of the fish, *Girardinus falcatus*, one selected to be lateralized and the other to be non-lateralized for preferred direction of turning at a barrier. Their efficiency of capturing shrimp was tested in the presence and in the absence of a predator fish. In the presence of the predator, the lateralized fish consumed the prey in half the time taken by the non-lateralized fish. No differences were present when the predator was absent. By measuring the eye used as the fish performed these tasks, the researchers were able to show that the lateralized fish monitored the predator with one eye and used the other eye for capturing the prey, whereas the non-lateralized fish swapped between tasks, using each eye to perform both prey capture and monitoring of the predator. Hence, efficient function is achieved by allocating each task to a different side of the brain, thus avoiding interference between functions.

Other experiments have shown that lateralized fish (*Girardinus falcatus*) are also better in spatial orientation than are nonlateralized fish, regardless of whether they are using geometric or nongeometric (featural or landmark) cues [Sovrano et al., 2005]. This result might suggest that lateralization is an advantage even when the individual fish is performing only a single function, but it is possible that attention to the salient spatial cues (e.g., geometric) in each test requires avoidance of distraction by the non-salient (e.g., nongeometric) cues. We know from research on chicks at least that geometric cues are processed by the right hemisphere and nongeometric by the left hemisphere [Tommasi et al., 2003] and, if it is the same for the fish, both hemispheres may be in use simultaneously, with conflict in the case of the poorly performing nonlateralized fish.

These experimental paradigms provide ways to determine the benefits of having a lateralized brain and lay a basis for asking some of the questions relevant to the evolution of lateralization. Previous research has shown that animals balance foraging against predator detection. An increased risk of predation is associated with a decrease in the efficiency of selection of prey, as shown in salmon [Metcalfe et al., 1987] and blue jays [Dukas & Kamil, 2000], and more difficulty in discriminating food from background is associated with reduced vigilance for predators, as shown in blackbirds [Lawerence, 1985]. These and other studies [see Dukas, 2004, for an overview] provide convincing evidence that animals have limited cognitive abilities for simultaneous predator detection and finding food, and there are contexts in which the balance is shifted towards performance of one task at the expense of the other. Here is where having a lateralized brain is advantageous. In fact, these are situations in which lateralization may well emerge as an evolutionary stable strategy.

LATERALIZATION IN INVERTEBRATES

Convincing evidence for lateralized function of the nervous system in invertebrate species is now coming to light. There are very few examples to indicate whether or not there is any similarity in the pattern of lateralization among invertebrate species, let alone to address the question of any possible homology of lateralization in invertebrates with lateralization in vertebrates. However, it is beginning to look as if the nervous system becomes lateralized rather readily or, at least, far more often than once thought. Probably it enhances neural capacity in invertebrates, as it does in vertebrates.

The few known examples of lateralization in invertebrates include both antisymmetric (individual level) and directional (population level) distributions. Byrne et al. [2002, 2004] have reported that wild-caught octopuses prefer to view a model crab held outside their aquarium by consistent use of one and the same eye. Since there is no consistent preference for the same eye to be used by the majority of individuals, the distribution of this form of lateralization is antisymmetrical. This means that almost all of the individual animals showed an eye preference and about half preferred to use the left eye and the other half the right eye. [Note that this bimodal distribution has nothing in common with fluctuating asymmetry, in which the frequency histogram conforms to a bell-shaped curve; see Møller & Swaddle, 1997; Rogers, 2002a.]

Directional lateralization has been found in some invertebrate species. The spitting spider, *Scytodes globula*, has a preference to use its anterior first and second legs on the left side to make probing touches of potential prey and leg loss in this species is more frequent on the left than on the right side [Ades & Ramires, 2002]. Heuts and Brunt [2005] have reported further evidence in support of the left side being more vulnerable to injury in spiders (18 families) and ants, suggesting that this may be a general directional bias in arthropod species. The latter researchers have remarked on the possibility that both vertebrates and invertebrates sustain more injuries on the left than the right side, as indeed has been reported for whitefish [Reist et al., 1987] and impala [Jarman, 1972], but we would need to know more details of the predator–prey interaction for each species before drawing the conclusion that they depend on similar causal mechanisms. As noted above, toads, chicks and dunnarts have been found to be more reactive to predators advancing on the left than on the right and this would lead to the logical prediction that they would sustain fewer injuries on their left sides, unless they have preference to orient the receptors on their left side towards potential predators. Also relevant to this discussion on side-bias and vulnerability to attack (or injury) is the finding that left-footed lizards sustain a higher level of tail injuries than do right-footed ones, the latter being more common in the population [Seligmann et al., 2003].

Bumblebees (*Bombus* spp.) also have lateralization, shown in direction of circling as they visit florets arranged in circles around a vertical inflorescence. Individual bees were shown to sample successive inflorescences by rotating in the same direction (clockwise or anticlockwise). Three out of the four species examined by Kells and Goulson [2001] exhibited directional biases (i.e., the majority of individual bees circled in the same direction). Given that two species had an anticlockwise preference and one preferred to circle clockwise, it seems that the directional bias is a characteristic of the bees rather than being secondary to an asymmetry in the structure of the florets. It may even be associated with the recently demonstrated lateralization of response to olfactory stimuli in honey bees, *Apis mellifera* [Letzkus et al., 2006]. Honeybees were trained on the standard proboscis extension reflex paradigm and it was found that they could learn about odors when using their right antenna but were less able to do so when using their left antenna. This asymmetry is directionally biased.

Another example of lateralization is that seen anatomically in the central nervous system of the fruit fly: Pascual et al. [2004] found that individual flies with an asymmetrical brain structure were able to form long-term memories more readily than those with a symmetrical brain structure. This finding is unexpectedly consistent with the function of lateralization in enhancing cognitive capacity, including memory, as shown in the chick (above). It suggests that, regardless of when the species evolved or how large, or complex, its brain is, lateralization may occur and cognitive capacity may be enhanced. Indeed, as Monaghan and Pollman [2003] have demonstrated using computational, neural network models, bilateral processing emerges spontaneously for processing complex tasks. This suggests that there may be a dynamic processing system within a brain of any level of complexity that may lead to division of functions between the left and right sides of the nervous system when the latter is asked to perform functions that it finds complex.

Even though it would be wise to resist drawing hasty conclusions about similarity of lateralization in invertebrates and vertebrates, we cannot turn aside from the gauntlet thrown down by these examples of lateralized behavior in invertebrates and simply treat them as being irrelevant to our understanding of lateralization in vertebrates. The examples of lateralization in invertebrates expand the field and open up new possibilities for conducting controlled experiments.

HAND PREFERENCES VERSUS BRAIN LATERALIZATION

Much has been made of hand preferences in primates as an indicator of whether or not a species has brain lateralization, from Walker [1980] to MacNeilage et al. [1987] and Marchant and McGrew [1996], this extrapolation being either stated or implied. While this approach might be valid when hand preferences for performing certain complex tasks are considered, as in the case of tool using [Hopkins & Cantalupo, 2005], it now seems to be invalid to use hand preferences measured on simpler tasks (e.g., to pick up food and take it to the mouth) for this purpose. In many species of primate, and as in the case of rodents, the preferred hand/paw to pick up food items and take them to the mouth appears to be a quite separate measure than hemispheric lateralization. Hand preference may be determined by which hemisphere is being used for the task or in a particular context. In other words, rather than hand preference in such a simple task being a unitary marker of brain lateralization, the hand preferred to reach for or pick up food seems to reflect the hemisphere most likely to be active when that individual is in a relaxed state and when there is no strong demand for fine manipulation or visually guided reaching. It follows, therefore, that the index of preferred use of one hand over the other in simple tasks is an inaccurate indicator of the presence (or absence) of hemispheric specialization.

Perhaps, the best way to explain how I have arrived at this conclusion is to give an example. Consider hand preferences and other types of lateralization in common marmosets. Several studies have reported that approximately half of common marmosets prefer to pick up pieces of food and hold them in the left hand while taking them to the mouth for chewing [Rothe, 1973; Box, 1977; Hook & Rogers, 2000], whereas the other half prefers to use the right hand for this purpose. This means that, on this measure, marmosets show a pattern of preferences similar to that originally reported for mice [Collins, 1985]; the majority of individuals have significant hand preferences but the population is antisymmetric and not directionally lateralized. If we were to finish our investigation of lateralization with this measure and conclude that the common marmoset as a species lacks brain lateralization, we would be incorrect. First, other measures of lateralization reveal that marmosets do show directional lateralization, as in the case of eye preferences for viewing stimuli through a peephole, asymmetry of facial expressions and head turning to listen to a vocalization played from behind the marmosets. Only one marmoset in a group of 21 failed to show a strong right-eye preference for viewing pieces of food [Hook-Costigan & Rogers, 1995]. Marmosets also show a group bias to express fear more strongly on the left side of their face [Hook-Costigan & Rogers, 1998], as has been shown also in rhesus macaques [Hauser, 1993] and humans [Davidson et al., 1990; Davidson, 1999]. Furthermore, marmosets display a left ear preference when they turn their head to listen to the tsik (mobbing) call or the phee (long-distance contact) call played behind them when they are sitting on a platform and feeding from a dispenser [Shuster & Rogers, unpublished data]. All of these measures show a directional or group bias and all were found to be independent of the hand preferences that the same individuals had shown for picking up food. Hence, they indicate that marmosets do have hemispheric specialization, for these functions at least.

Studies of rodents also demonstrate that paw preference is not a direct indication of the presence of brain lateralization. The early research of Denenberg [1981] used the technique of ablating one or the other hemisphere to reveal hemispheric specialization in the rat and the results showed clear evidence of a population bias of specialization of the right hemisphere for controlling affective behavior. Ehret [1987] was able to show that female mice have hemispheric specialization for recognition of the pups' ultrasounds, as demonstrated by plugging the left or right ear and looking at the effect of this on the maternal mouse's retrieval of her pups. These directional biases in cortical function are not reflected in a similar pattern of paw preferences. As mentioned above, studies on hand preferences in rodents performed at about the same time found that they display paw preferences when they are tested on a task requiring them to reach repeatedly into a tube to grasp pieces of food, but half of the population preferred to use the left paw and half the right paw. Although more recent studies on different populations/strains of rodents and using more sensitive techniques have reported directional biases in paw preference [in rats, Tang & Verstynen, 2002; in mice, Maarouf et al., 1999 and Waters & Denenberg, 1994], the general picture is one in which the distribution of paw preferences in the population is not a useful indicator of the existence of a population bias in the brain itself.

It is, of course, perfectly understandable, on ethical grounds, ease of study and as a matter of cost, why researchers have opted to study hand preferences in primates, rather than look for lateralization of sensory systems or within the brain itself. Also, hand preferences are relatively easy to study in natural populations. The problem is, as far as I can see, many researchers have tried to claim that hand preferences are the basis on which we should decide whether or not a species is lateralized rather than simply studying the hand preferences in their own right. When a directional bias is found, as in the case of tool using in chimpanzees [Hopkins & Cantalupo, 2005], it does, in fact, demonstrate a directional bias likely to be present in the brain but the converse does not hold: the absence of a directional bias in the frequency distribution of hand (or paw) preferences does not prove that the species lacks brain asymmetry.

Whether or not hand preferences have a directional bias would seem to depend on the demands of the task, not only the difficulty of the task itself, as recognized some time ago by Vauclair and Fagot [1993], but also on the visual and other sensory demands of the task. Hand preferences also depend on the postural demands, consistent with the hypothesis of MacNeilage et al. [1987], which states that the right hand and right side of the body is stronger and used by lower primates to support the body while reaching is performed by the left hand, but, as the primates evolved an upright posture, the right hand was increasingly freed from this constraint and could be used to manipulate objects. In other words, in the absence of postural demands, a specific hand will be used only if it is obligatory for a specific hemisphere to process inputs and control outputs in a particular task, in the same way that birds choose to use a particular eye to view certain types of stimuli [Dharmaretnam & Andrew, 1994].

Complexity or difficulty of the task may be why toads show a directional bias in paw preference [i.e., to use the right paw to wipe an offending object from the head or snout, or to pivot around when righting the body, Bisazza et al., 1996; Rogers, 2002b], whereas some strains of rodents and some primate species do not show directional lateralization when they are simply asked to reach for food. The toad may find the task of removing objects from its head or snout difficult to perform and demanding of the specialized neural circuitry of one side of the brain. In fact, to perform this task the toad may have to use the differently specialized circuitry on each side of the brain, since to raise one forelimb to wipe an object from the head is a skilled maneuver when, in parallel, the toad has to balance a rather rigid body on the other forelimb. This may be so demanding that it requires use of laterally specialized control systems of the forebrain.

If difficulty of the motor task is the factor demanding directional lateralization of limb use, we might expect to find directional biases of limb use in lower vertebrates on *seemingly* simple tasks, whereas higher vertebrates show them only on complex tasks requiring fine manipulation. This may explain why directionally biased limb preferences have been found in toads [Bisazza et al, 1996], lizards [Seligmann et al., 2003], chicks [Dharmaretnam et al., 2002; Rogers & Workman, 1993] and parrots [Rogers, 1980; Harris, 1989] on reaching and holding tasks that may be much more difficult for them than is first apparent, whereas forelimb preferences in mammals (rodents and primates) are exhibited only when task complexity places sufficient demands for use of the lateralized neural circuitry [e.g., for tool using in chimpanzees; Hopkins & Cantalupo, 2005].

ASSOCIATION BETWEEN HAND PREFERENCE AND EXPRESSED BEHAVIOR

Now let us consider the marmosets' hand preferences again and ask what they do tell us. Some experiments have shown that whether an individual has a left- or right-hand preference for picking up food and taking it to the mouth is linked to the behavior shown in novel settings. When marmosets that are left-handed for picking up food are taken from the home cage in a small nesting box and placed in the box with the door open in a novel room with novel objects, they are more reluctant to leave the box than are right-handed marmosets [Cameron & Rogers, 1999]. Once they have emerged from the box, the left-handers move around the room and interact with the novel objects less than right-handers. This result repeated a previous study by Hopkins and Bennett [1994] showing that left-handed chimpanzees interact with novel toys less than right-handed chimpanzees. These results have recently been confirmed by Braccini and Caine [2006] in similar tests using Geoffroy's marmosets (*Callithrix geoffroyi*): the right-handed marmosets were more likely to approach and sniff at novel foods and froze for longer on hearing the calls of a predator (a hawk).

Hence, hand preference for picking up food is associated with temperament or, more specifically, approach/withdrawal. The parsimonious explanation for this association is that preferred use of the left hand to pick up food indicates a balance towards using the right hemisphere (consistent with the crossed control of the forelimbs and hands), and vice versa. If primates have a general pattern of hemispheric specialization similar to that of humans, as I consider very likely, these findings can be explained on the Davison model [Davidson & Hugdahl, 1995] of approach being under left-hemisphere control and withdrawal under right-hemisphere control. In line with this, a body of evidence shows that the right hemisphere is specialized for fear expression [summarized in Rogers, 2002a]. It is, therefore, not surprising that a left-handed primate expresses more fear and a lesser likelihood of approaching novel stimuli.

In other words, the primate's preferred hand on the simple task of picking up food pieces and taking them to the mouth tells us something important about the hemisphere that individual will use in particular contexts. Here, it is important to note that individual marmosets maintain the same hand preference for performing this task across most of their lifespan [Hook & Rogers, 2000]: hence, the behavioral characteristics associated with hand preference are likely to be consistent aspects of individual behavior. Therefore, hand preference can be used to predict approach/withdrawal and perhaps other characteristics of temperament.

These behavioral associations with the hand preferred to pick up food in a relaxed context include social interactions. Westergaard et al. [2003, 2004] have shown that left- and right-hand preferring rhesus macaques exhibit different levels of dominance/subordinance and aggression. In their study, males with a right-hand preference received higher levels of grooming, were less likely to be submissive and received fewer aggressive interactions than males with a left-hand preference [Westergaard et al., 2003]. The opposite was the case for females [Westergaard et al., 2004].

Indirectly, the association between hand preference and particular behavioral characteristics of the individual indicates that the species (population) has hemispheric specialization (i.e., the majority of individuals having a lateralized brain in the same direction). In other words, if left-handers express one particular constellation of behavior and right-handers express another, then the population must have the same direction of hemispheric specialization. Note that to measure hand preferences alone would provide no indication of whether or not the population or species has a lateralized brain: a species may be comprised of equal numbers of left- and right-hand preferring individuals and still be strongly directionally biased in brain function. Only by investigating whether there is or there is not an association between preferred hand use and expressed patterns of behavior can we decide whether or not the species has a lateralized brain. This presents researchers studying hand preferences in primates in the wild with a problem if they wish to discover whether the species has hemispheric specialization, since to look only at hand preferences cannot lead to a conclusive answer.

We might now ask why some species, or populations, of primates are comprised of approximately equal numbers of left- and right-handed individuals for simple reaching and holding of food (i.e., antisymmetric), whereas others have a directional bias for left- or right-handedness. Hook [2004] examined the evidence for hand preferences in primates and found evidence of antisymmetry and directional asymmetry of various degrees in different species and groups, tested on various tasks and in differing laboratory contexts, and in some cases in the natural environment [and add the recent study of hand preferences in wild indri by Rigamonti et al., 2005]. It would seem that the expressed hand preferences in a population depend on the contextual demands for use of the left or right hemisphere. If expression of the functions of the right hemisphere is beneficial for a given species in a given context (i.e., to be cautious, or reactive, in the face of potential danger instead of being proactive), then we might expect to see directional bias for left-handedness and vice versa. In other environments, it might well be advantageous to maintain equal numbers of left- and right-handed individuals, with their associated temperaments, in the population since this would maintain a potential for rapid change under changing conditions. Hence, hand preferences of primate species in the wild may well depend on the ecological niche of the species and vary with environmental change.

A recent study of hand preferences in humans may suggest that similar processes operate even in our species, albeit while still maintaining overall right handedness. Faurie and Raymond [2004] have shown that the frequency of left-handedness is strongly and positively correlated with the incidence of homicides across traditional societies. Left-handedness ranges from 3 to 27% across the least to most violent and warlike of human societies. Having mentioned this, I recognize that there might well be alternative explanations for this association between hand preference and levels of violence in different societies (e.g., levels of reporting of crime, variation in type of violence, etc.). In interpreting these results, one also needs to take into consideration differences between societies in the degree to which they train against left-hand use. The finding of Faurie and Raymond [2004] is interesting in the light of my discussion above but preliminary and requiring further and broader investigation.

STRENGTH OF HAND PREFERENCE

I have considered the problem that direction of the distribution of hand preferences alone cannot tell us whether or not a species has a lateralized brain, but what can we say about *strength* of hand preference? In my opinion, that is an equally unreliable measure for deciding anything about species-level lateralization. Certainly, the strength of hand preference must indicate the degree of bias between activity of the hemispheres while the animal is performing a given task, those with a stronger preference having a greater bias towards the use of one hemisphere or the other than those with a weaker bias, but the individual with a weak, or no, hand preference may have a strongly lateralized brain. It may merely be more inclined to bring both hemispheres to bear on the task as opposed to one or the other hemisphere. Alternatively, individuals with a strong preference may allocate separate functions to each hemisphere and carry them out in parallel, whereas those with a weak preference divide the processing between hemispheres to a lesser degree.

A study completed recently in my laboratory supports the above reasoning [Piddington & Rogers, unpublished data; Rogers, 2006]. The hand preferences of marmosets to pick up food and take it to the mouth were determined and 11 females with a spread of strengths of hand preferences were tested singly on a test requiring them to detect a model predator (three different "predators" were used on repeated tests) while they were foraging. We found a significant negative correlation between latency to detect the predator and strength of hand preference, whereas the direction of hand preference had no effect on this result. Marmosets with weaker hand preferences took longer than those with stronger hand preferences to detect the predator. One could say that they were less efficient than the marmosets with strong hand preferences in performing the two tasks at the same time. In fact, when the marmosets were tested only with the predator and not required to forage at the same time, there was no relationship between strength of hand preference and latency to detect the predator.

Here, we are reminded not only of the results obtained with chicks (above) but also of a similar finding by McGrew and Marchant [1999]: they found that wild chimpanzees with stronger hand preferences displayed when they fished for termites from a mound by inserting a probe into the holes were more efficient in eating the termites than chimpanzees with weaker hand preferences. Termite fishing requires coordinated use of both limbs/hands, one to insert the probe and the other to act as a surface across, which to wipe the probe with its clinging termites before the chimpanzee scoops them up with its lips. Using each hand consistently for these functions is clearly the most efficient way of performing the task. The explanation for our result with marmosets has to be somewhat different since it depends on coordinated cognitive, not specifically motor, tasks (foraging and detecting a predator). If foraging demands use of neural circuits in the left hemisphere and predator detection on circuits in the right hemisphere (as we know for other species; see above), we might argue that the simultaneous performance of both aspects of the test is more efficient in individuals that have a stronger separation between functions of the two hemispheres. These separate functions might be well coordinated between the hemispheres (via the corpus callosum) but performed in parallel (i.e., strongly lateralized).

Strength of hand preference in our marmosets was also associated with their reactivity to the model predator. Those with a stronger hand preference not only responded to the model predator after a shorter latency but also exhibited higher levels of mobbing behavior than marmosets with weaker hand preferences. The strongly lateralized marmosets were confrontational and proactive, behaving in ways known to lower the levels of their stress hormone, cortisol, [Cross & Rogers, 2006], whereas the weakly lateralized marmosets were reactive/passive copers. Westergaard et al. [2001] have reported a similar behavioral result; they found a significant correlation between strength of hand preference and stress reactivity in rhesus macaques.

Hence, although somewhat limited so far, the evidence from research on primates is consistent with that on chicks, and also fish: it shows that having a strongly lateralized brain enables the individual animal to carry out foraging and predator detection simultaneously. Non-lateralized or weakly lateralized individuals cannot conduct these two tasks efficiently and would be at a selective disadvantage in the natural environment. In fact, this relationship between strength of limb preference and behavior might apply more broadly to animals that do not have hands but use their paws to manipulate or stabilize objects, as shown by a recent study relating paw preference to reactivity to fear-inducing sounds in dogs. Branson and Rogers [2006] determined the strength of paw preference in dogs to hold a cylindrical object, a Kong, while they obtained food from it and found that those dogs without a significant preference for one paw over the other were more likely to show excessive fear responses to fireworks and thunderstorms than either left- or right-pawed dogs. This might be explained according to the above findings in chicks and marmosets in paradigms requiring attention to two tasks simultaneously, if inhibition of response to fear-inducing sounds requires shifting attention to other less-threatening stimuli, since less lateralized dogs might be expected to be less able to do this. In fact, chicks that are not lateralized show higher levels of fear, expressed as distress calls, than lateralized chicks when they are tested with the task requiring simultaneous food search and detection of a model predator [Dharmaretnam & Rogers, 2005].

WHY DOES DIRECTIONAL LATERALIZATION EXIST?

The section on "Advantage of being lateralized" discusses the cognitive advantage of having a lateralized brain and this may well explain why individuals are lateralized, but it provides no explanation for those examples of lateralization that show a directional bias. In fact, it is reasonable to assume that there would be certain disadvantages that ensue from most individuals in the group (or species) being lateralized in the same direction. For example, predators might take advantage of their prey's left eye bias for detecting and reacting to their presence. Also, prey may be overlooked if they happen to be on the animal's left side. Presumably, therefore, some clear advantage of directional lateralization must overcome such disadvantages.

Vallortigara and I have proposed that social interactions among members of the same species are enhanced by directional lateralization [Vallortigara & Rogers, 2005]. For example, an individual might avoid being attacked by another by approaching the latter on its right side and so avoid its more antagonistic left side (right hemisphere). We have argued that directional lateralization is an advantage when one lateralized individual has to interact with another lateralized individual. Hence, directional lateralization should be present in social species only. One study that compared shoaling and non-shoaling species of fish supports this hypothesis: Bisazza et al. [2000] found that shoaling species showed directional bias for turning, whereas non-shoaling species had antisymmetrical distributions of turning preferences (individuals were lateralized but not in the same direction).

Of course, one could argue that embryological constraints limit certain forms of lateralization to develop as directional biases but many of the examples that I have discussed show sufficient flexibility that this seems to be unlikely. Added to this, experience in early life can modulate the presence or absence of lateralization, even when certain underlying asymmetries (e.g., orientation of the avian embryo in the egg) are genetically fixed. Therefore, it is likely that, when directional biases occur, they do so as a result of positive selection, as a stable strategy.

This hypothesis might be tested within one species by comparing social behavior in groups comprised of individuals lateralized in the same direction and groups comprised of either symmetrical individuals or ones in which half are lateralized on one direction and half in the other. Comparison of groups of chicks exposed to light before hatching with groups of dark-incubated chicks is one way to do this and, using this technique with a number of small groups of chicks, Rogers and Workman [1989] found that groups of chicks exposed to light before hatching form more stable social hierarchies than groups of chicks hatched from eggs incubated in the dark. The hierarchy was measured in terms of access to a localized source of food each day over the first two weeks of life. Chicks ranked as being at the top of the social hierarchy gain the most access to the food and those at the bottom of the hierarchy gain the least access to food. Since we know that the rank order determined in this way matches rank orders determined using other measures of competition between chicks [Rogers & Astiningsih, 1991], this reflects an important aspect of the social structure in chicks. In the groups comprised of light-exposed chicks, the same chicks remained at the top of the social hierarchy and the same ones remained at the bottom across the days of testing. In the groups of dark-incubated chicks, the chicks ranked at the bottom of the hierarchy varied across days: a hierarchy was present but it was less rigid and less stable from day to day. Of course, whether it is advantageous for young chicks to form more or less stable hierarchies may depend on a range of conditions in the wider social environment of the flock and the environmental conditions. Hens forced to leave their nests more frequently, either to search for food or because they are driven off, would be more likely to have offspring with directional lateralization (light exposed) and these offspring would form more rigid social hierarchies. As a consequence, survival of the highest-ranking chicks might be ensured at the expense of the lowest-ranking ones. This presents itself as a potentially successful response to low availability of food or conditions of social stress. I realize that this is just one possible scenario but this one, and others, can now be tested experimentally.

If enhancement of social interactions is the explanation for directional (group-level) lateralization, it is possible that individuals will be lateralized in the same direction as other individuals for those functions involved in social interactions but not for those functions that have no direct role in social behavior. This may explain why, for example, marmosets display directional lateralization for facial expressions, eye used in viewing, and ear used to attend to vocalizations, but not in hand preferences for picking up food. The hypothesis can be tested on different species and in differing social arrangements.

CONCLUSION

Among the vertebrate species, there are many examples of directional lateralization, as well as lateralization at the level of the individual but not biased in the same direction for the majority of individuals in the group (or species). A number of the directional lateralities known for a range of vertebrate species have the same pattern, which suggests that they are homologous. In fact, I have suggested that the fundamental pattern of lateralization is homologous and evolved with the first vertebrates. The recent evidence for lateralization in invertebrates is intriguing, and it offers an instructive comparison to lateralization in vertebrates, but insufficient examples are available to draw any conclusions about its homology or homoplasy with lateralization in vertebrates.

Evidence now available from vertebrates demonstrates a clear advantage in having a lateralized brain, in that lateralized individuals are able to perform more than one task simultaneously, whereas non-lateralized ones cannot do so. Likely, the same advantage of increased neural/cognitive capacity achieved by lateralization applies to both vertebrates and invertebrates. What emerges clearly here is the fact that the hypothesis that lateralization enhances the brain's capacity, once applied only to humans, pertains very broadly to organisms. Increasing the size of the brain (or cortex) is, of course, one solution to demands for enhancing cognitive capacity but allocating one set of functions to one hemisphere and another set to the other hemisphere is a common solution to the problem and one adopted both between and within species. Also within a species with the genetic potential for lateralization, ontogenetic processes can enhance or diminish the expression of lateralization to result in what seems to be ecologically relevant outcomes.

Regardless of whether the various types of lateralization in vertebrates are homologous or not, lateralization appears to be an evolutionary stable strategy [see Vallortigara & Rogers, 2005, for more on this; also Vallortigara, 2006]. Saying so does not mean that genetic determinants of lateralization are primary; as investigation of the development of lateralization in model species has shown, the influences of genes, hormones, experience and stimulation from the environment all interact to produce the expressed outcome of lateralization for any given function or collection of associated functions. So far, the interactive influences of these factors have been examined in two model systems only and during ontogeny only. The stage has been set for investigating not only the ontogeny of lateralization in other species but also looking to see whether rather similar interactions may occur in adulthood and also during ageing.

Strength of lateralization, including the absolutes of its presence or absence, has been shown to correlate with certain aspects of cognitive capacity, but this alone does not explain those examples of lateralization for which the majority of individuals within the group/species are lateralized in the same direction. An advantage of directional lateralization in social interactions has been proposed as the explanation for the latter. This hypothesis asks us to consider the social/solitary behavior of not only the species as a whole but also within the individual (i.e., whether the lateralized function(s) being investigated are used in social interactions or not). Until now, we have looked for consistencies of lateralization as species traits and for common patterns of lateralization between species. Now it might be worthwhile to consider the variations themselves and to study the processes that may modulate the expression of lateralization, as possible adaptations to environmental demands or constraints.

The discovery of lateralization in a wide range of vertebrate species and the availability of new techniques to examine its expression and development has provided us with a strong and broad base for comparison with lateralization in humans. By seeing lateralization of the human brain, not as an isolated phenomenon at the pinnacle of an assumed evolutionary process, I believe we are now in a position to advance understanding of human cerebral lateralization, and its changes with development and ageing.

REFERENCES

Ades, C.; Ramires, N. Asymmetry of leg use during prey handling in the spider *Scytodes globula* (Scytodidae). Journal of Insect Behavior 15: 563–570, 2002.

Andrew, R. J. Origins and evolution of lateralization. Pp. 94–125 in L. J. Rogers; R. J. Andrew, eds. Comparative Vertebrate Lateralization. Cambridge University Press, New York, 2002.

Andrew, R. J.; Johnston, A. N. B.; Robins, A.; Rogers, L. J. Light experience and the development of behavioural lateralization in chicks II. Choice of familiar versus unfamiliar model social partner. Behavioural Brain Research 155: 67–76, 2004.

Andrew, R. J.; Rogers, L. J. The nature of lateralization in tetrapods. Pp. 94–125 in L. J. Rogers; R. J. Andrew, eds. Comparative Vertebrate Lateralization. Cambridge University Press, New York, 2002.

Andrew, R. J.; Tommasi, L.; Ford, N. Motor control by vision and the evolution of cerebral lateralisation. Brain and Language 73: 220–235, 2000.

Bisazza, A.; Cantalupo, C.; Capocchiano, M.; Vallortigara, G. Population lateralization and social behavior: A study with sixteen species of fish. Laterality 5: 269–284, 2000.

Bisazza, A.; Cantalupo, C.; Robins, A.; Rogers, L. J.; Vallortigara, G. Right pawedness in toads. Nature 379: 408, 1996.

Bisazza, A.; Rogers, L. J.; Vallortigara, G. The origins of cerebral asymmetry: A review of evidence of behavioral and brain lateralization in fishes, reptiles and amphibians. Neuroscience and Biobehavioral Reviews 22: 411–426, 1998.

Box, H. Observations on spontaneous hand use in the common marmoset (*Callithrix jacchus*). Primates 18: 395–400, 1977.

Braccini, S.N., Caine, N.G. The link between hand preference, response to novelty, and fearfulness in captive marmosets (*Callithrix geoffroyi*). Abstracts of the American Society of Primatologists, August 2006, number 51, 2006.

Bradshaw, J. L.; Rogers, L. J. The Evolution of Lateral Asymmetries, Language, Tool Use, and Intellect. San Diego, Academic Press, 1993.

Branson, N.; Rogers, L. J. Relationship between paw preference strength and noise phobia in *Canis familiaris*. Journal of Comparative Psychology 120: 176–183, 2006.

Broad, K. D.; Mimmack, M. L.; Kendrick, K. M. Is right hemisphere specialization for face discrimination specific to humans? European Journal of Neuroscience 12: 731–741, 2000.

Brown, C.; Gardner, C.; Braithwaite, V. A. Population variation in lateralized eye use in the poeciliid *Brachyraphis episcope*. Proceedings of the Royal Society London B 271(Suppl.): S455–S457, 2004.

Byrne, R. A.; Kuba, M.; Griebel, U. Lateral asymmetry of eye use in *Octopus vulgaris*. Animal Behaviour 64: 461–468, 2002.

Byrne, R. A.; Kuba, M.; Meisel, D. V. Lateralized eye use in *Octopus vulgaris* shows antisymmetrical distribution. Animal Behaviour 68: 1107–1114, 2004.

Cameron, R.; Rogers, L. J. Hand preference of the common marmoset (*Callithrix jacchus*): Problem solving and responses in a novel setting. Journal of Comparative Psychology 113: 149–157, 1999.

Casey, M. B. Developmental systems, evolutionarily stable strategies, and population laterality. The Behavioral and Brain Sciences 28: 592–593, 2005.

Casey, M. B.; Karpinski, S. The development of postnatal turning bias is influenced by prenatal visual experience in domestic chicks (*Gallus gallus*). The Psychological Record 49: 67–74, 1999.

Casey, M. B.; Lickliter, R. Prenatal visual experience influences the development of turning bias in bobwhite quail chicks (*Colinus virginianus*). Developmental Psychobiology 32: 327–338, 1998.

Casey, M. B.; Martino, C. Asymmetrical hatching behaviours influence the development of postnatal laterality in domestic chicks (*Gallus gallus*). Developmental Psychobiology 34: 1–12, 2000.

Casey, M. B.; Sleigh, M. Cross-species investigations of prenatal experience, hatching behavior, and postnatal behavioral laterality. Developmental Psychobiology 35: 84–91, 2001.

Casperd, J. M.; Dunbar, R. I. M. Asymmetries in the visual processing of emotional cues during agonistic interactions in gelada baboons. Behavioural Processes 37: 57–65, 1996.

Collins, R. L. On the inheritance of direction and degree of asymmetry. Pp. 41–71 in S. D. Glick, ed. Cerebral Lateralization in Nonhuman Species. Academic Press, New York, 1985.

Costa, A. P.da; Leigh, A. E.; Man, M -S.; Kendrick, K. M. Face pictures reduce behavioural, autonomic, endocrine and neural indices of stress and fear in sheep. Proceedings of the Royal Society B, 2004. First Cite doi:10.1098/rspb.2004.2831.

Corballis, M. C. The Lopsided Brain: Evolution of the Generative Mind. Oxford University Press, Oxford, 1991.

Corballis, M. C. The trade-off between symmetry and asymmetry. The Behavioral and Brain Sciences 28: 594–595, 2005.

Cowell, P. E.; Denenberg, V. H. Development of laterality and the role of the corpus callosum in rodents and humans. Pp. 274–305 in L. J. Rogers; R. J. Andrew, eds. Comparative Vertebrate Lateralization. Cambridge University Press, New York, 2002.

Cowell, P. E.; Waters, N. E.; Denenberg, V. H. The effects of early environment on the development of functional laterality in Morris maze performance. Laterality 2: 221–232, 1997.

Cross, N. J.; Rogers, L. J. Mobbing vocalizations as a coping response in the common marmoset. Hormones and Behavior 49: 237–245, 2006.

Crow, T. J. The search for a critical event. Pp. 197–216 in T. J. Crow, ed. The speciation of modern *Homo sapiens*. Oxford University Press, Oxford, 2002.

Crow, T. J. The cerebral torque and directional asymmetry for hand use are correlates of the capacity for language in *Homo sapiens*. The Behavioral Brain Sciences 28: 595–596, 2005.

Csermely, D. Lateralization in birds of prey: Adaptive and phylogenetic considerations. Behavioral Processes 67: 511–520, 2004.

Dadda, M.; Bisazza, A. Does brain asymmetry allow efficient performance of simultaneous tasks? Animal Behaviour 72: 523–529, 2006.

Davidson, R. J. The functional neuroanatomy of emotion and affective style. Trends in Cognitive Science 3: 11–21, 1999.

Davidson, R. J.; Ekman, P.; Saron, C. D.; Senulis, J. A.; Friesen, W. V. Approach-withdrawal and cerebral asymmetry: Emotional expression and brain physiology I. Journal of Personality and Social Psychology 58: 330–341, 1990.

Davidson, R. J.; Hugdahl, K. Brain Asymmetry. MIT Press, Cambridge, 1995.

Deckel, A. W. Laterality of aggressive responses in *Anolis*. Journal of Experimental Zoology 272: 194–200, 1995.

Denenberg, V. H. Hemispheric lateralization in animals and the effects of early experience. Behavioral and Brain Sciences 4: 1–49, 1981.

Denenberg, V. H. Commentary: Is maternal stimulation the mediator of handling effect in infancy? Developmental Psychobiology 34: 1–3, 1999.

Denenberg, V. H. Behavioral symmetry and reverse asymmetry in the chick and rat. The Behavioral and Brain Sciences 28: 597, 2005.

Denenberg, V. H.; Fitch, R. H.; Schrott, L. M.; Cowell, P. E.; Waters, N. S. Corpus callosum: Interactive effects of infantile handling and testosterone in the rat. Behavioral Neuroscience 105: 562–566, 1991.

Denenberg, V. H.; Garbanati, J. A.; Sherman, G. F.; Yutzey, D. A.; Kaplan, R. Infantile stimulation induces brain lateralization in rats. Science 2201: 1115–1150, 1978.

Deng, C.; Rogers, L. J. Social recognition and approach in the chick: Lateralization and effect of visual experience. Animal Behaviour 63: 697–706, 2002a.

Deng, C.; Rogers, L. J. Factors affecting the development of lateralization in chicks. Pp. 206–246 in L. J. Rogers; R. J. Andrew, eds. Comparative Vertebrate Lateralization. New York, Cambridge University Press, 2002b.

Dharmaretnam, M.; Andrew, R. J. Age- and stimulus-specific use of right and left eyes by the domestic chick. Animal Behaviour 48: 1395–1406, 1994.

Dharmaretnam, M.; Rogers, L. J. Hemispheric specialization and dual processing in strongly versus weakly lateralized chicks. Behavioural Brain Research 162: 62–70, 2005.

Dharmaretnam, M.; Vijitha, V.; Priyadharshini, K.; Hashini, T.; Vathany, K. Ground scratching and preferred leg use in domestic chicks: Changes in motor control in first two weeks post-hatching. Laterality 7: 371–380, 2002.

Dukas, R. Causes and consequences of limited attention. Brain, Behavior and Evolution 63: 197–210, 2004.

Dukas, R.; Kamil, A. C. The cost of limited attention in blue jays. Behavioral Ecology 11: 502–506, 2000.

Dunaif-Hattis, J. Doubling the Brain. New York, Peter Lang, 1984.

Ehret, G. Left hemisphere advantage in the mouse brain for recognizing ultrasonic communication calls. Nature 325: 249–251, 1987.

Evans, C. S.; Evans, L.; Marler, P. On the meaning of alarm calls: Functional references in an avian vocal system. Animal Behaviour 46: 23–28, 1993.

Faurie, C.; Raymond, M. Handedness, homicide and negative frequency-dependent selection. Proceedings of the Royal Society B, published online, 2004. First Cite doi:10.1098/rspb.2004.2926.

Fitch, R. H.; Cowell, P. E.; Schrott, L. M.; Denenberg, D. H. Corpus callosum: Ovarian hormones and feminization. Brain Research 542: 313–317, 1991.

Güntürkün, O. The ontogeny of visual lateralization in pigeons. German Journal of Psychology 17: 276–287, 1993.

Güntürkün, O. Asymmetry pays: Visual lateralization improves discrimination success in pigeons. Current Biology 10: 1079–1081, 2000.

Güven, M.; Elalmis, D. D.; Binokay, S.; Tan, U. Population-level right-paw preference in rats assessed by a new computerized food-reaching test. International Journal of Neuroscience 113: 1691–1705, 2003.

Halpern, M. E.; Guntürkun, O.; Hopkins, W. D.; Rogers, L. J. Lateralization of the vertebrate brain: Taking the side of model systems. The Journal of Neuroscience 9: 10351–10357, 2005.

Harris, L. Footedness in parrots: Three centuries of research, theory, and mere surmise. Canadian Journal of Physiology 43: 369–396, 1989.

Hauser, M. D. Right hemisphere dominance for the production of facial expressions in monkeys. Science 261: 475–477, 1993.

Hauser, M. D.; Akre, K. Asymmetries in the timing of facial and vocal expressions by rhesus monkeys: Implications for hemispheric specialization. Animal Behaviour 61: 391–400, 2001.

Heuts, B. A.; Brunt, T. Behavioral left-right asymmetry extends to arthropods. The Behavioral and Brain Sciences 28: 601–602, 2005.

Hews, D. K.; Castellano, M.; Hara, E. Aggression in females is also lateralized: Left-eye bias during aggressive courtship rejection in lizards. Animal Behaviour 68: 1201–1207, 2004.

Hews, D. K.; Worthington, R. A. Fighting from the right side of the brain: Left visual field preference during aggression in free-ranging male tree lizards (*Urosaurus ornatus*). Brain, Behavior and Evolution 58: 356–361, 2001.

Hook, M. A. The evolution of lateralized motor functions. Pp. 325–370 in L. J. Rogers; G. Kaplan, eds. Comparative Vertebrate Cognition: Are Primates Superior to Non-Primates? Kluwer Academic, New York, 2004.

Hook, M. A.; Rogers, L. J. Development of hand preferences in marmosets (*Callithrix jacchus*) and effects of ageing. Journal of Comparative Psychology 114: 263–271, 2000.

Hook-Costigan, M. A.; Rogers, L. J. Hand, mouth and eye preferences in the common marmoset (*Callithrix jacchus*). Folia Primatologica 64: 180–191, 1995.

Hook-Costigan, M. A.; Rogers, L. J. Lateralized use of the mouth in production of vocalizations by marmosets. Neuropsychologia 36: 1265–1273, 1998.

Hopkins, W. D.; Bennett, A. Handedness and approach-avoidance behaviour in chimpanzees. Journal of Experimental Psychology: Animal Behavior Processes 20: 413–418, 1994.

Hopkins, W. D.; Cantalupo, C. Individual and setting differences in the hand preferences of chimpanzees (*Pan troglodytes*): A critical analysis and some alternative explanations. Laterality 10: 65–80, 2005.

Howard, K. J.; Rogers, L. J.; Boura, A. L. A. Functional lateralisation of the chicken fore-brain revealed by use of intracranial glutamate. Brain Research 188: 369–382, 1980.

Jarman, P. J. The development of the dermal shield in impala. Journal of Zoology 166: 349–356, 1972.

Kells, A. R.; Goulson, D. Evidence of handedness in bumblebees. Journal of Insect Behavior 14: 47–55, 2001.

Lawerence, E. S. Vigilance during 'easy' and 'difficult' foraging tasks. Animal Behaviour 33: 1373–1375, 1985.

Letzkus, P.; Ribi, W. A.; Wood, J. T.; Zhu, H.; Zhang, S-W.; Srinivasan, M. V. Lateralization of olfactory learning in the honeybee *Apis mellifera*. Current Biology 16: 1471–1476, 2006.

Levy, J. The mammalian brain and the adaptive advantage of cerebral asymmetry. Annals of the New York Academy of Sciences 299: 264–272, 1977.

Lippolis, G.; Bisazza, A.; Rogers, J.; Vallortigara, G. Lateralization of predator avoidance responses in three species of toads. Laterality 7: 163–183, 2002.

Lippolis, G.; Westman, W.; McAllan, B. M.; Rogers, L. J. Lateralization of escape responses in the striped-faced dunnart, *Sminthopsis macroura* (Dasyuridae: Marsupialia). Laterality 10: 457–470, 2005.

Lyons, D. M.; Afarian, H.; Schatzberg, A. F.; Sawyer-Glover, A.; Moseley, M. E. Experience-dependence asymmetric variation in primate prefrontal morphology. Behavioural Brain Research 136: 51–59, 2002.

Maarouf, F. D. L.; Roubertoux, P. L.; Carlier, M. Is mitochondrial DNA involved in mouse behavioural laterality? Behavior Genetics 29: 311–318, 1999.

MacNeilage, P. F.; Studdert-Kennedy, M. G.; Lindblom, B. Primate handedness reconsidered. The Behavioral and Brain Sciences 10: 247–263, 1987.

Marchant, L. F.; McGrew, W. C. Laterality of limb function in wild chimpanzees of Gombe National Park: Comprehensive study of spontaneous activities. Journal of Human Evolution 30: 427–443, 1996.

McGrew, W. C.; Marchant, L. F. Laterality of hand use pays off in foraging success in wild chimpanzees. Primates 40: 509–513, 1999.

McManus, I. C. Unity in the wild variety of nature, or just variety? The Behavioral and Brain Sciences 28: 606–608, 2005.

Metcalfe, N. B.; Huntingford, F. A.; Thorpe, J. E. Predation risk impairs diet selection in juvenile salmon. Animal Behaviour 35: 931–933, 1987.

Mikløsi, A.; Andrew, R. J. Right eye use associated with decision to bite in zebra fish. Behavioural Brain Research 105: 199–205, 1999.

Møller, A. P.; Swaddle, J. P. Asymmetry, Developmental Stability, and Evolution. Oxford University Press, Oxford, 1997.

Monaghan, P.; Pollman, S. Division of labour between the hemispheres for complex but not simple tasks: An implemented connectionist model. Journal of Experimental Psychology: General 132: 379–399, 2003.

Pascual, A.; Huang, K-L.; Nevue, J.; Préat, T. Brain asymmetry and long-term memory. Nature 427: 605–606, 2004.

Reist, J. D.; Bodaly, R. A.; Fudge, K. J.; Cash, K. J.; Stevens, T. V. External scarring of whitefish, *Coregonus nasus* and *C. clupeaformis* complex, from the western Northwest Territories, Canada. Canadian Journal of Zoology 65: 1230–1239, 1987.

Rigamonti, M. M.; Spiezio, C.; Poli, M. D.; Fazio, F. Laterality of manual function in foraging and positional behavior in wild indri (*Indri indri*). American Journal of Primatology 65: 27–38, 2005.

Robins, A.; Lipollis, G.; Bisazza, A.; Vallortigara, G.; Rogers, L. J. Lateralization of agonistic responses and hind-limb use in toads. Animal Behaviour 56: 875–881, 1998.

Robins, A.; Rogers, L. J. Lateralised prey catching responses in the toad (*Bufo marinus*): Analysis of complex visual stimuli. Animal Behaviour 68: 567–575, 2004.

Robins, R.; Rogers, L. J. Lateralized visual and motor responses in the Green tree frog (*Litoria caerulea*). Animal Behaviour 72: 843–852, 2006.

Rogers, L. J. Lateralisation in the avian brain. Bird Behaviour 2: 1–12, 1980.

Rogers, L. J. Light experience and asymmetry of brain function in chickens. Nature 297: 223–225, 1982.

Rogers, L. J. Light input and the reversal of functional lateralization in the chicken brain. Behavioural Brain Research 38: 211–221, 1990.

Rogers, L. J. The Development of Brain and Behaviour in the Chicken. Wallingford, CAB International, 1995.

Rogers, L. J. Early experiential effects on laterality: Research on chicks has relevance to other species. Laterality 2: 199–219, 1997.

Rogers, L. J. Evolution of hemispheric specialisation: Advantages and disadvantages. Brain and Language 73: 236–253, 2000.

Rogers, L. J. Lateralization in vertebrates: Its early evolution, general pattern and development. Pp. 107–161 in P. J. B. Slater; J. Rosenblatt; C. Snowdon; T. Roper, eds. Advances in the Study of Behavior, Vol. 31. Academic Press, San Diego, 2002a.

Rogers, L. J. Lateralized brain function in anurans: Comparison to lateralization in other vertebrates. Laterality 7: 219–239, 2002b.

Rogers, L. J. Cognitive and social advantages of a lateralized brain. Pp. 129–139 in Y. B. Malashichev; W. Deckel, eds. Behavioral and Morphological Asymmetries in Vertebrates. New York, Landes Bioscience, 2006.

Rogers, L. J.; Andrew, R. J. (eds). Comparative Vertebrate Lateralization. Cambridge University Press, New York, 2002.

Rogers, L. J.; Andrew, R. J.; Burne, T. H. J. Light exposure of the embryo and development of behavioural lateralisation in chicks: I Olfactory responses. Behavioural Brain Research 97: 195–200, 1998.

Rogers, L. J.; Andrew, R. J.; Johnston, A. N. B. Light experience and the development of behavioural lateralisation in chicks. III. Learning to distinguish pebbles from grains. Behavioural Brain Research 177: 61–69, 2007.

Rogers, L. J.; Anson, J. M. Lateralisation of function in the chicken fore-brain. Pharmacology, Biochemistry and Behavior 10: 679–686, 1979.

Rogers, L. J.; Astiningsih, K. Social hierarchies in very young chicks. British Poultry Science 32: 47–56, 1991.

Rogers, L. J.; Bolden, S. W. Light-dependent development and asymmetry of visual projections. Neuroscience Letters 121: 63–67, 1991.

Rogers, L. J.; Deng, C. Light experience and lateralization of the two visual pathways in the chick. Behavioural Brain Research 98: 277–287, 1999.

Rogers, L. J.; Deng, C. Corticosterone treatment of the chick embryo affects light-stimulated development of the thalamofugal visual pathway. Behavioural Brain Research 159: 63–71, 2005.

Rogers, L. J.; Rajendra, S. Modulation of the development of light-initiated asymmetry in chick thalamofugal visual projections by oestradiol. Experimental Brain Research 93: 89–94, 1993.

Rogers, L. J.; Sink, H. S. Transient asymmetry in the projections of the rostral thalamus to the visual hyperstriatum of the chicken, and reversal of its direction by light exposure. Experimental Brain Research 70: 378–384, 1988.

Rogers, L. J.; Workman, L. Light exposure during incubation affects competitive behaviour in domestic chicks. Applied Animal Behaviour Science 23: 187–198, 1989.

Rogers, L. J.; Workman, L. Footedness in birds. Animal Behaviour 45: 409–411, 1993.

Rogers, L. J.; Zappia, J. V.; Bullock, S. P. Testosterone and eye-brain asymmetry for copulation in chickens. Experientia 41: 1447–1449, 1985.

Rogers, L. J.; Zucca, P.; Vallortigara, G. Advantage of having a lateralized brain. Proceedings of the Royal Society London B 271(Suppl.): S420–S422, Biology Letters, Published on line May 2004, 2004.

Rothe, H. Handedness in the common marmosets (*Callithrix jacchus*). American Journal of Physical Anthropology 64: 417–433, 1973.

Seligmann, H.; Beiles, H.; Werner, Y. L. More injuries in left-footed individual lizards and Sphenodon. Journal of Zoology, London 260: 129–144, 2003.

Schwarz, I. M.; Rogers, L. J. Testosterone: A role in the development of brain asymmetry in the chick. Neuroscience Letters 146: 167–170, 1992.

Sherman, G. F.; Garbanati, J. A.; Rosen, G. D.; Hoffman, M.; Yutsey, D. A.; Denenberg, V. H. Lateralization of spatial preference in the female rat. Life Sciences 33: 189–193, 1983.

Sherman, G. F.; Garbanati, J. A.; Rosen, G. D.; Yutsey, D. A.; Denenberg, V. H. Brain and behavioral asymmetries for spatial preference in rats. Brain Research 192: 61–67, 1980.

Sovrano, V. A.; Dadda, M.; Bisazza, A. Lateralized fish perform better than nonlateralized fish in spatial reorientation tasks. Behavioural Brain Research 163: 122–127, 2005.

Sun, T.; Walsh, C. A. Molecular approaches to brain asymmetry and handedness. Nature Reviews 7: 655–662, 2006.

Tang, A. C.; Verstynen, T. Early life environment modulates 'handedness' in rats. Behavioural Brain Research 131: 1–7, 2002.

Tommasi, L.; Gagliardo, A.; Andrew, R. J.; Vallortigara, G. Separate processing mechanisms for encoding geometric and landmark information in the avian hippocampus. European Journal of Neuroscience 17: 1695–1702, 2003.

Tommasi, L.; Vallortigara, G. Footednesss in binocular and monocular chicks. Laterality 4: 89–95, 1999.

Tommasi, L.; Vallortigara, G. Hemispheric processing of landmark and geometric information in male and female domestic chicks (*Gallus gallus*). Behavioural Brain Research 155: 85–96, 2004.

Vallortigara, G. The evolutionary psychology of left and right: Costs and benefits of lateralization. Developmental Psychobiology 48: 418–427, 2006.

Vallortigara, G.; Andrew, R. J. Differential involvement of right and left hemisphere in individual recognition in the domestic chick. Behavioral Processes 33: 41–58, 1994.

Vallortigara, G.; Cozzutti, C.; Tommasi, L.; Rogers, L. J. How birds use their eyes: Opposite left-right specialisation for the lateral and frontal visual hemifield in the domestic chick. Current Biology 11: 29–33, 2001.

Vallortigara, G.; Rogers, L. J. Survival with an asymmetrical brain: Advantages and disadvantages of cerebral lateralization. The Behavioral Brain Sciences 28: 575–633, 2005.

Vauclair, J.; Fagot, J. Manual specialization in gorillas and baboons. Pp. 193–205 in J. P. Ward; W. D. Hopkins, eds. Primate Laterality: Current Behavioral Evidence of Primate Asymmetries. New York, Springer, 1993.

Walker, S. F. Lateralization of functions in the vertebrate brain: A review. British Journal of Psychology 71: 329–367, 1980.

Waters, N.; Denenberg, V. H. Analysis of two measures of paw preference in large population of inbred mice. Behavioural Brain Research 63: 195–204, 1994.

Weiss, D. J.; Ghazanfar, A. A.; Miller, C. T.; Hauser, M. D. Specialized processing of primate facial and vocal expressions: Evidence for cerebral asymmetries. Pp. 480–530 in L. J. Rogers; R. J. Andrew, eds. Comparative Vertebrate Lateralization. New York, Cambridge University Press, 2002.

Westergaard, G. C.; Chavanne, I. D.; Houser, L.; Cleveland, A.; Snoy, P. J.; Suomi, S. J.; Higley, J. D. Biobehavioural correlates of hand preference in free-ranging female primates. Laterality 9: 267–285, 2004.

Westergaard, G. C.; Chavanne, I. D.; Lussier, I. D.; Houser, L.; Cleveland, A.; Suomi, S. J.; Higley, J. D. Left-handedness is correlated with CSF monoamine metabolite and plasma cortisol concentrations, and with impaired sociality, in free-ranging adult male rhesus macaques (*Macaca mulatta*). Laterality 8: 169–187, 2003.

Westergaard, G. C.; Lussier, I. D.; Suomi, S. J.; Higley, J. D. Stress correlates of hand preference in rhesus macaques (*Macaca mulatta*). Developmental Psychobiology 38: 110–115, 2001.

Zappia, J. V.; Rogers, L. J. Light experience during development affects asymmetry of fore-brain function in chickens. Developmental Brain Research 11: 93–106, 1983.

This series of pictures depicts Reid, a chimpanzee from the Yerkes National Primate Research Center, throwing a poly-vinyl chloride pipe to observers located in a tower about 10 meters above him. Note that Reid stands bipedally as he prepares to throw, highlighting the role of posture on hand use in nonhuman primates. (Photograph by William Hopkins.)

The Evolution of Hemispheric Specialization in Primates
W.D. Hopkins (Editor)

PRESENT STATUS OF THE POSTURAL ORIGINS THEORY

Peter F. MacNeilage
Department of Psychology, University of Texas at Austin, Austin, TX, USA

ABSTRACT

In 1987, MacNeilage et al. argued that evidence of population-level handedness was evident in nonhuman primates, a claim that stood in strong contrast to the prevailing scientific view. Based on this paper, MacNeilage et al. went on to propose the postural origin of handedness theory. Since 1987, a plethora of studies have been conducted in nonhuman animals, notably primates. This chapter updates the extant data on handedness in nonhuman primates as it relates to the Postural Origins theory of handedness and argues that many aspects of the original theory are still supported by the new findings.

Keywords: posture, handedness, language, frame content theory

INTRODUCTION

The Postural Origins (PO) theory was originally an attempt to explain the origin, and subsequent course, of handedness in primates [MacNeilage et al., 1987, 1988, 1991; MacNeilage, 1991, 1993]. Later, an attempt was made to embed primate handedness and other primate specializations in a conception of the evolution of hemispheric specializations in vertebrates in general [MacNeilage, 1998a]. In this paper, I will concentrate on the implications for the theory of scores of studies of handedness in nonhuman primates conducted since the theory was first formulated in 1987, many of which were designed to test the theory. I will conclude that while a number of individual studies have failed to find hand preferences in particular groups of primates, and a few positive findings remain problematical, the original conception of primate handedness has been supported in its main outlines, and it remains the only conception of the evolution of handedness in *all* primates. In addition, I will add a few further comments on the broader conception of vertebrate-wide hemispheric specializations, according to which there is a left-hemisphere specialization for routine action control complementary to a right-hemisphere specialization for emergency reactions.

My colleagues Lindblom and Studdert-Kennedy and I originally became interested in the question of primate handedness in the context of another evolutionary theory regarding complex action systems – the "Frame/Content" theory of evolution of speech [MacNeilage, 1998b, 2007; MacNeilage & Davis,

ISSN: 1936-8526 DOI: 10.1016/S1936 8526(07)05003-8

2005; MacNeilage et al., 1984]. According to this theory, speech evolved a mode of organization whereby "content" elements – consonants and vowels – could be programed into syllable-structure "frames." An initial conception regarding the origin of this mode of organization [since abandoned – see MacNeilage, 1998b, for an alternative view] was that it could have evolved from a particular form of bimanual coordination – the kind in which an object, such as a plastic container full of pills, is held in one hand while the lid of the container is removed and a pill is taken by the other hand. In such a mode, the support hand provides the frame, and operations of the other hand constitute content elements.

As we knew that a number of nonhuman primates had good bimanual coordination, we turned our attention to the primate handedness literature, expecting that it would throw light on the phylogeny of human handedness and its relation to speech evolution. To our surprise, we found that nobody thought that other primate populations had any sign of a human-like right-hand preference – that is, a tendency to prefer the right hand for manual actions in general. The accepted authority on this question was J. M. Warren. In a number of review papers [e.g., Warren, 1977, 1980; Warren & Nonneman, 1977], he concluded unequivocally that handedness in other primates differed from human handedness in two ways. First, right- and left-handers were about equal in number, but there were also many animals with no hand preference. Second, certain animals showed a tendency to prefer different hands for different tasks.

Primate handedness seemed, then, to be a dead issue. But in looking into this question, we found that Warren's obituaries were a bit premature. We found some instances of statistically significant population-level tendencies toward hand preference that were not included in the Warren reviews. He had never reviewed more than 13 studies at a time. We reviewed 45, and published our results in the peer-review journal *Behavioral and Brain Sciences* [MacNeilage et al., 1987]. Before summarizing our review and the PO theory that arose from it, I will make a few brief orienting remarks about primates, focusing on the more unfamiliar prosimians – the most direct descendents of the earliest forms [for a primate taxonomy, see Purvis, 1995].

A defining characteristic of the 234 species constituting the primate order [Rowe, 1996] is that, with a few exceptions [including, notably, ourselves], they are *arboreal* mammals. They live in the trees. The first primates may have begun to evolve in the explosive mammalian radiation that followed the extinction of the dinosaurs about 65 million years ago [mya]. The original true primates – called "prosimians" – evolved as early as 50 mya, descending from an animal something like a modern squirrel or marmot. In what could be described as a three-dimensional experiment, these animals came to live entirely in the trees.

Prosimians are visually distinguishable from other primates by their pointed, dog-like faces. Most of the 50 species of modern prosimians [Rowe, 1996] live in central Africa and Madagascar, but some live in Southeast Asia as well. About 40 mya a distinct "simian" [monkey and ape] taxon evolved from prosimian ancestors, giving rise to two major groups of monkeys – New World monkeys, living in the Americas, and Old World monkeys, living in Eurasia and Africa [Byrne, 2000]. Apes perhaps began to evolve at about 30 mya and a split between the lesser apes – gibbons and siamangs – and a line leading to the great apes – a group now comprising orangutans, gorillas, chimpanzees, and hominids – may have originated at about 20 mya. One of the most important findings of the past couple of decades is that, according to DNA dating evidence, we and our nearest relations, chimpanzees, may have shared a common ancestor as recently as 4–6 mya [Byrne, 2000].

THE 1987 REVIEW: DATA AND THEORY – SUMMARY OF FINDINGS

In the 1987 review of primate handedness studies, we found over a dozen statistically significant hand preferences. Perhaps most surprisingly, we found some evidence of a *left*-hand preference when reaching for food in *prosimians*. In one study [Subramoniam, 1957] it was reported informally that all 8 of a group of slender lorises consistently picked up small insects with their left hand. In another study, Sanford et al. [1984] found that of 25 bushbabies reaching for food from their typical vertical posture [they cling to vertical supports], 12 made over 80% of the reaches with their left hand, and the preferences of the other 13 were distributed rather evenly across other preference levels.

In monkeys, we also found some signs of a left-hand food-reaching preference. But there was also evidence that some monkeys – in one case, a group with a left-hand reaching preference [Beck & Barton, 1972] – preferred the *right* hand for tasks that involved manipulation – that is, hand skill. There was very little data on hand preferences in the great apes.

POSTURAL CONSIDERATIONS

On the basis of this evidence, my colleagues and I formulated, in 1987, the beginnings of what later became the "Postural Origins" theory of the evolution of hemispheric specializations in primates [MacNeilage, 1991]. Posture received central status because of our conviction that hand movements are typically made in the context of a support configuration of all or much of the whole body, and that such support must be an important variable in an arboreal habitat. The connotation of "posture" that we had in mind was "the position or bearing of the body whether characteristic or assumed for a special purpose" [*Merriam-Webster's Collegiate Dictionary*, 11th ed.]. The term applies equally to static and dynamic aspects of action control. This connotation of "posture" should be distinguished from the one more common in neuropsychology, which has to do specifically with the maintenance of the position of the body relative to gravity. This latter usage of the term is more like that of the everyday term "balance."

The postural implications of hand use in our sense of the term become particularly clear when contemplating the *origin* of primates. The arboreal three-dimensional experiment of ancestral primates had profound implications for posture. As Cartmill has pointed out, in contrast to the relatively homogeneous support provided for an animal in terrestrial, aquatic, or aerial media, the supports provided by an arboreal habitat are "[1] discontinuous, [2] limited and variable in width, [3] mobile, and [4] oriented at all possible angles to the gravity vector" [Cartmill, 1974, p 45]. Under these circumstances, practically every action must involve unique postural demands affecting all or most of the body.

The first task of the theory was an attempt to account for the somewhat paradoxical finding of left-handedness in prosimians, a finding that has been replicated numerous times since, as shown below. But some further background on prosimians is necessary before we can get to the theory.

PROSIMIANS: A LEFT HAND–RIGHT HEMISPHERE PREHENSION SPECIALIZATION?

The original prosimians are considered to have had a locomotor style known as "vertical clinging and leaping" [Fleagle, 1988]. At rest, they clung to vertical supports, and they moved around by leaping from one such support to the next. They have generally been considered unimanual predators [though see later discussion]. They caught live prey by grabbing it with one hand. Some modern small prosimians – for example, the bushbabies studied by Sanford et al. [1984] – are vertical clingers and leapers and unimanual predators, and in these traits are considered to "represent a conservative primate pattern carried forward into the present relatively unchanged" [Eisenberg, 1981, p 161].

Unimanual predation is an extremely momentous adaptation when put in the context of the evolution of predation in vertebrates. For about half a billion years before this adaptation evolved, the business end of vertebrate predation was bilaterally symmetrical. Animals caught prey with a bilaterally symmetrical structure, the mouth or tongue. Now, for the first time, predation became *obligatorily asymmetrical*. On each individual occasion, one side of the body was occupied with the reaching movement while the other side provided the primary postural support. A somewhat important though limited analogy to this dual-purpose function can be seen in the design of cranes. The arm of the crane extends in space to pick up objects and move them around. But cranes also have a heavy metal block at the base in back of the cab area, which provides a necessary counterweight to the arm. In addition, the reaching movement of these animals was necessarily highly specialized. Present-day prosimians, including bushbabies, can reach for food with a movement faster than the human eye can see [Bishop, 1964]. Such movements are considered to be "ballistic" in the sense that once they are initiated, their trajectory cannot be modified. The question that needs to be asked is: Given the inevitability of the action asymmetry, did these animals tend to build a single specialized crane involving one particular active arm and support from the other side of the body, or did they build two, thus indulging in some redundancy?

The momentous nature of the adoption of a totally arboreal status is indicated by the fact that it was accompanied by two major anatomical adaptations. The first was *frontal vision*. The eyes moved from the sides of the face, as in squirrels, to the front of the face. The result was a big increase in the overlap between the fields of vision of the two eyes. This allowed better depth perception, accruing from the fact that each eye had a slightly different view of the same object, up to a distance of several feet. The second adaptation was the *prehensile hand* – the ability to enclose an object within a single forelimb extremity. It has often been thought that these two adaptations may have evolved to support locomotion in an arboreal habitat. To move from one branch to another, it is useful to have precise knowledge of how far away the target branch is and also to be able to grab it on arrival. But Cartmill [1974] has suggested that the two adaptations evolved primarily for predation. He pointed out, for example, that aerial predators, such as eagles and owls, also have frontal vision, while many other birds that move from branch to branch do not. And they have claws that are prehensile – not a common accompaniment to aerial or arboreal life.

In the light of the extreme specialization involved in unimanual predation, its necessarily asymmetric nature, and the evidence for a current leftwards bias in prosimian reaching, my colleagues and I suggested that these animals chose the one-crane option rather than the two-crane option. And we suggested that the arrangement was accompanied by two complementary hemispheric specializations – a specialization of the left hand–right hemisphere for unimanual predatory prehension, and a specialization of the right side of the body, controlled by the left hemisphere, for postural support [MacNeilage et al., 1987].

MONKEYS: A RIGHT SIDE–LEFT HEMISPHERE OPERATIONAL SPECIALIZATION?

As I mentioned, there was also evidence of left-hand reaching preferences in monkeys. This was found in marmosets by Box [1977], in a group of 10 stumptail macaques by Beck and Barton [1972], in 3 field studies of Japanese macaques, and in early stages of 4 studies of rhesus macaques by Ettlinger and his colleagues [for references to the latter 7 studies, see MacNeilage et al., 1987]. But Ettlinger and his colleagues also found a number of trends toward *right*-hand preference when the rhesus macaques had a chance to practice performing a task in which the lid of a container had to be removed to obtain food. In addition, as mentioned earlier, Beck and Barton found extremely strong right-hand preferences in the same group of stumptail macaques who favored the left hand for reaching, when they were faced with a different task, one that involved manipulation – for example, moving a hasp on a puzzle box with a single finger.

These studies suggested the possibility that monkeys may retain some traces of the left-hand reaching preference presumably present in their prosimian ancestors, but may have also developed some human-like tendency to prefer the right hand for tasks involving manual skill. Before I discuss how we tried to explain *this* development, I need to make some general comments on the evolution of monkeys.

One important thing that happened in the evolution of monkeys was the abandonment not only of vertical clinging and leaping as the primary locomotor adaptation but also of predation as a primary feeding strategy. The universal readoption of basic mammalian quadrupedalism as the primary mode of locomotion freed the forelimbs from the postural demands associated with vertical clinging. A wide range of feeding strategies evolved – from, at one extreme, almost completely folivorous (foliage-eating) animals, through those that are also frugivorous (fruit-eating), to those that are basically omnivorous and thus continue to practice at least some predation. Feeding strategies of these animals are correlated with their habitats, which range from entirely arboreal at various levels (from close to the ground to tree tops) to terrestrial, as with some baboons. To the degree that the now quadrupedal animals moved toward a terrestrial habitat, the overall body postural demands that accompanied hand action decreased.

Along with these developments in diet and habitat came a spectrum of associated developments in the anatomy and function of the hand. While some animals remain like the typical prosimians – whole-hand grabbers – others develop hand-internal control, i.e., the ability to control the different digits independently. Some even become capable of a human-like precision grip done in a human-like way, by opposing the thumb to the index finger. The ability for hand-internal control was often put to use for food processing, which included the frame/content mode – the form of bimanual coordination in which an object is held in one hand while the other operates on it. Animals take shells off nuts, peel fruit, remove sharp spines from caterpillars, etc. Some of the more omnivorous animals developed a new feeding strategy – invasive foraging. They basically took the environment apart to find food, breaking up rotten wood, or tearing off palm fronds to expose food in the interstice of the frond and the trunk, etc.

We suggested that the right-hand preference in monkeys may have evolved in the context of some of these adaptations. What might have happened was this: "With the abandonment of vertical clinging and the advent of more omnivory and more invasive foraging and manual food processing in many higher primates, the right side of the body, with its greater physical strength and its postural heritage of on-line control, may have become the operative side" [MacNeilage, 1998c, p 230]. A particular consequence of this may have been that the more the animal became free from the postural constraints of positioning itself in the arboreal habitat, the more the right hand would be favored in object manipulation.

GREAT APES

In our original review, we found only five studies of chimpanzees, gorillas, and orangutans in which there were more than a couple of animals. An early study by Finch [1941] showed no evidence of population-level hand preference in a group of 30 chimpanzees. Preilowski and Leder [1984] showed a similar lack of evidence in 6 lowland gorillas. Schaller [1963] reported a right-hand preference for initiating chest beating in 59 of 72 gorillas. Fischer et al. [1982] found right-hand preference in three tasks for all four of the chimpanzees studied. Dimond and Harries [1984] reported a left-hand preference for face touching in all three great ape taxa, which was significant for chimpanzees and gorillas. Thus, overall we did not find enough evidence of handedness in great apes to place them in an evolutionary perspective. But their intermediate evolutionary status between monkeys and humans, present-day representatives of which showed some signs of right-handedness, and right-handed humans, led us to suggest that great apes might have a handedness pattern intermediate between those of monkeys and humans.

PRIMATE HANDEDNESS SINCE 1987

There were 40 peer responses to our review paper in *Behavioral and Brain Sciences*. They were spread over three issues of the journal, indicating a considerable interest in the question. The main response was one of incredulity. People were simply unwilling to believe that other primates had handedness, and attempted to bolster their conclusion with a variety of critical comments. But despite this early skepticism, our paper has, in the words of Parnell [2001], become "widely accepted as having breathed new life into this area of study" (p 365). Over a hundred studies have been conducted since our review, and, as already noted, many have tended to bear out our original hypotheses. Let us now review this more recent work.

PROSIMIANS: VERIFICATION OF THE LEFT-HAND PREFERENCE

A large number of studies in Jeanette Ward's laboratory at Memphis State University have replicated the finding that prosimians are predominantly left-handed. These studies have involved not only vertical clingers and leapers such as the bushbaby and the sifaka, but also a number of quadrupedal species, particularly lemurs. Most of the studies have been of simple reaching. In a review of 8 studies, Ward et al. [1993] found a total of 143 examples of left preference, 107 of right preference, and 52 of no preference. In addition, a study of 22 lemurs by Masataka [1989] showed 20 of them to prefer the left hand for 4 different actions. Mason et al. [1995] also found a group of 15 sifakas to be predominantly left-handed, though not significantly so. In a study of 40 black lemurs, Colquhoun [1994] found that 48 of 53 instances of reaching into tree holes by black lemurs, usually a drinking maneuver, involved the left hand. Finally, in a study by Rigamonti et al. [2005], a population of 16 wild indris [a large vertical clinger and leaper] was found to favor the left hand over the right for reaching, though again not significantly so. There were 5 significant left-hand preferences and 2 significant right-hand preferences.

One further aspect of the Rigamonti et al. study should be mentioned. They also tested the prediction from the PO theory that at rest, the right arm should be typically positioned higher than the left because the right side is providing more postural support. They found that 4 animals showed a significant right-hand preference and no animal had a significant left-hand preference. If one is permitted to put these two findings together – left reaching and right hand supporting during resting clinging – there is a 9:2 ratio of support for the PO theory, which is significant on a test of the binomial distribution [$p = 0.033$]. In an apparently related finding, Hook and Rogers [2002] found a tendency in monkeys [marmosets] to favor the right side of the body to initiate leaping and to favor the right hand and foot in initiation of terminal contact. Finally, one extremely strong and apparently posturally based left-hand preference should be noted. Forsythe et al. [1988] reported that although a group of 5 ruffed lemurs did not show particularly strong left-hand reaching preferences while reaching for food placed on the ground, members of the group made 514 out of 515 extended reaches for food floating in a moat with their left hands, with the aid of a counterposed right side.

Thus, there is now little doubt that modern prosimians tend to be left-handed. No study of prosimians has shown a significant right-hand preference. Consequently, the part of the PO theory that says that the original prosimians may have had a left hand–right hemisphere specialization for unimanual predation, though still speculative, remains viable.

One fact worthy of comment here is that this well-established result of prosimian left-handedness is virtually never mentioned in studies of handedness in monkeys or apes, and certainly not by those wedded to the anthropocentric perspective that handedness began in humans [e.g., Annett, 2002; Crow, 2004]. But it will ultimately need to be dealt with if we are to understand human handedness in the context of descent with modification. I am trying to provide a first step in this direction.

LEFT-HAND PREHENSION IN MONKEYS

The evidence we initially presented for a left-hand preference for prehension of food objects on the ground from a characteristically quadrupedal stance in monkeys was typically rather weak. But there have been a few confirmations of that tendency. First, consider Old World monkeys. Observing about a 2:1 left-to-right ratio in the well-known Koshima population of Japanese macaques, Watanabe and Kawai [1993] concluded that "feral monkeys display a general left hand preference in picking up wheat grains, washing wheat grains and washing sweet potatoes" (p 191). Westergaard et al. [1998] found a significant left-hand bias for quadrupedal reaching in rhesus macaques. As to New World monkeys, Laska [1996] found a significant [9:1] bias towards a left-hand preference in spider monkeys. And after reviewing a number of studies of prehension in marmosets, Hook-Costigan and Rogers [1996] concluded that "there may be a non-significant skew towards left hand preferences" (p 178).

In contrast to these findings of left-hand preferences for quadrupedal reaching in prosimians and monkeys, there have been no findings of left-hand preferences for quadrupedal reaching in the great apes, supporting our speculation that these taxa might be more closely related to humans than monkeys are.

A LEFT-HAND BALLISTIC SPECIALIZATION IN MONKEYS?

One study suggests that the specialized left-hand ballistic reaching capability that we postulated in prosimians may also be present in monkeys. King and Landau [1993] found that, while a group of 16 squirrel monkeys showed no overall hand preference for quadrupedal reaching for objects on the cage floor, a significant majority of them (13/16) showed a predominant left-hand preference for the presumably ballistic movement necessary for catching goldfish in a bowl or in a pool.

RIGHT-HAND PREFERENCES IN CAPUCHIN (*CEBUS*) MONKEYS

Capuchin monkeys are interesting from the point of view of evolution of primate handedness. Though primarily arboreal, they are considered to be the most versatile tool users among monkeys [Visalberghi & Fragaszy, 2002] – and have been observed to use tools in the wild [e.g., Fernandes, 1991]. Hook-Costigan and Rogers [1996] concluded that "Overall, there appears to be a tendency toward increasing right hand preferences in feeding in *Cebus apella* with 3 out of 6 studies reporting significant group biases" (p 183). The studies showing right-hand preferences were by Masataka [1990], Gomperts and Costello [1991], and Parr et al. [1997]. The other three studies reported non-significant findings. Capuchin monkeys have also been found to be right-handed for bimanual coordination in one study [Spinozzi et al., 1998], although there was a non-significant finding in another [Westergaard & Suomi, 1996]. In addition, in their review of hand preferences in New World primates, Hook-Costigan and Rogers [1996] also found a consensus of evidence for a right-hand preference in cebus tool use.

In contrast to these findings, capuchin monkeys have been found to prefer the left hand in searching for and grasping food objects that were not visible, thus implicating haptic [tactile] perception [Lacreuse & Fragaszy, 1996; Parr et al., 1997]. As Hook-Costigan and Rogers [1996] point out, this result perhaps implicates a human-like right hemisphere spatial specialization.

Hook-Costigan and Rogers [1996] conclude that, "like humans, capuchins [*Cebus apella*] appear to be right handed for most tasks...." (p 198). Even more importantly, they conclude that "tool use and right handedness may have evolved before bipedalism, and well before the apes and, indeed, humans evolved" (p 195).

It should be added that another New World monkey, the tamarin, has been twice reported to favor the right hand for picking up and holding objects [Diamond & McGrew, 1994; King, 1995]. Hook-Costigan and Rogers [1996] note that these animals have a manipulative feeding strategy but to a lesser degree than capuchins.

POSTURAL INFLUENCES ON HAND PREFERENCE

Beyond its more general function in rekindling interest in primate handedness, the PO theory has played a heuristic role in emphasizing the importance of posture in primate hand preferences. For example, Hook-Costigan and Rogers [1996], in a review of 32 studies of hand preferences in New World monkeys, concluded that "the posture adopted during hand use appears to be the most influential variable on [*sic*] handedness ..." (p 200).

Westergaard et al. [1998] made a review of comparisons of quadrupedal and bipedal reaching in 9 nonhuman primate species. The bipedal condition considered here involved the placement of the non-reaching hand against the vertical wall of the cage, apparently as an aid to balance. Their findings were as follows:

> In the present analysis, 3 of 9 non-human species [chimpanzees, gorillas, and tufted capuchins] exhibited significant population-level right-hand preferences for bipedal reaching, although none of these species exhibited a significant right-hand preference for quadrupedal reaching. A significant shift toward greater use of the right hand for bipedal versus quadrupedal reaching was also noted in rhesus macaques [from a population-level left-hand preference, noted earlier, to a lack of population-level bias], and trends toward greater use of the right hand were noted for orangutans and bonobos. Trends toward greater use of the left hand for bipedal versus quadrupedal reaching were noted for gibbons, squirrel monkeys, and bush babies.

The authors concluded that "Highly dextrous [manipulative] primates, such as great apes, tufted capuchins and rhesus monkeys [but not less dextrous species such as bushbabies, squirrel monkeys and gibbons (see Torigoe, 1985)] are predisposed towards greater use of the right hand when in bipedal posture" (p 62).

Some further comments can be added. In two of the three instances in which bipedal posture evoked a left-hand preference – galagos and gibbons – their vertical orientation was the one that these highly arboreal taxa normally favor. Galagos are vertical clingers and leapers, and gibbons are brachiators; that is, they hang vertically by hand from horizontal supports, move by swinging from hand to hand, and feed by unimanual prehension. In both cases, a left-hand prehension specialization in bipedal (vertical) posture would be expected from the PO theory. Apparently, this preference is nullified when they are in their less customary quadrupedal posture. The trend toward a left-hand preference in the bipedal condition in squirrel monkeys is part of a complicated picture. Although these animals tended toward a left-hand preference when they were fed in a location that involved the free hand being placed against the side of the cage, there was a very strong right-hand preference (21/7) when the animals were required to cling to the edge of the cage with one hand and their feet while reaching for food [King & Landau, 1993]. In the case of the appearance or enhancement of a *right*-hand preference in the bipedal condition in monkeys and apes, perhaps the postural instability associated with the adoption of the uncharacteristic bipedal posture *induced* the favored asymmetrical mode.

A number of authors have suggested that the trend toward human right-handedness was either initially invoked (if one is anthropocentrically inclined) or strengthened (apparently the more likely alternative) by habitual bipedalism associated with locomotion. It is generally agreed that this condition is extremely posturally demanding. In the picturesque metaphor of primate locomotion specialist Alan Walker; "From an engineering perspective, bipedalism is a ridiculous answer to the need for locomotion, posing problems akin to balancing an apple on top of a moving pencil" [Walker & Shipman, 1996, p 199]. This characterization certainly suggests that bipedalism may have severely constrained the performance of tasks requiring whole body asymmetry in hominids, perhaps forcing them into the mode of asymmetry more supported by their phylogeny. Here, again the one-crane metaphor seems more appropriate than the two-crane metaphor.

It should be noted parenthetically that in two of the great ape species studied here, Shafer has shown *general-purpose* handedness in captive groups of animals. She tabulated instances of hand use in 47 gorillas [Shafer, 1993] and 14 bonobos [Shafer, 1997] across a broad spectrum of acts – touch self, touch other, hit, throw, eat, manipulate small and large object, dig/sift, gestures, and a miscellaneous category. She found more right-hand than left-hand use for all categories in both species. A total of 26 gorillas were significantly right-handed and 10 were significantly left-handed. For bonobos, 11 were significantly right-handed and 2 were significantly left-handed.

BIMANUAL COORDINATION

Another aspect of handedness that involves integration of action of both sides of the body, though often without much overall bodily postural demand (e.g., while sitting), is bimanual coordination. Hopkins has invented an ingenious "Tube" test of bimanual coordination that presents essentially the same problem for many different species, thus facilitating their comparison [see Hopkins, 1995]. The test consists of a six-inch length of PVC pipe, an inch in diameter, in one end of which the experimenter can place a food reward, such as peanut butter. The animal must stabilize the pipe by holding it with one hand while extracting the food with the other. Population-level right-hand preferences for food extraction on this test have been found for several monkey and ape species.

This preference was found in rhesus monkeys by Westergaard and Suomi [1996]. It has also been found in capuchin monkeys by Spinozzi et al. [1998], though not by Westergaard and Suomi [1996]. Most recently, in a large-scale study of 104 baboons, Vauclair et al. [2005] found that although there was no overall hand preference for simple reaching, there was a significant excess of right-hand over left-hand preferences in the Tube test (52/33).

In the great apes, Hopkins et al. [2003] found a significant trend toward right-hand preferences in chimpanzees [replicating Hopkins' [1995] earlier finding] and a marginally significant trend in gorillas. But they found a significant *left*-hand preference in a group of 19 orangutans. Most recently, Hopkins and his colleagues have administered the Tube test to an astonishing total of 467 chimpanzees from three captive populations [Hopkins et al., 2004]. They found a significant tendency toward right-hand preferences that was relatively uniform across the 3 groups. There were 272 right preferences, 162 left preferences, and 33 without significant preferences. Thus, almost 60% of these animals had a right-hand preference, and the number of animals without preference was remarkably small.

These findings of right-hand preferences for bimanual coordination in monkeys and apes are of particular interest because bimanual-coordination tasks have the methodological advantage of forcing an animal to reveal which of the two hands it favors for the more demanding action, thus revealing the more specialized role of the contralateral hemisphere. The finding of so many right-hand preferences in monkeys and apes with relatively well-developed manipulative capabilities is important evidence for the PO theory.

RIGHT-HAND PREFERENCES FOR THROWING AND GESTURING

The chimpanzee right-hand preference apparently extends to social and communicative behavior. Chimpanzees like to throw things – at lab visitors, among other targets. Hopkins et al. [1993] have shown a significant tendency toward a right-hand preference for throwing in a group of 36 chimpanzees. Twenty-four animals were right preferent, 9 were left preferent, and 3 were ambipreferent.

With respect to gestures, Hopkins and Wesley [2002] have replicated an earlier study by Hopkins and Leavens [1998] to the effect that the right hand is preferred significantly more often than the left for food begs in chimpanzees. Hopkins and deWaal [1995] found a significant tendency toward use of the right hand for communicative gestures in bonobos. In addition, Hopkins and Cantero [2003] showed that the right-hand preference for gestural communication in chimpanzees is significantly enhanced when accompanied by a vocalization. The authors conclude that, "the lateralization of manual and speech systems of communication may date back as far as 5 million years ago" (p 55). In a remarkable recent finding with profound implications for the phylogeny of human communication as well as handedness, Meguerditchian and Vauclair [2006] found a significant right-hand preference for the communicative gesture of ground-patting in a group of 60 baboons.

HANDEDNESS IN CAPTIVITY AND IN THE WILD

The instances of right-hand preferences in captive populations reviewed here have, for the most part, not been paralleled by similar findings in the wild. Some researchers have concluded from this that the findings of population-level trends in captive animals are artifacts [e.g., McGrew & Marchant, 1997], although no satisfactory explanation for the nature of the artifact has been advanced. Imitation is an obvious possibility that can be raised, but that is certainly not the explanation, for a number of reasons. Probably the modal situation for gaining information used for manual movements would be a face-to-face one. Bekkering and Prinz [2002] summarize a number of experiments in which young *children* imitated manual movements of an adult facing them. One example of a task involved adults touching one of their own ears with either their left or right hand, making either a contralateral or an ipsilateral movement. They concluded that, "young children always moved to the correct goal such as an object or a particular ear to reach for, but widely ignored the agent (a particular hand to move with or the movement path ipsi or contralateral to the object)" (p 563). In addition, it appears that monkeys do not imitate. Visalberghi and Fragaszy [2002] came to this conclusion following a sustained program of research on the capuchin monkey, the monkey in which the strongest right-hand preferences have been observed. And they go on to say, "As capuchins do not imitate, we have no strong grounds at present for expecting that any other species of monkey will be able to do so" (p 493).

Another reason to believe that right-handedness in captivity is not derived from humans is the finding of Hopkins et al. [2004], in their study of 467 chimpanzees, that the tendency towards a right-hand preference was independent of whether the chimpanzees were reared by humans or not.

It is more likely that the circumstances of testing primates in captivity, such as the paucity of arboreal postural demands, and invariant modes of presentation of food and other objects on horizontal surfaces, afford an opportunity for animals to adopt stereotyped manual responses, and that their choice of hands under these circumstances indicates an intrinsic laterality. In addition, Hopkins and Cantalupo [2005] have noted that the characteristically small samples of animals used in studies in the wild reduce the statistical power of analyses.

Most importantly, Lonsdorf and Hopkins [2005] have now found strong evidence for two different kinds of population-level handedness in wild chimpanzees. In a sample of 17 animals fishing for termites, they showed that there were significantly more left-handed (12) than right-handed (4) or ambiguously handed (1) animals. In addition, when combining their results with those of other studies of 37 wild animals, a similar pattern with similar statistical significance was obtained. Most interestingly, in the light of the claim that hand preferences of captive animals are artifactual, they found that the pattern of hand preference in 29 captive animals fishing in artificial termite mounds did not significantly differ from the pattern observed in the wild.

Lonsdorf and Hopkins also pooled available data from two other instances of chimpanzee tool use in the wild – nut-cracking ($N = 63$) and wadge-dipping ($N = 16$). The latter task, also called leaf sponging, involves bunching up leaves and dipping them into cavities in tree trunks to obtain water. They found that the predominance of right-hand preference on these tasks resulted in a significantly different preference pattern from that found for termite-fishing, and that the right-hand preference tendency was significant for nut-cracking and approached significance for wadge-dipping. Another conspicuous exception to the typical null findings in the wild was a tendency toward right-hand preference in bimanual coordination tasks in wild gorillas that would have been significant if subject to a one-tailed statistical test, a test that would have been appropriate for an evaluation of the prediction of the PO theory [Byrne & Byrne, 1991].

The finding of right-hand preference tendencies for nut-cracking, wadge-dipping, and bimanual coordination in great apes in the wild is consistent with the PO theory. All tasks involve the coordinated use of both sides of the body. Nut-cracking involves considerable manual skill as well as the postural demands of force application. It is important to crack the nut but not obliterate it. Wadge-dipping also includes whole-body postural demands.

On the face of it, the finding of a left-hand preference for termite-fishing might seem contrary to the PO theory. But it is more likely that some factors not specifically dealt with in the theory are involved. A possibility consistent with the theory is that a human-like right-hemisphere spatial specialization is being used to conceptualize the relation between the termite tool and the non-visible internal space of the mound. In addition, it is well known that there is a right-hemisphere specialization for tactile perception in humans [Harada et al., 2004] and this sometimes shows up in a left-hand advantage in discrimination and identification tasks [e.g., Fagot et al., 1997]. Lacreuse and Fragaszy have repeatedly found a left-hand preference for object-directed haptic (touch-based) actions in capuchins in the absence of visual information [Lacreuse & Fragaszy, 1996, 1997, 1999]. A left-hand preference in haptic tasks has also been shown for spider monkeys [Laska, 1996] and rhesus monkeys [Fagot et al., 1991]. The result for the capuchins is most indicative of a right-hemisphere spatial specialization because, as noted earlier, they are typically right-handed for manipulative tasks. Lacreuse and Fragaszy [1999] concluded that, "left hand preferences for the haptic tasks may reflect a right hemisphere specialization to integrate the spatial and motor components of action" (p 65).

In contrast to these findings in monkeys, Lacreuse et al. [1999] found a *right*-hand preference for chimpanzees retrieving grapes from an opaque, partially occluded, water-filled bucket. But it is not clear from the description of this study how much visual information was available at the point of entry of the hand.

Another possibility is that a vigilance factor may be involved in termite-fishing because of the subtlety of the tactile difference between whether termites are on the tool or not. Heilman [1995] summarizes a number of studies suggesting a right-hemisphere advantage for vigilance in humans, though the specific question of tactile vigilance was not addressed. Unfortunately, little information is available on hand preferences for termite-fishing and nut-cracking in the same animals. The prediction of the PO theory for animals participating in both tasks is that left-hand preference for the former will tend to co-occur with right-hand preference for the latter.

SUMMARY

In summary, a large body of evidence now supports the hypothesis, arising from the PO theory, that prosimians show a generalized left-hand preference. Similarly, a large body of evidence supports the hypothesis that there tends to be a right-hand preference in monkeys and apes that is sensitive to postural variables and particularly marked in those with relatively well-developed manipulative skills. I conclude from this that the PO theory remains a viable conception of how handedness evolved in all primates, humans included. There is at present no alternative theory that has implications for the entire primate order.

THE QUESTION OF NON-SIGNIFICANT FINDINGS

Having said this, it is only fair to acknowledge that practically every finding of handedness is accompanied by one and sometimes more null findings. Some of these findings arise from studies with small numbers of animals, but this is not true of all such studies. Perhaps the main reason to continue to try to uphold the PO theory in the presence of these null findings is the evidence that there has almost certainly been a *phylogenetic flip-flop in asymmetry* – left and not right for prosimian prehension and (except for orangutans – see below) right but not left for skilled tasks in higher primates. The implications of this take us far beyond handedness as such.

HANDEDNESS RESULTS PROBLEMATICAL TO THE THEORY

One implication of the theory is that the use of the right hand in monkeys relative to the left should be greater to the degree that the animals make use of a terrestrial habitat rather than an arboreal one. However, findings that capuchin monkeys and tamarins, which are relatively arboreal species, tend to favor right hand for food reaching, and that macaque monkeys, which are more terrestrial in habitat, tend to favor the left hand, are not consistent with this assumption. One could perhaps attribute the right-hand preferences to the manual versatility of capuchin monkeys and tamarins, despite their arboreal habitat. But although macaque monkeys are also rather manually versatile, they nevertheless show some preference for left-hand prehension.

Orangutans are at present a rather problematical case for the theory. Earlier, after noting the Westergaard et al. [1998] trend toward a right-hand preference for bipedal reaching, I argued that, as in the case of other similar results, the demand for a bipedal posture in orangutans may have induced a basic preference to reveal itself. But the left-hand preference for bimanual coordination, which is perhaps the single primate handedness finding least compatible with the PO theory, calls this interpretation into question in their case. In addition, Rogers and Kaplan [1996] found a significant left-hand preference for manipulative face touching in orangutans [confirming the trend for simple face touching noted by Dimond and Harries [1984] and for scratching]. And they did not find a preference for reaching while hanging in a vertical position, a situation which could perhaps be analogous to the one for which Westergaard, Kuhn, and Suomi found a tendency toward a right hand preference. Thus, there is an overall tendency for orangutans to be left-handed, and though they are almost totally arboreal, they do have a respectable capacity for manual manipulation, which, according to the PO theory, should potentiate right-hand preferences.

CONTINUITY IN THE LEFT-HEMISPHERE SPECIALIZATION

Hook-Costigan and Rogers [1996] interpret the co-existence of predominant left-hand preferences in prosimians with predominant right-hand preferences in New World monkeys as possible evidence for "an evolutionary shift in hemisphere dominance" (p 201). This may not be the best interpretation of these results. From the postural origins point of view, rather than equate direction of hand preference with hemispheric dominance, we should note that hemispheric specialization may, in important ways, remain the same. According to the PO theory, the left hemisphere retains an important role in action control throughout primate evolution. But the role changes from being one of whole-body involvement in general, to one of also supporting a right-hand operational preference. And, as we will see later, the right hemisphere, rather than changing specializations, retains some vestige of the ballistic reaching specialization in humans. The retention of the left-hemisphere specialization is indicated by one particular line of evidence available in all major primate taxonomic groups – the tendency to initiate locomotion (quadrupedal and/or bipedal as appropriate) with the right side of the body. Instances of this have been observed in prosimians [Forsythe & Ward, 1987], monkeys [Hook & Rogers, 2002, mentioned earlier], four species of great apes [Heestand, 1986], bonobos [Hopkins & deWaal, 1995], and humans without marching experience [Seltzer et al., 1990]. Humans subject to marching experience tended to lead off with the left foot, as, by convention, this ritual demands. Hopkins et al. [1997] also found a significant tendency for neonatal chimpanzees to initiate crawling on the right side.

The PO theory requires an additional role for the *left* side of the body, in the transition from prosimians to monkeys, just as it does for the right side. But this role is new rather than being an extension of the previous capability. As the right side of the body becomes increasingly dominant for operations on the environment, the left side becomes increasingly important for background postural support. Consistent with this hypothesis, Spinazzola et al. [2003] have recently found that left-hemispheric damage in humans impairs whole-body praxis (skill), while right-hemispheric damage impairs whole-body posture.

EXTENSION OF THE POSTURAL ORIGINS THEORY: VERTEBRATE HEMISPHERIC SPECIALIZATIONS

In 1998, I suggested that the PO theory of primate handedness evolution might be expanded into a conception of the evolution of hemispheric specializations in vertebrates in general [see MacNeilage, 1998a, for a more detailed discussion]. I raised the possibility that there may be two complementary vertebrate-wide hemispheric specializations; a left-hemisphere specialization for whole-body control under routine circumstances, and a right-hemisphere specialization for emergency reactions under, in the extreme, life-threatening circumstances. Here I will briefly discuss these possibilities, adding some evidence that I was not aware of in 1998.

A LEFT-HEMISPHERE ROUTINE-ACTION SPECIALIZATION

One implication of the origin of a left-hemisphere routine-action control specialization in early vertebrates is that this already-existing left-hemisphere action specialization may have been put to use in the form of the right-side dominance associated with the clinging and leaping motor adaptation characteristic of everyday early prosimian life. This would have made the left hand more available for prehension of food objects. This suggestion is compatible with the recently stated hypothesis that the earliest primates may not have been predatory vertical clingers and leapers but instead quadrupedal small-branch feeders capitalizing primarily on plant life in the outer canopies of trees [Bloch & Boyer, 2002]. If so, then the left-hemisphere action-control capacity favoring right-sided postural support may have triggered the asymmetric reaching adaptation favoring the hand, which is the non-dominant hand for vertical clinging and leaping – the left hand – before the manual-predation specialization in vertical clingers and leapers, and its accompanying ballistic reaching capacity, evolved.

In arguing for a generalized role of the left hemisphere for routine body control, I am allying myself with the Russian psychologist Bianki [1988], who concluded that in all vertebrates "the left hemisphere mainly controls motor activity" (p 147). He reached this conclusion after reviewing an extremely large body of evidence, mainly about rats, accumulated by numerous Russian researchers. One important basis for his conclusion was the finding that anesthetizing the left hemisphere decreased general activity levels in rats while anesthetizing the right hemisphere increased them. He concluded that, "it is the left hemisphere that dominates in activating motor function whereas the right hemisphere inhibits it" (p 140). This conjecture regarding rats is in remarkable correspondence with the recent rather unequivocal conclusion that there is an inhibitory specialization in the right frontal cortex of humans [Aron et al., 2004].

Evidence supporting the existence of this left-hemisphere dominance for routine action in another major vertebrate taxon comes from work on birds, which have the convenient anatomical property of having almost entirely contralateral projections from the eyes to the brain. A number of studies [e.g., Andrew et al., 2000] report that chickens have better visual discrimination in pecking for food when guided by their right eye [and left hemisphere] than when guided by their left eye. In a study showing the implications of this finding for activity levels, Güntürkün and Hoferichter [1985] found that cutting the *left* forebrain bundle, which is the main outflow from the forebrain, markedly reduces the activity level of the birds when pecking at a target key to obtain food, while cutting the right forebrain bundle has no such effect.

Findings in motor development in humans and chimpanzees support the conclusion that they share a basic right-sided posture/action preference. Hepper et al. [1990] have shown that over 90% of human fetuses suck their right thumb in utero. Hepper et al. [2005] recently found that all 60 infants who sucked their right thumb in utero turned out to be right-handed. There is also a well-known rightward supine orientation preference in human neonates, correlated with subsequent handedness [Michel, 2002]. Güntürkün [2003] has found, by means of surreptitious observation of parting rituals in airport terminals that this rightward orientation preference tends to be shared by both participants in the act of kissing.

This orientation/action preference has precursors in chimpanzees. Bard et al. [1990] have found a significant tendency for newborn chimpanzees to bring their right hand to the mouth. Hopkins and Bard [1995] have found a significant rightward orientation preference in supine neonatal chimpanzees. Fagot and Bard [1995] have shown significantly stronger grips for the right hand and foot in neonatal chimpanzees. Imitation is obviously not a factor in these human-like neonatal rightward preferences of chimpanzees.

A LEFT-HEMISPHERE VOCAL-COMMUNICATION SPECIALIZATION

It is well known that there is a widespread left-hemisphere specialization for vocal communicative *action* in vertebrates ranging from fish [Fine et al., 1996] through amphibians [frogs – Bauer, 1993], songbirds [Arnold & Bottjer, 1985], non-primate mammals [Mongolian gerbils – Holman & Hutchinson, 1991], to monkeys [Hook-Costigan & Rogers, 1998]. It has been known for a long time that monkeys have a left-hemisphere *perceptual* specialization for vocal communication [Hauser & Andersson, 1994; Heffner & Heffner, 1984; Petersen et al., 1978; Poremba et al., 2004]. Recently, evidence for a left-hemisphere perceptual specialization has also been found in seals [Boye et al., 2005]. In the absence of an obvious reason why *communicative* action would be specialized in the left hemisphere independent of any other action, I suggested that this is further evidence for a left-hemisphere specialization for routine action control in general [MacNeilage, 1998a]. Certainly, any vocal-action specialization must have evolved in the context of postural configurations of the body rather than simply as a "talking head" specialization. The implication is that rather than the human specialization for vocal action in humans being derived from the left-hemisphere manual-control specialization, as is often proposed, they are both offshoots of a more general left-hemisphere specialization for the control of the whole body under routine circumstances. Thus, I would give the finding of Hopkins and his colleagues of a right-hand gestural preference in chimpanzees associated with vocalizations mentioned earlier, a hybrid explanation in terms of both the left-hemisphere communicative specialization and the right-hand operational specialization. The chimpanzee's right-hand throwing preference of Hopkins et al. [1993] would be more attributable to the latter specialization alone.

A POSTURAL LINK WITH THE LEFT-HEMISPHERE SPEECH SPECIALIZATION? EVIDENCE FROM HUMANS

Strong evidence for the relation between the vocal-action specialization and the specialization for whole-body postural configurations comes from an unlikely source – humans. There are now four studies showing that the hemispheric specialization for speech is more closely related to the hemisphere controlling whole-body action than to the hemisphere specialized for the control of manual skill [Searleman, 1980; Maki, 1990; Day & MacNeilage, 1996; Elias & Bryden, 1998]. These are all dichotic listening studies with the wildly counter-intuitive finding that in the 10% of the human population with mixed-limb preferences [5% right-handed left-footers and 5% left-handed right-footers], *language lateralization is more closely related to the foot preference than to the hand preference.* Foot preference is obviously related to a contralateral specialization for whole-body action control, because any selective use of the foot when in a bipedal stance must involve postural control of the whole body. A prediction arising from these findings is that, even though there are some dissociation between foot preference and manual-skill lateralization in humans, foot preference for manipulative acts in monkeys and apes will tend to be concordant with right-hand preferences. Both of these are presumed indicants of a left-hemisphere specialization for routine action control.

A RIGHT-HEMISPHERE AFFECTIVE SPECIALIZATION?

In a further broadening of the PO theory, I have suggested that complementary to the left-hemisphere routine action-control specialization there may have evolved a right-hemisphere specialization "for apprehension of the world in situations with survival-related risk" [MacNeilage, 1998a]. Further discussion of this possibility, echoed in the work of a number of other conceptions of right-hemispheric specializations [e.g., Liotti & Tucker, 1995; Robinson & Downhill, 1995; Wittling, 1995] must be limited here, but its implications for primate handedness deserve mention.

Considering fear in particular, Vallortigara and Rogers [2005] conclude that "A variety of different classes ... appear to be more reactive to predators seen in their left rather than right hemifield." In fact this phenomenon can be found in all major vertebrate taxonomic groups – fish, amphibians, reptiles, birds, and [primate and non-primate] mammals. According to Vallortigara and Rogers, these findings "suggest that predator escape and associated fear responses are controlled by the right side of the brain." (p 575). Fear also seems to have implications for hand use in monkeys. Hatta and Koike [1991] have found extremely strong left-hand preferences in adult female monkeys when removing infants from threatening situations, though not in routine actions that involve picking up food. Several studies by Westergaard and his colleagues [e.g., Westergaard et al., 2001] have shown the influence of affective variables on characteristic hand preference in monkeys with more left-hand use associated with negative affect.

A somewhat analogous dichotomy to the routine/emergency dichotomy of hemispheric function has been suggested by Davidson and his colleagues [e.g., Davidson, 1995]. They suggest that the left hemisphere is specialized for approach behavior while the right hemisphere is specialized for avoidance. This dichotomy has been shown to exist in human infants as well as adults, and there is also some evidence for it in monkeys. Again, with respect to handedness, Hopkins and Bennett [1994] found that right-handed chimpanzees approached and touched novel objects significantly more quickly than left-handed chimpanzees, suggesting a relationship between handedness and each of the two poles of the affective approach-withdrawal continuum. Cameron and Rogers [1999] obtained a similar result in marmosets.

From a PO standpoint, what inferences can be drawn from these various findings showing a tendency of negative affect to be associated with left-hand preference in other primates? It is possible that these instances of left-hand preference could be masking the basic evolutionary move toward right-handedness that has been postulated by the PO theory. They raise the possibility that, in instances of overall null findings in monkeys and apes, animals with right- and left-hand preferences may be showing these preferences for different reasons. While the animals with right-hand preferences may be revealing a basic evolutionary trend culminating in characteristic right-handedness, animals with left-hand preferences may be showing either situationally induced [e.g., Hatta and Koike's result] or characteristic preferences for using that hand for affective reasons [Hopkins & Bennett's 1994 result]. The likelihood that within group, left- and right-hand preferences are not simply a result of mirror-image lateralities comes from findings suggesting that the right-handers tend to show behavior more related to human-like manual skill than left-handers. For example, Costello and Fragaszy [1988] found more use of the prototypically human precision grip in capuchin monkeys that preferred the right hand than in those that preferred the left. Hopkins et al. [2002] found the same result in a group of 140 chimpanzees, and also found that the right-hand-preferent animals made fewer errors.

A HUMAN ATTENTIONAL DICHOTOMY

The hypothesis of evolution of a routine/emergency complementarity of the left and right hemispheres in vertebrates, though formulated from phylogenetic evidence, receives some support from studies of *human* performance on laboratory tasks amenable to brain imaging methodology. On the basis of such studies, Fox et al. [2006] conclude that "two attention systems exist in the human brain: a bilateral dorsal attentional system involved in top-down orienting of attention and a right lateralized ventral attentional system involved in reorienting attention in response to salient sensory stimuli." (p 10,046). Of course a top-down mode of organization would be characteristic under routine life circumstances. In contrast, the bottom-up stimulus-dependent mode would be invoked for what Corbetta and Shulman [2002] describe as "the detection of behaviorally relevant stimuli, particularly when they are salient and unexpected." (p 201). Corbetta and Shulman suggest that the ventral system "works as a 'circuit-breaker' for the dorsal system, directing attention to salient events."

The right-hemisphere lateralization of the ventral system in humans is what would be expected from the aspect of the PO theory that suggests that the right hemisphere has evolved a specialization for emergency reactions. In addition, it is possible that the top-down system, while found to be bilaterally symmetrical in what Fox et al. [2006] call "conventional task-response studies" (p 10,046) could be predominantly left lateralized under routine non-laboratory interactions with the environment, as expected from the PO theory, particularly if Gazzaniga's hypothesis that the left hemisphere, plays a role as the human brain's "interpreter" [Gazzaniga, 2000] is correct.

A HUMAN RESIDUE OF THE LEFT-HAND ACTION SPECIALIZATION?

According to the PO theory, the right hemisphere in prosimians, as in humans and perhaps other vertebrates, would be specialized for apprehension of unexpected stimuli, but prosimians [and apparently squirrel monkeys] also have a specialized left-handed response in predation situations, which could certainly be regarded as threatening. There is some evidence for the continued existence of such a specialized response capacity in humans under laboratory conditions. Carson et al. [1995] note that in reaction time tasks in which "movement preparation is not permitted in advance of the imperative stimulus," a number of studies have been "remarkably consistent" in showing a manual asymmetry favoring the left hand (p 151). Advantages cited ranges from 7.5 to 21 ms. Carson et al. [1995] replicated and extended these findings and concluded that,

"A putative mechanism for the left hand advantage for reaction time is a greater degree of right hemisphere engagement in the integration and feedforward of information about the position and orientation of an effector, relative to a target and the environment, prior to initiation of rapid movement (p 152)". Additional evidence of a possible link between the specialized right-hemisphere mechanism suggested for prosimians and modern human perceptuomotor capabilities is provided by the finding of a left visual field/right-hemisphere advantage for left hand responses to perceived movement trajectories in humans. Boulinguez et al. [2003] showed this advantage in a reaction time task involving prediction of coincidence between viewed ballistic trajectories of dots on a screen and a target dot. The possibility that these capacities reveal a residue of a formerly adaptive function is suggested by the fact that they do not seem to be useful to modern human function except perhaps for the left jab in boxing, and in fielding in baseball and cricket!

THE RIGHT-HEMISPHERE SPATIAL SPECIALIZATION

The best known specialization of the human right hemisphere is for spatial function [DeRenzi, 1982]. It is most clearly revealed in the left-sided neglect that often follows right hemisphere lesions. My colleagues and I suggested that a precursor to this specialization, perhaps used for the targeting phase preceding ballistic reaching, may have associated with the left-hand preference for unimanual predation in prosimians [MacNeilage et al., 1987]. In support of this possibility, Vallortigara and Rogers [2005] have pointed out that such a specialization "has been largely documented in birds ... and in mammals ...." (p 589). Corbetta and Shulman [2002] observe that in humans the anatomical correlates of neglect, in addition to being characteristically in the right hemisphere, are more associated with the anatomical loci of the ventral attentional system (particularly the parieto-temporal junction and inferior frontal cortex) than with the dorsal system. Thus, it appears that the spatial and bottom-up attentional systems are functionally related in humans, and this relationship could well have already existed in prosimians.

THE GLOBAL AND LOCAL SPECIALIZATIONS

In humans [Sergent, 1982] and in chimpanzees [Hopkins et al., 1997], there is a right-hemisphere specialization for processing global aspects of stimuli and a complementary left-hemisphere specialization for processing local aspects of stimuli. This is revealed with stimuli in which the three lines of a large letter (say "H") are formed by series of a small letter (say "L"). It is found in normal subjects that, while the left visual field (right hemisphere) can most readily detect the (global) "H," the right visual field (left hemisphere) can more readily detect the small (local) "L" [Sergent, 1982]. In a related result, patients with left-hemisphere lesions tend to successfully reproduce the global stimulus, but simply with a set of straight lines. In contrast, patients with right-hemisphere lesions tend to produce a cluster of instances of the local stimulus but without the global spatial organization called for by the large letter [Delis et al., 1986]. The ability to rapidly synthesize global patterns in the right hemisphere could be an aspect of its rapid emergency response capacity. To the contrary an ability to characterize the environment in detail, while in a top-down mode in the service of subsequent action, could be a valuable component of a left-hemisphere routine action capacity.

HUMAN HANDEDNESS/LANGUAGE AS A SALTATION

The biological definition of *saltation* given in *Merriam-Webster's Collegiate Dictionary* [11th ed.] is: "the origin of a new species or a higher taxon in essentially a single evolutionary step that in some esp. former theories is held to be due to a major mutation or to unknown causes." Despite the steadily accumulating evidence of primate homologs to human hemispheric specialization [e.g., Rogers & Andrew, 2002; Vallortigara & Rogers, 2005] and, in particular, anthropoid primate homologs to human handedness and left-hemispheric precursors to language, some theorists, notably Annett [2002] and Crow [2004], continue to see only a saltational genetic step to human handedness/language. In fact, this is probably still the most common view in the scientific community.

One might ask from this standpoint why language laterality is more related to foot preference than to hand preference, and why the left hand is faster at responses to unexpected events than the right hand. More generally, we must remember that, as Darwin [1859] said in *The Origin of Species* [1859–1952], "natura non facit saltum" (nature does not make jumps) (p 92). The saltational view of evolution common in the early 20th century supposedly resulting in the creation of what Goldschmidt [1940, p. 183], a proponent of this view, called "hopeful monsters," has been passé for a long time in evolutionary biology [Mayr, 1984]. And no evidence for a mutation resulting in a species-specific genetic basis for human handedness and language, together or apart, has yet been unearthed. Perhaps, more disturbing is that nothing is said by Annett or Crow about *exactly* what this supposed gene actually *did* to produce handedness/language. We are simply asked to believe that whatever these theorists believe human handedness and language to consist of (and neither of them have any particular expertise in the language sciences), a gene did it. Despite the natural human tendency to be anthropocentric, encouraged particularly by Descartes and Western religion, the time for this conception of the evolution of human left-hemispheric specializations has obviously passed.

CONCLUSION

Positive findings of primate handedness since 1987 have tended to uphold the PO theory. It is now clear that prosimians tend to have a left-hand preference for prehension, and that a right-hand preference for acts requiring skill is widespread among monkeys and apes. And although there have been many null findings regarding primate hand preferences, there have been very few positive findings counter to the theory. Results support the claim of the PO theory that there is a tendency, perhaps vertebrate-wide, for the left hemisphere to be specialized for action control under routine, non-threatening circumstances. Human right-handedness and left-hemisphere language control are considered to be separate offshoots of this more basic specialization. In contrast, the right hemisphere may be specialized for emergency reactions. Goldberg and Podell [1995] have made a similar suggestion regarding the organization of the left and right hemispheres of humans but at the cognitive level. They conclude that, "The right hemisphere appears to be critical for processing novel cognitive situations. The left hemisphere appears to be central to the processes mediated by well routinized representations and strategies" (p 93).

Obviously a number of aspects of the theory remain speculative, particularly in its expanded form. But I would prefer to believe Harrison and Byrne's [2000] conclusion that, over half a generation after its original formulation, the theory "even today ... remains a useful heuristic framework in which to study laterality in nonhuman primates" (p 13). Perhaps the most important contribution of the theory has been its role in encouraging the development of a sophisticated ethology of hand use in primates, bent on understanding the relation between hand preference and use of hands in various ecological contexts. This development will surely be valuable for the understanding of how humans evolved by descent with modification.

ACKNOWLEDGMENTS

This paper was prepared with the support of grant HD 27733-12 from the Public Health Service. I thank John Trimble for help in preparing the manuscript.

REFERENCES

Andrew, R. J.; Tommasi, L.; Ford, N. Motor control by vision and the evolution of cerebral lateralization. Brain and Language 73: 220–235, 2000.

Annett, M. Handedness and Brain Asymmetry. New York, Taylor and Francis Inc, 2002.

Arnold, A. P.; Bottjer, S. W. Cerebral lateralization in birds. Pp. 11–39 in S. D. Glick, ed. Cerebral Lateralization in Nonhuman Species. Orlando, Academic Press, 1985.

Aron, A. R.; Robins, T. W.; Poldrack, R. A. Inhibition and the right frontal cortex. Trends in Cognitive Science 8: 170–177, 2004.

Bard, K.; Hopkins, W. D.; Fort, C. Lateral bias in infant chimpanzees (*Pan troglodytes*). Journal of Comparative Psychology 104: 309–321, 1990.

Bauer, R. H. Lateralization of neural control for vocalization by the frog (*Rana pipiens*). Psychobiology 21: 243–248, 1993.

Beck, C. H.; Barton, R. L. Deviation and laterality of hand preference in monkeys. Cortex 8: 339–363, 1972.

Bekkering, H.; Prinz, W. Goal representations in imitative actions. Pp. 555–572 in K. Dautenhahn; C. L. Nehaniv, eds. Imitation in Animals and Artifacts. Cambridge, MA, MIT Press, 2002.

Bianki, V. L. The Right and Left Hemispheres of the Animal Brain. New York, Gordon and Breach, 1988.

Bishop, A. Use of the hand in lower primates. Pp. 133–255 in J. Buettner-Janush, ed. Evolutionary and Genetic Behavior of Primates. New York, Academic Press, 1964.

Bloch, J. I.; Boyer, D. M. Grasping primate origins. Science 298: 1606–1610, 2002.

Boulinguez, P.; Ferrois, M.; Graumer, G. Hemispheric specialization for trajectory perception. Brain Research: Cognitive Brain Research 16: 219–225, 2003.

Box, H. O. Observations on spontaneous hand use in the common marmoset (*Callithrix jacchus*). Primates 18: 395–400, 1977.

Boye, M.; Güntürkün, O.; Vauclair, J. Right ear advantage for conspecific calls in adults and subadults, but not infants, California sea lions (*Zalophus californianus*): Hemispheric specialization for communication? European Journal of Neuroscience 21: 1727–1732, 2005.

Byrne, R. W. The evolution of primate cognition. Cognitive Science 24: 543–570, 2000.

Byrne, R. W.; Byrne, J. E. Hand preferences in the skilled gathering tasks of mountain gorillas (*Gorilla g. beringei*). Cortex 27: 521–546, 1991.

Cameron, R.; Rogers, L. J. Hand preference of the common marmoset (*Callithrix jacchus*): Problem solving and responses in a novel setting. Journal of Comparative Psychology 113: 149–157, 1999.

Carson, R. G.; Chua, R.; Goodman, D.; Byblow, W. D. The preparation of aiming movements. Brain and Cognition 28: 133–154, 1995.

Cartmill, M. Pads and claws in arboreal locomotion. Pp. 45–83 in F. A. Jenkins, ed. Primate Locomotion. New York, Academic Press, 1974.

Colquhoun, I. C. Drinking behavior and left handed preference among *Eulemur macaco macaco* at Ambato Massif, Madagascar. American Journal of Physical Anthropology 18(Supplement): 69, 1994.

Corbetta, M.; Shulman, G. L. Control of goal-directed and stimulus-driven attention in the brain. Nature Reviews Neuroscience 3: 201–215, 2002.

Costello, M. B.; Fragaszy, D. M. Prehension in cebus and saimiri 1: Grip type and hand preference. American Journal of Primatology 15: 235–245, 1988.

Crow, T. J. Directional asymmetry is the key to the origin of modern *Homo sapiens* (the Broca-Annett axiom): A reply to Rogers' review of *The Speciation of modern Homo sapiens*. Laterality 9: 233–242, 2004.

Darwin, C. The origin of species. London, John Murray, 1859. (Reprinted in Great Books of the Western World No. 49, Chicago, Encyclopedia Brittanica, 1952.)

Davidson, R. J. Cerebral asymmetry, emotion and affective style. Pp. 361–388 in R. J. Davidson; K. Hugdahl, eds. Brain Asymmetry. Cambridge, MA, MIT Press, 1995.

Day, L. B.; MacNeilage, P. F. Postural asymmetries and language lateralization in humans (*Homo sapiens*). Journal of Comparative Psychology 110: 88–96, 1996.

Delis, D.; Robertson, L.; Efron, R. Hemispheric specialization of memory for visual hierarchical stimuli. Neuropsychologia 24: 205–214, 1986.

DeRenzi, E. Disorders of Space Exploration and Cognition. New York, Wiley, 1982.

Diamond, A.; McGrew, W. C. True handedness in the cotton-top tamarin (*Saguinus oedipus*)? Primates 35: 69–77, 1994.

Dimond, S.; Harries, R. Face touching in monkeys, apes and man: Evolutionary origins and cerebral asymmetry. Neuropsychologia 7: 161–166, 1984.

Eisenberg, J. F. The Mammalian Radiations: An Analysis of Trends in Evolution, Adaptation and Behavior. Chicago, University of Chicago Press, 1981.

Elias, L. J.; Bryden, M. P. Footedness is a better predictor of language lateralization than handedness. Laterality 3: 41–52, 1998.

Fagot, J.; Bard, K. A. Asymmetric grasping response in neonate chimpanzees (*Pan troglodytes*). Infant Behavior and Development 18: 253–255, 1995.

Fagot, J.; Drea, C.; Wallen, K. Asymmetrical hand use in rhesus monkeys (*Macaca mulatta*) in tactually and visually regulated tasks. Journal of Comparative Psychology 105: 260–268, 1991.

Fagot, J.; Lacreuse, A.; Vauclair, J. Role of sensory and post-sensory factors on hemispheric asymmetries in tactile perception. Pp. 469–494 in S. Christman, ed. Cerebral Asymmetries in Sensory and Perceptual Processing. Amsterdam, Elsevier, 1997.

Fernandes, E. B. M. Tool use and predation of oysters (*Crassostrea rhizophorae*) by the tufted capuchin, *Cebus apella apella*, in brackish water mangrove swamp. Primates 32: 529–531, 1991.

Fine, M. L.; McElroy, D.; Rafi, J.; King, C. B.; Loesser, K. B.; Newton, S. Lateralization of pectoral stridulation sound production in the channel catfish. Physiology and Behavior 60: 753–757, 1996.

Finch, G. Chimpanzee handedness. Science 94: 117–118, 1941.

Fischer, R. B.; Meunier, G. F.; White, P. J. Evidence of laterality in the lowland gorilla. Perceptual and Motor Skills 54: 1093–1094, 1982.

Fleagle, J. C. Primate Adaptation and Evolution. New York, Academic Press, 1988.

Forsythe, C.; Milliken, G. W.; Stafford, D. K.; Ward, J. P. Posturally related variations in the hand preferences of the ruffed lemur (*Varecia variegata variegata*). Journal of Comparative Psychology 102: 248–250, 1988.

Forsythe, C.; Ward, J. P. The lateralized behavior of the ruffed lemur. American Journal of Primatology 12: 342, 1987.

Fox, M. D.; Corbetta, M.; Snyder, A. Z.; Vincent, J. L.; Raichle, M. E. Spontaneous neuronal activity distinguishes human dorsal and ventral attention systems. Proceedings of the National Academy of Sciences 103: 10,046–10,051, 2006.

Gazzaniga, M. S. Cerebral specialization and interhemispheric communication: Does the corpus callosum enable the human condition? Brain 123: 1293–1326, 2000.

Goldberg, E.; Podell, K. Lateralization in the frontal lobes. Pp. 85–96 in H. H. Jasper; S. Riggio; P. S. Goldman-Rakic, eds. Epilepsy and the Functional Anatomy of the Frontal Lobe. New York, Raven Press, 1995.

Goldschmidt, R. The Material Basis of Evolution. New Haven, Yale University Press, 1940.

Gomperts, S.; Costello, M. Preliminary results of hand use in free-ranging capuchins (*Cebus capucinus*). American Journal of Primatology 24: 104, 1991.

Güntürkün, O. Human behavior: Adult persistence of head-turning asymmetry. Nature 421: 711, 2003.

Güntürkün, O.; Hoferichter, H. H. Neglect after section of a left telencephalotectal tract in pigeons. Behavioral Neuroscience 18: 1–9, 1985.

Harada, T.; Daito, D. N.; Kashikura, K-I.; Sato, T.; Yonekura, Y.; Honda, M.; Sadato, N. Asymmetrical neural substrates of tactile discrimination in humans: A functional magnetic imaging study. The Journal of Neuroscience 24: 7524–7530, 2004.

Harrison, K. E.; Byrne, R. W. Hand preference in unimanual and bimanual feeding by wild vervet monkeys (*Cercopithecus aethiops*). Journal of Comparative Psychology 114: 13–21, 2000.

Hatta, T.; Koike, M. Left-hand preference in frightened mother monkeys in taking up their babies. Neuropsychologia 29: 207–209, 1991.

Hauser, M.; Andersson, K. Left hemisphere dominance for processing vocalizations in adult, but not infant, rhesus monkeys: Field experiments. Proceedings of the National Academy of Sciences, USA. 91(9): 3946–3948, 1994.

Heestand, J. Behavioral Lateralization in Four Species of Apes? Unpublished Ph.D. Dissertation, University of Washington, Seattle, 1986.

Heffner, H. E.; Heffner, R. S. Temporal lobe lesions and the perception of species-specific vocalizations by macaques. Science 226: 75–76, 1984.

Hepper, P. G.; Shahidullah, S.; White, R. Origins of fetal handedness. Nature 347: 431, 1990.

Hepper, P. G.; Wells, D. L.; Lynch, C. Prenatal thumb sucking is related to postnatal handedness. Neuropsychologia 43: 313–315, 2005.

Heilman, K. Attentional asymmetries. Pp. 217–234 in R. J. Davidson; K. Hugdahl, eds. Brain Asymmetry. Cambridge, MA, MIT Press, 1995.

Hook, M. A., Rogers, L. J. Leading limb preferences in marmosets. 7: 145–162, 2002.

Hook-Costigan, M. A.; Rogers, L. J. Hand preferences in new world primates. International Journal of Comparative Psychology 9: 173–207, 1996.

Hook-Costigan, M. A.; Rogers, L. J. Lateralized use of the mouth in production of vocalizations by marmosets. Neuropsychologia 36: 1265–1273, 1998.

Holman, S. D.; Hutchinson, J. B. Lateralized action of androgen and development of behavior and brain sex differences. Brain Research Bulletin 27: 261–265, 1991.

Hopkins, W. D. Hand preferences for a coordinated bimanual task in 110 chimpanzees (*Pan troglodytes*): Cross-sectional analysis. Journal of Comparative Psychology 109: 291–297, 1995.

Hopkins, W. D.; Bard, K. Asymmetries in spontaneous head turning in infant chimpanzees. Behavioral Neuroscience 109: 808–812, 1995.

Hopkins, W. D.; Bard, K. A.; Griner, K. M. Locomotor adaptation and leading limb asymmetries in neonatal chimpanzees (*Pan troglodytes*). International Journal of Primatology 18: 104–114, 1997.

Hopkins, W. D.; Bard, K. A.; Jones, A.; Bales, S. L. Chimpanzee hand preference in throwing and infant cradling: Implications for the origin of human handedness. Current Anthropology 34: 786–790, 1993.

Hopkins, W. D.; Bennett, A. J. Handedness and approach-avoidance behavior in chimpanzees (*Pan troglodytes*). Journal of Experimental Psychology: Animal Behavior Processes 20: 413–418, 1994.

Hopkins, W. D.; Cantalupo, C. Individual and setting differences in the hand preferences of chimpanzees (*Pan troglodytes*): A critical analysis and some alternative explanations. Laterality 10: 65–80, 2005.

Hopkins, W. D.; Cantalupo, C.; Wesley, M. J.; Hostetter, A. B.; Pilcher, D. L. Grip morphology and hand use in chimpanzees (*Pan troglodytes*): Evidence of a left hemisphere specialization in motor skill. Journal of Experimental Psychology: General 131: 412–423, 2002.

Hopkins, W. D.; Cantero, M. From hand to mouth in the evolution of language: The influence of vocal behavior on lateralized hand use in manual gestures by chimpanzees (*Pan troglodytes*). Developmental Science 6: 55–61, 2003.

Hopkins, W. D.; deWaal, F. B. M. Addendum to "Behavioral laterality in captive bonobos (*Pan paniscus*)": Replication and extension. International Journal of Primatology 16: 261–276, 1995.

Hopkins, W. D.; Leavens, D. A. Hand use and gestural communication in chimpanzees (*Pan troglodytes*). Journal of Comparative Psychology 112: 95–99, 1998.

Hopkins, W. D.; Stoinski, T. S.; Lukas, K. E.; Ross, S. R.; Wesley, M. J. Comparative assessment of handedness for a coordinated bimanual task in chimpanzees (*Pan troglodytes*), gorillas (*Gorilla gorilla*), and orangutans (*Pongo pygmaeus*). Journal of Comparative Psychology 117: 302–308, 2003.

Hopkins, W. D.; Wesley, M. J.; Izard, M. K.; Hook, M.; Schapiro, S. J. Chimpanzees are predominantly right-handed: Replication in three colonies of apes. Behavioral Neuroscience 118: 659–663, 2004.

Hopkins, W. D.; Wesley, M. J. Gestural communication in chimpanzees (*Pan troglodytes*): The influence of experimenter position on gesture type and hand preference. Laterality 7: 19–30, 2002.

King, J. E. Laterality in hand preferences and reaching accuracy of cotton-top tamarins. (*Saguinus oedipus*). Journal of Comparative Psychology 107: 403–410, 1995.

King, J. E.; Landau, V. I. Manual preferences in varieties of reaching in squirrel monkeys. Pp. 107–124 in J. P. Ward; W. D. Hopkins, eds. Primate laterality: Current evidence of primate asymmetries. New York, Springer, 1993.

Lacreuse, A.; Fragaszy, D. M. Hand preference for a haptic searching task by tufted capuchins (*Cebus apella*). International Journal of Primatology 17: 613–632, 1996.

Lacreuse, A.; Fragaszy, D. M. Manual exploratory procedures and asymmetries for a haptic search task: A comparison between capuchins (*Cebus apella*) and humans. Laterality 2: 247–266, 1997.

Lacreuse, A.; Fragaszy, D. M. Left hand preferences in capuchins (*Cebus apella*): Role of spatial demands in manual activity. Laterality 4: 65–78, 1999.

Lacreuse, A.; Parr, L. A.; Smith, H. M.; Hopkins, W. D. Hand preferences for a haptic task in chimpanzees (*Pan troglodytes*). International Journal of Primatology 20: 867–881, 1999.

Laska, M. Manual laterality in spider monkeys (*Ateles geoffroyi*) solving visually and tactually guided food-reaching tasks. Cortex 32: 717–726, 1996.

Liotti, M.; Tucker, D. M. Emotion in asymmetric corticolimbic networks. Pp. 389–424 in R. J. Davidson; K. Hugdahl, eds. Brain Asymmetry. Cambridge, MA, MIT Press, 1995.

Lonsdorf, E. V.; Hopkins, W. D. Wild chimpanzees show population-level handedness for tool use. Proceedings of the National Academy of Sciences 102: 634–638, 2005.

MacNeilage, P. F. The postural origins theory of neurobiological asymmetries in primates. Pp. 165–188 in N. Krasnegor; D. Rumbaugh; M. G. Studdert-Kennedy; R. Schiefelbusch, eds. The Biological Foundations of Language Development. Hillsdale, NJ, Erlbaum, 1991.

MacNeilage, P. F. Implication of primate functional asymmetries for the evolution of cerebral hemispheric specializations. Pp. 319–341 in J. P. Ward;

W. D. Hopkins, eds. Primate Laterality: Current Evidence of Primate Asymmetries. New York, Springer, 1993.

MacNeilage, P. F. Towards a unified view of cerebral hemispheric specializations in vertebrates. Pp. 167–183 in A. D. Milner, ed. Comparative Neuropsychology. Oxford, Oxford University Press, 1998a.

MacNeilage, P. F. The frame/content theory of evolution of speech production. Behavioral and Brain Sciences 21: 499–511, 1998b.

MacNeilage, P. F. Evolution of the mechanism of language output: Comparative neurobiology of vocal and manual communication. Pp. 222–241 in J. R. Hurford; M. G. Studdert-Kennedy; C. Knight, eds. Approaches to the Evolution of Language: Social and Cognitive Bases. Cambridge, Cambridge University Press, 1998c.

MacNeilage, P. F. The frame/content theory. in B. L. Davis; K. Zajdo, eds. The Development of the Syllable: The Frame/Content Theory and Beyond. Mahwah, NJ, Erlbaum, 31–55, 2007.

MacNeilage, P. F.; Davis, B. L. The frame/content theory of evolution of speech: A comparison with a gestural origins alternative. Journal of Interaction Studies 6: 173–199, 2005.

MacNeilage, P. F.; Studdert-Kennedy, M. G.; Lindblom, B. Functional precursors to language and its lateralization. American Journal of Physiology R (Regulatory, Integrative and Comparative Physiology 15) 246: 912–915, 1984.

MacNeilage, P. F.; Studdert-Kennedy, M. G.; Lindblom, B. Primate handedness reconsidered. Behavioral and Brain Sciences 10: 247–303, 1987.

MacNeilage, P. F.; Studdert-Kennedy, M. G.; Lindblom, B. Primate handedness: A foot in the door. Behavioral and Brain Sciences 11: 748–758, 1988.

MacNeilage, P. F.; Studdert-Kennedy, M. G.; Lindblom, B. Primate handedness: The other theory, the other hand and the other attitude. Behavioral and Brain Sciences 14: 344–349, 1991.

Maki, S. An Experimental Approach to the Postural Origins Theory of Neurobiological Asymmetries in Primates. Unpublished Ph.D. Dissertation, University of Texas at Austin, 1990.

Masataka, N. Population level asymmetries of hand preference in lemurs. Behaviour 110: 1–4, 1989.

Masataka, N. Handedness of capuchin monkeys. Folia Primatologia 55: 189–192, 1990.

Mason, A. M.; Wolfe, L. D.; Johnson, J. C. Hand preference in the sifaka (*Propithecus, verreauxi coquereli*) during feeding in captivity. Primates 36: 275–280, 1995.

Mayr, E. The Growth of Biological Thought. Cambridge, MA, Bellknap, 1984.

McGrew, W. C.; Marchant, L. F. On the other hand: Current issues in and meta-analysis of the behavioral laterality of hand function in nonhuman primates. Yearbook of Physical Anthropology 40: 201–232, 1997.

Meguerditchian, A.; Vauclair, J. Baboons communicate with their right hand. Behavioral Brain Research 171: 170–174, 2006.

Michel, G. F. Development of infant handedness. Pp. 1665–1685 in D. J. Lewkowics; R. Lickliter, eds. Conceptions of Development: Lessons from the Laboratory. New York, Psychology Press, 2002.

Parnell, R. J. Hand preference for food processing in wild western lowland gorillas (*Gorilla gorilla gorilla*). Journal of Comparative Psychology 115: 365–375, 2001.

Parr, L. A.; Hopkins, W. D.; De Waal, F. B. M. Haptic discrimination in capuchin monkeys (*Cebus apella*): Evidence of manual specialization. Neuropsychologia 35: 143–152, 1997.

Petersen, M.; Beecher, B.; Zoloth, S.; Moody, D.; Stebbins, W. Neural lateralization of species-specific vocalizations by Japanese macaques (*Macaca fuscata*). Science 202: 324–327, 1978.

Poremba, A.; Malloy, M.; Saunders, R. C.; Carson, R. E.; Hersovitch, P.; Mishkin, M. Species-specific calls evoke asymmetric activity in the monkey's temporal poles. Nature 427: 448–451, 2004.

Preilowski, B.; Leder, F. Comparative studies of laterality: Hand use of a captive group of lowland gorillas in A. Albert, ed. Proceedings of the 34th Congress of the German Society for Psychology, Vienna, Austria, 1984.

Purvis, A. A composite estimate of primate phylogeny. Philosophical Transactions of the Royal Society of London, Series B, 348: 405–421, 1995.

Rigamonti, M. M.; Spiezio, C.; Poli, M. D.; Feruccio, F. Laterality of manual function in foraging and positional behavior in wild indri (*Indri indri*). American Journal of Primatology 65: 27–38, 2005.

Robinson, R. J.; Downhill, C. Lateralization of psychopathology in response to focal brain injury. Pp. 693–712 in R. J. Davidson; K. Hugdahl, eds. Brain Asymmetry. Cambridge, MA, MIT Press, 1995.

Rogers, L. J.; Andrew, R. J. Comparative Vertebrate Lateralization. Cambridge, Cambridge University Press, 2002.

Rogers, L. J.; Kaplan, G. Hand preferences and other lateral biases in rehabilitated orang-utans, *Pongo pygmaeus pygmaeus*. Animal Behaviour 51: 13–25, 1996.

Rowe, N. The Pictorial Guide to Living Primates. East Hampton, NY, Pegonias Press, 1996.

Sanford, C.; Guin, K.; Ward, J. P. Posture and laterality in the bushbaby (*Galago senegalensis*). Brain, Behavior and Evolution 25: 217–224, 1984.

Schaller, G. B. The Mountain Gorilla. Chicago, Chicago University Press, 1963.

Searleman, A. Subject variables and cerebral organization for language. Cortex 16: 239–254, 1980.

Seltzer, C.; Forsythe, C.; Ward, J. P. Multiple measures of motor lateralization in human primates (*Homo sapiens*). Journal of Comparative Psychology 104: 159–166, 1990.

Sergent, J. The cerebral balance of power: Confrontation or cooperation. Journal of Experimental Psychology: Human Perception and Performance 8: 253–272, 1982.

Shafer, D. D. Patterns of handedness: Comparative study of nursery school children and captive gorillas. Pp. 267–284 in J. P. Ward; W. D. Hopkins, eds. Primate Laterality: Current Evidence of Primate Asymmetries. New York, Springer, 1993.

Shafer, D. D. Hand preference behaviors shared by two groups of captive bonobos. Primates 38: 303–313, 1997.

Spinazzola, L.; Cubelli, R.; Della Sala, S. Impairment of trunk lesions following left and right hemisphere lesions: Dissociation between apraxic errors and postural instability. Brain 126: 2,656–2,666, 2003.

Spinozzi, G.; Castorina, M. G.; Truppa, V. Hand preferences in unimanual and coordinated-bimanual tasks by tufted capuchin monkeys (*Cebus apella*). Journal of Comparative Psychology 112: 183–191, 1998.

Subramoniam, S. Some observations on the habits of the slender loris, *Loris tardigradus L.* Journal of the Bombay Natural History Society 54: 386–398, 1957.

Torigoe, T. Comparison of object manipulation among species of non-human primates. Primates 26: 182–194, 1985.

Vallortigara, G.; Rogers, L. J. Survival with an asymmetrical brain: Advantages and disadvantages of cerebral lateralization. Behavioral and Brain Sciences 28: 575–589, 2005.

Vauclair, J.; Meguerditchian, A.; Hopkins, W. D. Hand preferences for unimanual and coordinated bimanual tasks in baboons (*Papio anubis*). Brain Research Cognitive Brain Research 25: 210–216, 2005.

Visalberghi, E.; Fragaszy, D. "Do monkeys ape?" – Ten years after. Pp. 471–5000 in K. Dautenhahn; C. L. Nehaniv, eds. Imitation in Animals and Artifacts. Cambridge, MA, MIT Press, 2002.

Walker, A.; Shipman, P. The Wisdom of Bones. London, Weidenfeld and Nicolson, 1996.

Ward, J. P.; Milliken, G.; Stafford, D. K. Patterns of lateralized behavior in prosimians. Pp. 43–74 in J. P. Ward; W. D. Hopkins, eds. Primate Laterality: Current Evidence of Primate Asymmetries. New York, Springer, 1993.

Warren, J. M. Handedness and cerebral dominance in monkeys. Pp. 151–172 in S. Harnad; R. W. Doty; L. Goldstein; J. Jaynes; G. Krauthamer, eds. Lateralization in the nervous system. New York, Academic Press, 1977.

Warren, J. M. Handedness and laterality in humans and other animals. Physiological Psychology 8: 351–359, 1980.

Warren, J. M.; Nonneman, A. J. The search for cerebral dominance in monkeys. Pp. 732–744 in S. R. Harnad; H. D. Steklis; J. Lancaster, eds.

Origins and Evolution of Language and Speech. New York, New York Academy of Sciences, 1977.
Watanabe, K.; Kawai, M. Pp. 183–192 in J. P. Ward; W. D. Hopkins, eds. Primate Laterality: Current Evidence of Primate Asymmetries. New York, Springer, 1993.
Westergaard, G. C.; Champoux, M.; Suomi, S. J. Plasma cortisol is associated with handedness in infant rhesus monkeys. Developmental Psychobiology 38: 116–122, 2001.
Westergaard, G. C.; Kuhn, H. E.; Suomi, S. J. Bipedal posture and hand preference in humans and other primates. Journal of Comparative Psychology 112: 55–64, 1998.
Westergaard, G. C.; Suomi, S. J. Hand preference for a bimanual task in tufted capuchins (*Cebus apella*) and rhesus macaques (*Macaca mulatta*). Journal of Comparative Psychology 110: 406–411, 1996.
Wittling, W. Brain asymmetry in the control of autonomic – physiologic activity. Pp. 305–358 in R. J. Davidson; K. Hugdahl, eds. Brain Asymmetry. Cambridge, MA, MIT Press, 1995.

Stained images of Brodmann's area 45 in the left and right hemispheres of a human brain. Sections were stained using a modified Gallyas stain for Nissl substance. The image was taken using a Nikon E400 microscope and digitized with a resolution of 1.47 μm/pixel (Photograph by Natalie Schenker.)

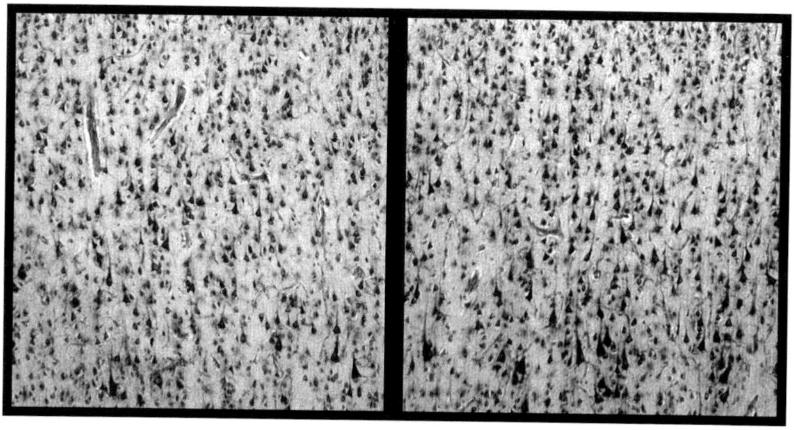

The Evolution of Hemispheric Specialization in Primates
W.D. Hopkins (Editor)

MICROSTRUCTURAL ASYMMETRIES OF THE CEREBRAL CORTEX IN HUMANS AND OTHER MAMMALS

Natalie M. Schenker, Chet C. Sherwood, Patrick R. Hof and Katerina Semendeferi

Department of Anthropology, University of California at San Diego, La Jolla, CA, USA (NMS, KS); Department of Anthropology, The George Washington University, Washington, DC, USA (CCS); Department of Neuroscience, Mount Sinai School of Medicine, New York, NY, USA (PRH); New York Consortium in Evolutionary Primatology, New York, NY, USA (PRH); Department of Anthropology, Graduate Program in Neurosciences, University of California at San Diego, La Jolla, CA, USA (KS)

ABSTRACT

The human brain shows marked gross anatomical and microstructural asymmetries that are presumably related to lateralized motor and cognitive functions. This chapter generally summarizes the extant data on gross morphological asymmetries in human and nonhuman mammal brains. In addition, the evidence of microstructural asymmetries, such as gray-level index, minicolumn width, and cellular organization, are presented. Although there are few studies of microstructrual asymmetries in nonhuman primates, it is argued that such studies are important for validating morphological asymmetries as well as for understanding the cellular basis for hemispheric specialization in primates, including humans.

Keywords: microstructural asymmetry, histological asymmetry, cytoarchitectonics, primates, minicolumn asymmetry, primate brain evolution

INTRODUCTION

The cerebral cortex can be parcellated into areas that differ in their cytoarchitecture [Brodmann, 1909], chemoarchitecture [Krubitzer & Huffman, 2000; Krubitzer & Kahn, 2003], connectivity [Barbas & Rempel-Clower, 1997], and distribution of receptors for neurotransmitter molecules [Zilles et al., 2002]. It is well known that such regional variation in cortical microstructure contributes to the many distinct functional specializations of the cortex. For example, differential activation of cortical areas known to vary in microstructure has been demonstrated in numerous studies in humans, using functional magnetic resonance imaging (fMRI), positron emission tomography (PET), and electroencephalography (EEG) [e.g., Azari et al., 2001; Beauregard et al., 2001;

ISSN: 1936-8526 DOI: 10.1016/S1936-8526(07)05004-X

Bookheimer et al., 1997; Dagher et al., 1999; Grabowski et al., 2001; Pesenti et al., 2001; Rowe & Passingham, 2001].

The cerebral hemispheres are also often differentially recruited in divergent functional activities. Thus, a region that is recognized as homotopic between hemispheres on the basis of topology and basic cytoarchitectural organization may participate in different information processing in each hemisphere. The best documented example of such cerebral lateralization is language, where 95% of humans are left hemisphere dominant for the production and comprehension of language [Branche et al., 1964; Ojemann, 1991; Petersen et al., 1988]. As there are structural differences between cortical areas that exhibit functional variation, interhemispheric functional asymmetry in the cerebral cortex arises, in part, from differences in the connectivity and microstructure of homotopic cortical areas [Hutsler & Galuske, 2003]. Therefore, important insights into the anatomical substrates of the lateralized functions of the cerebral cortex can be gained through the incorporation of microstructural data.

At present, however, the majority of studies concerning human brain asymmetries are at the level of macrostructure. For example, the human brain exhibits large-scale asymmetries in the protrusion of the frontal and occipital lobes, called petalias [Holloway & de Lacoste-Lareymondie, 1982]. Furthermore, it is known that regions associated with language function are also asymmetric in their gross anatomy. Still, it is not entirely clear how such large-scale asymmetries are reflected in the microstructure of the cerebral hemispheres. It has been suggested that volumetric differences are primarily caused by changes in the number of cells, rather than by changes in cell density [Galaburda et al., 1986; Rosen, 1996]. However, other types of microstructural asymmetries also exist. Asymmetries in cell size, columnar organization, the complexity of dendritic arbors, and chemoarchitectural organization have been reported [Buxhoeveden et al., 2001; Garcia et al., 2004; Hayes & Lewis, 1995; Scheibel et al., 1985]. Here, we review known structural asymmetries in the human cerebral cortex, with special attention to the relatively small literature on microstructural asymmetries in the brains of other mammals. We discuss the implications of current data to our understanding of the evolution of functional cortical asymmetries in humans.

Gross Anatomical Asymmetries

Pioneering research on the human brain by Paul Broca drew attention to the functional asymmetries of the human inferior frontal cortex through evidence that lesions to the left cerebral hemisphere tend to result in language impairments [Broca, 1861]. These initial observations inspired an explosion of studies concerning the distribution of gross anatomical asymmetries of the human brain as they relate to functional asymmetries. While early efforts were constrained to rely on small samples of postmortem brains, in recent years, the advent of magnetic resonance imaging (MRI) and voxel-based morphometry methods have allowed measurement of cerebral asymmetries in larger samples. With this proliferation of data, much has been learned regarding how gross anatomical asymmetries in the human cortex vary with age, sex, psychiatric, and neurological conditions. As a rule, these gross morphological data are interpreted to reflect corresponding changes in some aspect of underlying neural circuitry. Many asymmetries are measured using sulcal landmarks, which may [Brodmann, 1909] or may not [Amunts et al., 1999; Sherwood et al., 2003] reflect the boundaries of cortical areas as defined by microstructural details. Nonetheless, if we assume that gross morphological asymmetries emerge from some underlying factor that causes the displacement of tissue volumes, sulci, and gyri, then consideration of cortical asymmetry at the macroanatomical level yields suggestive evidence regarding regions where microstructural architecture may express asymmetry as well. Studies of this variety have primarily addressed the length of particular sulci as well as the volume of areas defined by sulcal boundaries [Tomaiuolo et al., 1999; White et al., 1997a].

Asymmetries in sulcal lengths and trajectories in humans have been described for the central sulcus and the Sylvian fissure. In one study, the length of the central sulcus at the cortical surface was found not to evince a consistent pattern of lateral asymmetry [White et al., 1997a]. The depth of the central sulcus, however, as measured on horizontal MR images, displays an interaction with sex and handedness [Amunts et al., 2000]. Asymmetry of the contralateral central sulcus was most pronounced in right-handed men, with a decreased degree of asymmetry in mixed- and left-handed men. In contrast, asymmetries in central sulcus depth did not differ across handedness in women. Likewise, the "hand knob" of primary cortex in chimpanzees does not display a significant population-level asymmetry [Hopkins & Cantalupo, 2004]. However, a correlation between the volumetric asymmetry of the hand knob in motor cortex and hand preference in a specific tool task was reported, with a tendency for leftward volumetric asymmetry to be associated with preference for using the right hand. Similarly, in male capuchin monkeys, asymmetries in the depth of the central sulcus are also correlated with hand dominance on a coordinated bimanual task [Phillips & Sherwood, 2005], suggesting that this neuroanatomical relationship to handedness may be more widespread among primates.

The length of the Sylvian fissure was found to be longer in the left hemisphere in humans [Blanton et al., 2001; Foundas et al., 1999], partly because of the difference in the shape of the sulcus between the hemispheres. The right sulcus typically contains an upward bend at the posterior end, whereas the left sulcus remains relatively horizontal [Foundas et al., 1999].

A few regions have been found to express volumetric asymmetry on the basis of gross morphological criteria. In humans, the cytoarchitectural subdivisions composing Broca's area, a region involved with language production, frequently lie within the morphological boundaries of the inferior frontal gyrus (IFG). Within the IFG, the ascending (vertical) ramus of the Sylvian fissure separates pars opercularis (Brodmann's area 44) from pars triangularis (area 45), and the anterior (horizontal) ramus separates pars triangularis from pars orbitalis (area 47). Numerous studies have investigated macrostructural asymmetry in Broca's area using these sulcal landmarks to subdivide the region. However, the results from these studies differ markedly, depending on methodology and anatomical definitions. While measures of the convexity of the cortical surface area of the frontal operculum (including the pars opercularis and the posterior portion of the pars triangularis) have not revealed significant population-level leftward dominance [Wada et al., 1975], asymmetries are significant when intrasulcal cortex is included [Falzi et al., 1982; Tomaiuolo et al., 1999]. Furthermore, some volumetric MRI-based studies have found that both pars triangularis and pars opercularis [Foundas et al., 1998, 2001] are leftward dominant, but others have not found volumetric asymmetry in pars opercularis [Knaus et al., 2006; Tomaiuolo et al., 1999]. Thus, a consensus has yet to be reached regarding macrostructural asymmetries in the human IFG. Nonetheless, population-level leftward asymmetry of fronto-orbital sulcal length, an external morphologic feature in this region of great apes, has been reported in chimpanzees [Cantalupo & Hopkins, 2001].

A multi-species MRI analysis of living ape brains found volumetric asymmetries in two structurally defined subdivisions of the frontal lobe: the dorsal sector (composed of most of the cortex on the lateral surface of the lobe) showed a rightward asymmetry, while the medial cortex (composed of the entire cortex on the medial surface of the lobe) showed a leftward asymmetry. The orbital sector shows no asymmetry [Schenker et al., 2005].

Gross morphological studies have also identified volumetric asymmetries of the planum temporale (PT, including the posterior part of Brodmann's area 22; also known as Tpt), a temporal lobe region involved in language processing, in both humans and chimpanzees using MRI [Emmorey et al., 2003; Hopkins & Cantalupo, 2004; Hopkins et al., 1998; Penhune et al., 1996] and postmortem specimens [Anderson et al., 1999; Gannon et al., 2001]. A third region, the angular gyrus in the inferior parietal lobe, recently confirmed as a region in a language circuit [Catani et al., 2005], also exhibits the same left greater than right volumetric asymmetry [Watkins et al., 2001]. Furthermore, the region shows reversed (right greater than left) volumetric asymmetry in schizophrenic patients [Buchanan et al., 2004; Niznikiewicz et al., 2000]. A nearby region, occupying the posterior bank of the posterior ascending branch of the Sylvian fissure, shows a significant interaction between handedness and sex in the analysis of volumetric asymmetry. Right-handed men and left-handed women show a strong rightward asymmetry. Right-handed women also have a rightward asymmetry (but not as strong), while left-handed men show a weak leftward asymmetry [Jancke et al., 1994]. A study of the temporal lobes in MR images of living ape brains reported limited evidence of hemispheric asymmetry in gyrification and surface area [Rilling & Seligman, 2002]. Another study found no evidence of asymmetry in images of postmortem chimpanzee brains [Zilles et al., 1996].

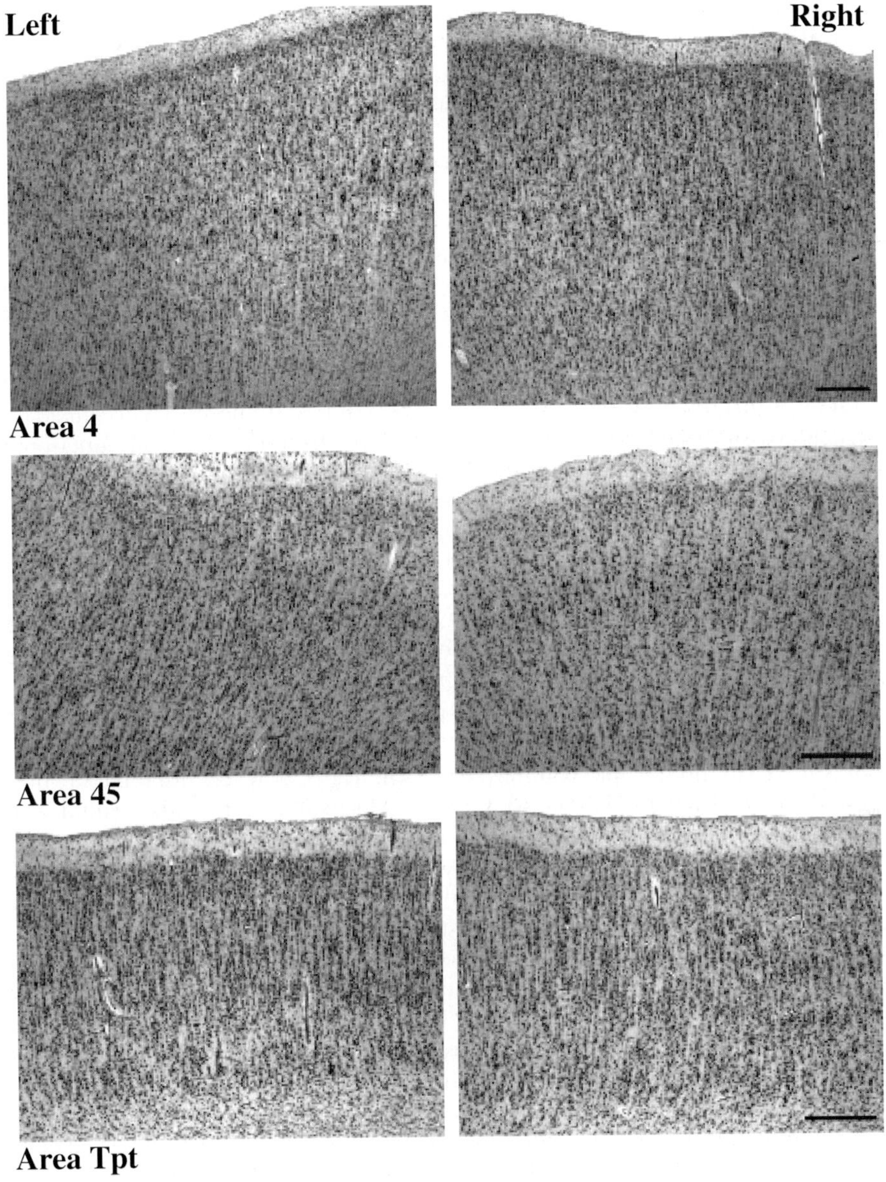

Figure 1. Representative micrographs of cortex from the left and the right hemispheres of a human brain in areas 4, 45, and Tpt. Notice the overall similarity between cortical areas in each hemisphere. The asymmetries that have been detected have relied on quantitative methods that are capable of measuring subtle variations in histological features such as cell sizes, cell densities, and the space between minicolumns.

Gross anatomical studies of asymmetry have focused predominantly on areas that demonstrate functional asymmetry. Such studies have revealed asymmetries in motor cortex related to hand preference in both humans and chimpanzees. Studies of language related areas (IFG, PT, angular gyrus) report a tendency for these regions to exhibit leftward asymmetry, matching the prevalence of left hemisphere dominance for language, particularly among right-handed individuals. However, repeated studies of the IFG, with varying conclusions, reveal the continuing lack of consensus on the presence and magnitude of such macrostructural asymmetries.

Microstructural Asymmetries

While gross anatomical analyses can provide an overview of where structural asymmetries may exist, microstructural studies are necessary to elucidate particular hemispheric specializations of neural wiring that underlie functional lateralization. In addition, although comparative microarray analyses of gene expression in the brain can reveal interesting differences between the transcriptomes of humans and other primates [Caceres et al., 2003; Enard et al., 2002; Uddin et al., 2004], this approach is relatively insensitive to subtle variation in gene expression levels among small populations of cells [Geschwind, 2000]. This is especially problematic in analyses of regional differences in the cerebral cortex because of the cellular heterogeneity of its composition and large degree of interindividual variation. In this regard, it is noteworthy that a recent study, looking for differentially expressed genes, was unable to distinguish among three regions known to differ in function: human Broca's area in the left hemisphere, its homotopic counterpart in the right, and left dorsolateral prefrontal cortex [Khaitovich et al., 2004]. This same study also could not detect greater differences in transcript levels between human and chimpanzee Broca's area as compared to several other cortical areas. Therefore, to reveal the correlates of functional cortical lateralization, it is necessary to examine interhemispheric differences in microstructural organization. Several microstructural studies of histological specimens, including investigations of regional volume, cell density, dendritic structure, and cell size, have been performed to investigate asymmetries in human brain areas that have well-established patterns of population-wide gross morphological asymmetry. Figure 1 shows interhemispheric comparisons of human cortex in three of these regions demonstrating that variation between the hemispheres is expressed in quite subtle details of histological architecture. Before reviewing the current evidence concerning histological asymmetries in the cerebral cortex, it is important to note that very limited sample size is a problem that plaques all such studies. Indeed, it is not uncommon for samples to be smaller than $n = 10$. Thus, interpretation of negative findings should be considered with caution and should not necessarily be taken as definitive evidence of lack of asymmetry in the larger population.

Volume. Volumetric studies of histological samples based on cytoarchitectural criteria have been conducted in several regions of cortex, including primary motor (Brodmann's area 4), primary visual (area 17), and language related cortices (areas 44, 45, and part of 22).

In a study of six postmortem human brains of unknown handedness, five had a leftward asymmetry in the total volume of primary motor cortex, while the sixth had a larger motor area in the right hemisphere [Zilles et al., 1996]. However, another study of the same region, using 20 postmortem brains, did not find a significant population-level asymmetry in the total volume of primary motor cortex, with a similar number of brains having a leftward asymmetry as had a rightward one [White et al., 1997b]. When only the hand representation area of primary motor cortex was measured, there was a population-level leftward asymmetry that approached significance, but 6 of 20 specimens exhibited a rightward asymmetry [White et al., 1997b].

The primary visual cortex of humans shows a rightward asymmetry at the population level, although some individual brains show a leftward asymmetry. In a study of 31 brains, the volume of this region of cortex exhibited a significant rightward asymmetry across the sample [Murphy, 1985]. The average asymmetry, regardless of direction, was 8%; 24 of the 31 postmortem brains exhibited a rightward asymmetry, while the remaining 7 had a leftward asymmetry. In another cytoarchitectural study, a similar right-hemispheric bias in the volume of primary visual cortex was found in 11 of 14 brains, with a mean asymmetry of 13.7% [Andrews et al., 1997].

Volumetric asymmetries have also been reported in both anterior and posterior language areas, involved in production and comprehension, respectively. The anterior language area consists of Brodmann's areas 44 and 45. Using computer-assisted observer-independent quantification of laminar cytoarchitectural variation, recent studies have been able to parcellate and measure the volume of areas 44 and 45 on the basis of the multivariate distances between their quantitative cytoarchitectural profiles [Amunts et al., 1999; Schleicher et al., 1999; Uylings et al., 2006]. When this method of volumetric measurement was applied to 10 adult human brains of both sexes, a robust leftward volumetric asymmetry was found in area 44 (all 10 brains had a larger area 44 in the left than in the right hemisphere). In area 45, the degree of asymmetry (without regard to direction) was similar to that found in area 44. All five female brains displayed leftward asymmetry, but there was no significant asymmetry detected among the five male brains [Amunts et al., 1999; Uylings et al., 2006]. In these studies, the location of cytoarchitectural boundaries among area 44, 45, and adjacent areas did not correspond to external sulcal landmarks. Of note, another study concerning the correspondence between the boundaries of area 44 in common chimpanzees as defined by cytoarchitecture, myeloarchitecture, and the distribution of nonphosphorylated neurofilament protein-immunoreactive neurons, also failed to reveal a correlation between the borders of this cortical area and external morphologic features [Sherwood et al., 2003].

The posterior language area includes part of Brodmann's area 22. A study of area Tpt (which comprises the posterior part of area 22) in four human brains found left hemisphere volumetric dominance in all four specimens based on qualitative assessment of cytoarchitectural boundaries [Galaburda & Sanides, 1980]. A correlation between the volume of Tpt and the surface area of the grossly defined planum temporale has also been reported [Galaburda et al., 1978]. One of the four brains in this study had a slight rightward asymmetry in the planum temporale, but this was the brain with the least asymmetric area Tpt.

Gray-Level Index (GLI). Asymmetries of cell density have been analyzed indirectly by measurement of GLI, which estimates the fraction of tissue volume that is occupied by Nissl-stained cell bodies versus neuropil space. In the region of hand representation in primary motor cortex (area 4), the right hemisphere was found to have, on average, greater GLI than the left hemisphere in 12 postmortem human brains [Amunts et al., 1996]. This means that in the right hemisphere a greater percentage of the total tissue volume was occupied by cell soma than the corresponding region in the left hemisphere. In contrast, the left hemisphere contained relatively more neuropil space, which is occupied primarily by dendrites, axons, and synapses. This asymmetry was not confined to a single layer, but was observed across the entire cortical depth. In young children, asymmetry of GLI is present in this region as a whole, although development of adult-like asymmetry in supragranular cortical layers (layers II and III), which are involved in corticocortical association projections, is delayed as compared to infragranular layers (layers V and VI), which are involved in projections to subcortical structures [Amunts et al., 1997b].

Similar analyses of asymmetries in GLI have been conducted in the IFG. Amunts et al. [1999] found GLI to be greater in left area 44 than in the corresponding area in the right hemisphere in all of the male ($n = 5$) and three of the female brains ($n = 5$) studied. Thus, there may be a sex difference in the presence of asymmetry in this region. No asymmetry or sex difference in GLI was observed in area 45 [Amunts et al., 1999]. However, in a subsequent study, using a larger sample that included the brains from the earlier study, Amunts et al. [2003] reported a significantly greater GLI in areas 44 and 45 in the *right* hemisphere when the two areas were analyzed together. Furthermore, they found that asymmetry increases with age, with infants showing little or no asymmetry. GLI decreased in both areas with age, primarily during early childhood. However, the decrease differed between hemispheres, meaning that the adult pattern of asymmetry did not appear until relatively late in development (age 5 for area 45 and age 11 for area 44).

Pyramidal Cell Somatodendritic Geometry. The size of pyramidal cell dendritic arbors, the number and complexity of their branches, and their spine density are known to vary among cytoarchitectonic areas in humans [Jacobs et al., 2001] and macaques [Elston, 2000; Elston & Rockland, 2002] as well as among species [Elston et al., 2001]. These parameters may provide a picture of the complexity of the integrative capacity of individual pyramidal cells.

Several studies of human cortex have focused specifically on measuring asymmetry in the dendritic arbors of pyramidal cells based on Golgi impregnations. Scheibel et al. [1985] analyzed the dendritic profiles of pyramidal cells in the orofacial region of primary motor cortex and found that, overall, the number of dendritic segments was greater in the left hemisphere in a sample of six right-handed adult human males. However, total dendritic length was slightly greater in the right hemisphere. They also found an interaction between hemisphere and order of the segment (which is the distance of the segment from the cell body in terms of the number of branching points). The number and total length of segments more proximal to the soma were greater in the right hemisphere, while higher-order segments were greater in the left hemisphere.

Scheibel et al. [1985] also analyzed the IFG and found longer total dendritic length in the left frontal opercular region. However, as in the motor cortex, this difference was mainly due to the length of higher-order segments (those further from the cell body) in the left hemisphere, as lower-order segments were longer in the right [Scheibel et al., 1985]. In contrast, another study examined asymmetries in only the largest pyramidal cells in layer III of area 45 in humans [Hayes & Lewis, 1996]. This population of pyramidal neurons, known as magnopyramidal cells, mainly furnishes long corticocortical association projections. Total dendritic length, dendritic complexity (numbers of branches and maximal branch order), and spine densities were found to be greater in the right hemisphere. Contrary to the authors' expectations, dendritic length was positively correlated with soma volumes of magnopyramidal cells only in the left hemisphere and not in the right [Hayes & Lewis, 1996]. This finding suggests that there are additional factors that significantly contribute to the size of cells and the size of their dendritic arbors. Furthermore, these two studies of pyramidal cell geometry in anterior language-related cortical areas indicate that there are differences in the somatodendritic geometry of different cell populations within a single region.

Studies of dendritic parameters in the planum temporale are somewhat more difficult to interpret. One study found a slight leftward asymmetry in total dendritic length averaged across 20 human individuals [Jacobs et al., 1993]. Individually, only 12 of the 20 brains showed a leftward asymmetry. A similar asymmetry in the number of dendritic spines was observed, however, no asymmetry in mean dendritic segment length was found. Furthermore, there was a significant negative correlation between age and total dendritic length, and accompanying the decrease in length was a decrease in asymmetry. That is, in younger individuals (<50 years), total dendritic length was significantly greater in the left than in the right hemisphere, but a clear asymmetry was not present in the older sample [Jacobs & Scheibel, 1993]. In contrast, another study found a right greater than left asymmetry in three dendritic parameters: total basal dendritic length, number of dendrite branches, and number of dendritic spines, in seven of nine individuals [Anderson & Rutledge, 1996]. But, like Jacobs and

Scheibel [1993], these authors also found a negative correlation between these three variables and age, with no corresponding difference in cell soma size.

Cell Columns and Connectivity. The isocortex (neocortex) is populated by vertically oriented aggregates of cells with strong vertical interconnections among layers, forming fundamental structural and functional units known as minicolumns [Douglas & Martin, 1992; Mountcastle, 1997]. The emergence of columnar organization is related to the migration of neuroblasts from the ventricular and subventricular zones into radial columns during development [Rakic, 1995]. Cellular minicolumns differ from pyramidal cell modules, which have also been identified as minicolumns [Rockland & Ichinohe, 2004]. Such modules are formed by a core of apical dendrites surrounded by neurons that do not necessarily align in vertical rows [Peters & Kara, 1987; Peters & Sethares, 1991; Rockland & Ichinohe, 2004]. In contrast, cellular minicolumns comprise single rows of neurons [Buxhoeveden & Casanova, 2002; Mountcastle, 1997]. The width of a cellular minicolumn is a measure of the size of the core region of the minicolumn, which contains the majority of the neurons and apical dendrites, and both myelinated and unmyelinated fibers [Buxhoeveden & Casanova, 2002; Mountcastle, 1997; Seldon, 1981]. A cell-poor region, containing dendritic arbors, unmyelinated axons, and synapses, surrounds each minicolumn. The size of the cell-poor area is quantified as the distance between minicolumns. The width of these columns has been investigated in multiple auditory areas in the temporal lobe, including von Economo and Koskina's areas TA, TB, and TC [Seldon, 1981]. Both the width and the distance between minicolumns were found to be greater in the left hemisphere than in the right throughout auditory cortex. In most auditory cortical areas, the length of basal dendrites was found to compensate for asymmetric differences in distance between columns, but *not* in the planum temporale. In the planum temporale, the tangential extent of dendrites was increased in the left hemisphere, but not to a degree that completely compensated for the differences in minicolumn spacing between hemispheres. Some of these findings have been replicated by recent studies of area Tpt, which found a greater width of minicolumns and a relatively larger volume of neuropil space in the left hemisphere of humans, but no such asymmetry in chimpanzees and rhesus macaques [Buxhoeveden & Casanova, 2000].

Interconnectivity among cell columns in posterior area 22 has also been studied using carbocyanine dye to anterogradely label axons and retrogradely label cells in postmortem human specimens [Galuske et al., 2000]. Labeled terminal axon arbors and labeled cells were found to be superimposed, forming regularly spaced clusters surrounding the injection site. The average size of the clusters was the same in the two hemispheres, but the distance *among* clusters was significantly larger in the left hemisphere. Such an asymmetry did not exist in primary auditory cortex. Galuske et al. [2000] suggest that the labeled clusters represent different subsystems of interconnected columns and that the greater distance between clusters in the left hemisphere implies the presence of more subsystems within area 22 in the left hemisphere than in the right. However, it is also possible that larger cell columns may account for the distance between clusters in the left hemisphere. This would mean that axons in the right and left hemispheres reach across the same number of columns and that subsystems in the left hemisphere are simply more spread out, but not more numerous.

Using postmortem specimens and gross morphological boundaries, Anderson et al. [1999] found a volumetric asymmetry in white matter within the posterior superior temporal gyrus corresponding to a part of area Tpt in humans and then examined the contribution of various microstructural factors to this asymmetry. Although no interhemispheric differences were found in the number of nonphosphorylated neurofilament protein-immunoreactive neurons, the relative volume of glial cells, or the diameters of axons, the axons in the left hemisphere had significantly thicker myelin sheaths than their counterparts in the right hemisphere. It would be interesting to know whether comparable asymmetries in myelination might explain interhemispheric asymmetries in neuropil space reported in various other cortical areas.

Cell Size and Number. Another parameter in which asymmetries have been investigated is the size of individual cells. Neuronal soma volume is determined by the biosynthetic and metabolic requirements of the entire cell, including its dendritic arbors and axon [Kaas, 2000], suggesting that differences in cell volume may represent changes in the thickness and ramifications of cells' neurites or differences in metabolic activity. Asymmetries in cell sizes have been found in several regions throughout human cortex. In primary visual cortex, the left hemisphere tends to have larger neurons than the right, an asymmetry that is markedly absent in dyslexic patients [Jenner et al., 1999]. However, this region has greater numbers of neurons in the right hemisphere in rhesus macaques [Suner & Rakic, 1996].

In frontal cortex, the presence and direction of asymmetry differs by the region analyzed. In dorsal area 4, no interhemispheric differences were found in the mean size of layer III magnopyramidal cells. In area 45, layer III magnopyramidal cells were larger in the left hemisphere, while in area 46 these cell types were larger in the right hemisphere. Within the right hemisphere, no difference in cell size was observed among the three areas. However, cells in left area 45 were significantly larger than those in either left area 46 or left area 4 [Hayes & Lewis, 1995].

Furthermore, asymmetry within area 45 is observable only in the largest pyramidal cells in layer IIIb [Hayes & Lewis, 1995]. If all pyramidal cells are sampled equally, there is no difference in mean cell size between area 45 in the right and left hemispheres. This suggests that the distribution of cell sizes differs between the two hemispheres. If thc mean size of all pyramidal cells is equal in the two hemispheres, then left area 45 must also have more small cells than right area 45 in order to counterbalance the larger magnopyramidal cells. Total neuron count in the IFG of humans may also be asymmetrical [Uylings et al., 2006].

In a study of 10 human brains, Uylings et al. [2006] found neuron numbers in area 44 to be greater in the left than right hemisphere in all the 10 brains, but the difference only reached significance within 5 male brains. In area 45, neuron numbers were leftwardly asymmetric in all 5 female brains, but only in 2 of 5 male brains. However, no asymmetry was found in neuron *density* in this study.

A study of pyramidal cells in the superior temporal lobe found greater numbers of magnopyramidal cells throughout auditory cortex in the left hemisphere than in the right, including primary and secondary auditory cortex, as well as regions within Wernicke's area, such as the planum temporale and the supramarginal gyrus. The largest interhemispheric differences were seen in the anterior regions (Heschl's gyrus and anterior planum temporale) and the magnitude decreased posteriorly [Hutsler & Galuske, 2003].

Cell size asymmetries have also been found in particular subpopulations of pyramidal cells. Hutsler and Gazzaniga [1996] found size asymmetries in acetylcholinesterase-enriched cells in lower layer III of several regions in the posterior superior temporal lobe, not restricted to putative language areas. Another study also reports greater size of acetylcholinesterase-rich layer III magnopyramidal cells that is restricted to left area 45 of humans [Garcia et al., 2004]. Acetylcholinesterase is an enzyme responsible for the deactivation of acetylcholine. Density of the labeled cells was symmetrical, but differed among cortical areas. Interestingly, however, a previous study of choline acetyltransferase, an enzyme that facilitates the formation of acetylcholine, found that the concentration of this enzyme was significantly greater in the left hemispheres than in the right hemispheres of four individuals [Amaducci et al., 1981]. Peak concentration seemed to be located within layers II and IV in both hemispheres.

DISCUSSION

Functional Anatomy

Microstructural asymmetries have been primarily reported in four regions of the human brain, including primary motor cortex, primary visual cortex, and both anterior (Broca's) and posterior (Wernicke's) language areas.

Motor control is one of the most conspicuously lateralized functions. Most humans exhibit a preference for using one hand rather than the other for most tasks, with the majority showing right-hand dominance. Functional studies of primary motor cortex show that the strongest activation is generally within the primary motor cortex contralateral to the movement [Rao et al., 1995]. Greater usage is known to be related to the size of cortical areas within motor cortex in both humans [Amunts et al., 1997a; Karni et al., 1995; Pascual-Leone et al., 1995] and squirrel monkeys [Nudo et al., 1996]. Thus, if volumetric asymmetry of motor cortex reflects population-level right-hand dominance, we might find a leftward asymmetry in the total size of the motor cortex. The presence of such an asymmetry may [Zilles et al., 1996] or may not [White et al., 1997b] exist.

However, it is likely that such an asymmetry exists within the hand representation region of motor cortex [White et al., 1997b]. Other parameters also display asymmetry in portions of primary motor cortex. Relative neuropil volume seems to be greater in the left hemisphere in the hand region of area 4 [Amunts et al., 1996], which is particularly interesting if the same region expresses a volumetric asymmetry. Thus, in humans greater macrostructural asymmetry of the hand representation in the left hemisphere may be due in part to elaboration of interconnections in the dominant hemisphere. The complexity of connectivity also differs between hemispheres in the orofacial region as represented by dendritic geometry [Scheibel et al., 1985], but no interhemispheric difference has been found in cell size [Hayes & Lewis, 1995].

Vision is also functionally lateralized, with one eye being dominant. As with motor cortex, visual cortex is more strongly activated contralaterally [Miki et al., 2000, 2001]. However, despite the higher prevalence of right-eye dominant individuals [Annett, 2000], visual cortex more frequently exhibits a rightward volumetric asymmetry [Murphy, 1985] suggesting that the anatomical asymmetry is unrelated to eye dominance. Murphy [1985] has suggested the rightward bias in volume might underlie right hemisphere/left visual field superiority for a number of visual tasks [see Kimura & Durnford, 1974]. An asymmetry in the reverse direction is observed in cell size; the left hemisphere tends to have larger neurons than the right, an asymmetry that is markedly absent in dyslexic patients [Jenner et al., 1999].

The finding of an asymmetry in visual cortex that relates to language is interesting because language is an aspect of cognition that has garnered significant attention regarding both functional and anatomical lateralization. Broca [1861] and Wernicke [1874] first identified regions in the left hemisphere of the brain that are crucial to proper language function. Since those classical studies, the functional lateralization of these regions (IFG and posterior superior temporal) has been confirmed via countless studies using the lesion method and/or functional imaging. Furthermore, asymmetries have been reported in at least one language area for each microstructural parameter reviewed here.

Volumetric asymmetries have been reported in both anterior and posterior language areas. Population-level asymmetries were found in areas 44 and Tpt [Amunts et al., 1999; Galaburda & Sanides, 1980]. Asymmetry was also reported at the individual level in area 45, but not at the population level [Amunts et al., 1999]. Volumetric asymmetry in the white matter of the posterior superior temporal lobe has been found to be related to the thickness of the myelin sheaths encompassing axons [Anderson et al., 1999]. There is a greater percentage of neuropil space per unit volume in the dominant hemisphere in the IFG [Amunts et al., 2003].

Reports regarding dendritic parameters differ among cell populations and among regions. In the IFG, total dendritic length is greater in the left hemisphere [Scheibel et al., 1985], but if only magnopyramidal cells are examined, then dendritic parameters are greater in the right hemisphere [Hayes & Lewis, 1996]. In the temporal language area, there seems to be substantial individual variation in dendritic parameters. Asymmetries with left- [Jacobs et al., 1993] or right-hemisphere [Anderson & Rutledge, 1996] dominance have both been reported.

Cell columns have been found to be larger in the left planum temporale than in the right [Buxhoeveden & Casanova, 2000]. This may correspond to reports that patterns of interconnectivity differ between the two hemispheres in this region [Galuske et al., 2000]. To our knowledge, cell columns and related interconnectivity have yet to be analyzed in the IFG.

The size of magnopyramidal cell soma in layer III displays asymmetry in area 45, such that this subpopulation of the largest pyramidal cells is larger in the left hemisphere. This is in contrast to a neighboring region with no known language involvement, area 46, in which the corresponding cells are larger in the right hemisphere [Hayes & Lewis, 1995]. Similar asymmetries have been observed in temporal auditory and language areas, with greater numbers of layer III magnopyramidal cells present in the left hemisphere [Hutsler & Galuske, 2003]. Also, acetylcholinesterase-enriched cells are larger in left auditory areas and area 45 than in the corresponding regions in the right hemisphere [Garcia et al., 2004; Hutsler & Gazzaniga, 1996].

Individual Variability and Population-Level Asymmetry

In all of the above studies that report individual values, there is strong evidence that individual variability is present. Thus, even where substantial asymmetry exists in individuals, it does not necessarily follow that population-level asymmetry is present. For example, in a sample of 10 human brains, 9 exhibited strong asymmetry in the volume of area 45 (>6%), but 4 favored the right hemisphere and 5 favored the left, meaning that no significant asymmetry was seen at the population level [Amunts et al., 1999].

Similarly, the presence of a population-level asymmetry does not rule out the presence of individual variability. In fact, the presence of substantial interindividual variability means that the population-level findings of a study can change with the addition of more individuals. This is evident by a comparison of Amunts and colleagues' [1999, 2003] studies on asymmetries in Broca's area. In the earlier study, the sample size was 10, which in histological studies is quite large. Nonetheless, the subsequent study included additional adult individuals (sample size of 16) and resulted in different conclusions. Furthermore, the differences between hemispheres in these two studies are quite small when compared with interindividual variability.

Asymmetry in Nonhumans

Very limited evidence is currently available to examine whether humanlike asymmetries are present in nonhuman species [Buxhoeveden et al., 2001;

Gannon et al., 2000; Kheck et al., 1999; Rosen et al., 1993; Sherwood et al., 2005]. In a direct comparison of the microstructure of area Tpt in humans, chimpanzees, and rhesus macaques, Buxhoeveden and colleagues [2001] reported that only humans have left dominant asymmetry of neuropil volume and minicolumn widths, suggesting that the microstructure of the dominant hemisphere has been reorganized in humans for its involvement in language function. Small samples of nonhuman species in this study, however, give reason to be cautious in concluding that such histological asymmetries are entirely absent. If population-level asymmetries are present, but to a lesser magnitude than found in humans, they would be more difficult to detect statistically in a small sample. In this regard, it is notable that another study that examined volumetric asymmetries of cytoarchitecturally defined area Tpt in long-tailed macaques, revealed significantly greater volume of this cortical area in the left hemisphere [Gannon et al., 2000]. Interestingly, asymmetries have also been described in the distribution of calcium-binding protein-immunoreactive inhibitory interneuron subtypes within area Tpt of macaques [Kheck et al., 1999]. Unfortunately, comparable data on interneuron distributions within area Tpt of humans does not exist. Taken together, these findings suggest that asymmetry of the size and some aspects of microcircuitry in area Tpt of humans may be an ancestral homology that is shared with other Old World primates. This interpretation is consistent with observations based on behavior, functional imaging, and lesion studies indicating that macaques are left hemisphere dominant for the processing of acoustic features in conspecific vocal calls [Hauser & Andersson, 1994; Heffner & Heffner, 1984; Petersen et al., 1978, 1984; Poremba et al., 2004].

It has been suggested that, given the conduction delays associated with interhemispheric transfer, functional and structural asymmetries evolve as an adaptation to preserve temporal fidelity in the processing of complex streams of serial information, such as the vocal calls of conspecifics and the performance of fine motor sequences [Ringo et al., 1994]. Thus, it might be expected that due to these network constraints, lateral asymmetries will emerge among any species that relies heavily on acoustic communication in its social interactions or that displays high dexterity of movements. Indeed, house mouse mothers exhibit a right ear preference in their orientation response to the ultrasonic distress calls of their pups [Ehret, 1987] and electrophysiological mapping reveals a greater extent of auditory cortex surface area in the left hemisphere compared to the right [Stiebler et al., 1997]. Furthermore, data from multiunit recordings in starlings show lateralization in the strength of neuronal activation in response to the presentation of species-specific songs, but not other artificial sounds, with interindividual variation in the dominant hemisphere [George et al., 2002]. Although the results of these studies would seem to suggest that humanlike asymmetry of auditory cortex is prevalent among vertebrates, a cytoarchitectural study of auditory cortical area Doppler-shifted constant frequency area (DSCF) in mustached bats did not find population-level asymmetries in neuronal densities, glial-neuron ratios, or the distribution of magnopyramidal cells

[Sherwood et al., 2005]. This is in spite of the fact that mustached bats use a complex repertoire of vocalizations, with heteroharmonic calls that can last a second or more [Kanwal et al., 1994], and some auditory cortical neurons display specialized response properties for social vocalizations [Esser et al., 1997; Ohlemiller et al., 1996].

While functional asymmetry may be a common feature in processing social vocal communication among vertebrates, the microstructural correlates of this phenomenon have yet to be fully elucidated. Evidence from mice and macaques suggest that the volumetric extent of auditory cortical areas may express left hemisphere dominance. However, most studies of the intrinsic microstructure of the cortex in nonhumans have failed to reveal patterns of asymmetries that are homologous with humans. Some suggest that the particular pattern of asymmetries present in the human cortex is the defining characteristic of the human species [Crow, 1998a, 1998b]. The argument is that there are species-level asymmetries in humans, showing a consistency in the direction of asymmetry across individuals that may be absent in other species. Furthermore, certain psychological disorders, such as schizophrenia, may be associated with abnormal or absent asymmetries [Buchanan et al., 2004; Crow, 2004; Irle et al., 2005; Niznikiewicz et al., 2000] of the cortex, suggesting that typical asymmetries are important for normal human brain function. However, one must consider that the absence of evidence for asymmetries in nonhuman primates is not evidence of absence. Given the paucity of comparative data concerning microstructural cortical asymmetries, it is premature to assume that such asymmetries are absent among nonhumans.

Future Directions

There is ample evidence for microstructural asymmetries in human isocortex. However, the evidence is neither consistent nor uniformly distributed. No parameter has been analyzed equivalently across many cortical areas, and repeated analyses of a single measure in one cortical area sometimes produce differing results. Furthermore, our knowledge of the presence or absence of asymmetries in nonhumans is even more limited than our knowledge of asymmetries in humans.

There are tremendous opportunities for additional research in this area. Much is not yet known regarding how asymmetry in a particular region reflects the function of that region, how much individual variability there is in asymmetry and how that variability is reflected in population-level asymmetry, nor how cortical asymmetries have evolved over time. Future studies are needed to fill this gap. Ongoing investigations include the analysis of cell columns in humans and their closest relatives (great apes and gibbons) in multiple regions and the mapping of individual cortical areas across species. Additional future studies should include increased investigations of both humans and nonhumans and examination of microstructural parameters that can be compared with existing studies. Further studies of humans that can replicate and expand upon existing findings will help to elucidate the functional anatomy of asymmetry, while additional comparative analyses will serve to illuminate the evolution of asymmetry and further shed light on possible correlations with known functional asymmetry.

ACKNOWLEDGMENTS

This manuscript was prepared while being supported by the National Science Foundation (BCS-0406605, BCS-0515484, and BCS-0549117), the National Institutes of Health (NS-42867), the Wenner-Gren Foundation, and the James S. McDonnell Foundation (22002078).

REFERENCES

Amaducci, L.; Sorbi, S.; Albanese, A.; Gainotti, G. Choline-acetyltransferase (ChAT) activity differs in right and left human temporal lobes. Neurology 31(7): 799–805, 1981.

Amunts, K.; Jancke, L.; Mohlberg, H.; Steinmetz, H.; Zilles, K. Interhemispheric asymmetry of the human motor cortex related to handedness and gender. Neuropsychologia 38(3): 304–312, 2000.

Amunts, K.; Schlaug, G.; Jancke, L.; Steinmetz, H.; Schleicher, A.; Dabringhaus, A.; Zilles, K. Motor cortex and hand motor skills: Structural compliance in the human brain. Human Brain Mapping 5(3): 206–215, 1997a.

Amunts, K.; Schlaug, G.; Schleicher, A.; Steinmetz, H.; Dabringhaus, A.; Roland, P. E.; Zilles, K. Asymmetry in the human motor cortex and handedness. Neuroimage 4(3): 216–222, 1996.

Amunts, K.; Schleicher, A.; Burgel, U.; Mohlberg, H.; Uylings, H. B. M.; Zilles, K. Broca's regions revisited: Cytoarchitecture and intersubject variability. Journal of Comparative Neurology 412: 319–341, 1999.

Amunts, K.; Schleicher, A.; Ditterich, A.; Zilles, K. Broca's region: Cytoarchitectonic asymmetry and developmental changes. Journal of Comparative Neurology 465(1): 72–89, 2003.

Amunts, K.; Schmidt-Passos, F.; Schleicher, A.; Zilles, K. Postnatal development of interhemispheric asymmetry in the cytoarchitecture of human area 4. Anatomy and Embryology 196(5): 393–402, 1997b.

Anderson, B.; Rutledge, V. Age and hemisphere effects on dendritic structure. Brain 119: 1983–1990, 1996.

Anderson, B.; Southern, B. D.; Powers, R. E. Anatomic asymmetries of the posterior superior temporal lobes: A postmortem study. Neuropsychiatry, Neuropsychology, and Behavioral Neurology 12(4): 247–254, 1999.

Andrews, T. J.; Halpern, S. D.; Purves, D. Correlated size variations in human visual cortex, lateral geniculate nucleus, and optic tract. Journal of Neuroscience 17(8): 2859–2868, 1997.

Annett, M. Predicting combinations of left and right asymmetries. Cortex 36(4): 485–505, 2000.

Azari, N. P.; Nickel, J.; Wunderlich, G.; Niedeggen, M.; Hefter, H.; Tellmann, L.; Herzog, H.; Stoerig, P.; Birnbacher, D.; Seitz, R. J. Neural correlates of religious experience. European Journal of Neuroscience 13(8): 1649–1652, 2001.

Barbas, H.; Rempel-Clower, N. Cortical structure predicts the pattern of corticocortical connections. Cerebral Cortex 7(7): 635–646, 1997.

Beauregard, M.; Levesque, J.; Bourgouin, P. Neural correlates of conscious self-regulation of emotion. Journal of Neuroscience 21(18): U11–U16, 2001.

Blanton, R. E.; Levitt, J. G.; Thompson, P. M.; Narr, K. L.; Capetillo-Cunliffe, L.; Nobel, A.; Singerman, J. D.; McCracken, J. T.; Toga, A. W. Mapping cortical asymmetry and complexity patterns in normal children. Psychiatry Research-Neuroimaging 107(1): 29–43, 2001.

Bookheimer, S. Y.; Zeffiro, T. A.; Blaxton, T.; Malow, B. A.; Gaillard, W. D.; Sato, S.; Kufta, C.; Fedio, P.; Theodore, W. H. A direct comparison of PET activation and electrocortical stimulation mapping for language localization. Neurology 48(4): 1056–1065, 1997.

Branche, C.; Milner, B.; Rasmussen, T. Intracarotid sodium amytal for the lateralization of cerebral speech dominance. Journal of Neurosurgery 21: 399–405, 1964.

Broca, P. Remarques sur le siège de la faculté du langage articulé, suivies d'une observation d'aphémie. Bulletin de la Société Anatomique Paris 36: 330–357, 1861.

Brodmann, K. Vergleichende Lokalisationslehre der Grosshirnrinde in ihren Prinzipien dargestellt auf Grund des Zellenbaues (Garey L. J. translator). Leipzig: Johann Ambrosius Barth, 1909.

Buchanan, R. W.; Francis, A.; Arango, C.; Miller, K.; Lefkowitz, D. M.; McMahon, R. P.; Barta, P. E.; Pearlson, G. D. Morphometric assessment of the heteromodal association cortex in schizophrenia. American Journal of Psychiatry 161(2): 322–331, 2004.

Buxhoeveden, D.; Casanova, M. Comparative lateralization patterns of minicolumns in the language area of human, chimpanzee, and rhesus monkey brains. Laterality 5(4): 315–330, 2000.

Buxhoeveden, D.; Switala, A.; Litaker, M.; Roy, E.; Casanova, M. Lateralization of minicolumns in human planum temporale is absent in nonhuman primate cortex. Brain Behavior and Evolution 57: 349–358, 2001.

Buxhoeveden, D. P.; Casanova, M. F. The minicolumn hypothesis in neuroscience. Brain 125: 935–951, 2002.

Caceres, M.; Lachuer, J.; Zapala, M.; Redmond, J.; Kudo, L.; Geschwind, D.; Lockhart, D.; Preuss, T.; Barlow, C. Elevated gene expression levels distinguish human from non-human primate brains. Proceedings of the

National Academy of Sciences of the United States of America 100: 13030–13035, 2003.

Cantalupo, C.; Hopkins, W. D. Asymmetric Broca's area in great apes. Nature 414: 505, 2001.

Catani, M.; Jones, D.; ffytche, D. Perisylvian language networks of the human brain. Annals of Neurology 57: 8–16, 2005.

Crow, T. J. Sexual selection, timing and the descent of man: A theory of the genetic origins of language. Cahiers De Psychologie Cognitive-Current Psychology of Cognition 17(6): 1079–1114, 1998a.

Crow, T. J. Why cerebral asymmetry is the key to the origin of *Homo sapiens*: How to find the gene or eliminate the theory. Cahiers De Psychologie Cognitive-Current Psychology of Cognition 17(6): 1237–1277, 1998b.

Crow, T. J. Cerebral asymmetry and the lateralization of language: Core deficits in schizophrenia as pointers to the gene. Current Opinion in Psychiatry 17(2): 97–106, 2004.

Dagher, A.; Owen, A. M.; Boecker, H.; Brooks, D. J. Mapping the network for planning: A correlational PET activation study with the Tower of London task. Brain 122: 1973–1987, 1999.

Douglas, R.; Martin, K. Exploring cortical microcircuits: A combined anatomical, physiological, and computational approach. Pp. 381–412 in T. McKenna; J. Davies; S. Zornetzer, eds. Single neuron computation. San Diego, Academic Press, 1992.

Ehret, G. Left hemisphere advantage in the mouse brain for recognizing ultrasonic communication calls. Nature 325: 249–251, 1987.

Elston, G. N. Pyramidal cells of the frontal lobe: All the more spinous to think with. Journal of Neuroscience 20(18): RC95, 2000.

Elston, G. N.; Benavides-Piccione, R.; DeFelipe, J. The pyramidal cell in cognition: A comparative study in human and monkey. Journal of Neuroscience 21(17): RC163, 2001.

Elston, G. N.; Rockland, K. S. The pyramidal cell of the sensorimotor cortex of the macaque monkey: Phenotypic variation. Cerebral Cortex 12(10): 1071–1078, 2002.

Emmorey, K.; Allen, J. S.; Bruss, J.; Schenker, N.; Damasio, H. A morphometric analysis of auditory brain regions in congenitally deaf adults. Proceedings of the National Academy of Sciences of the United States of America 100(17): 10049–10054, 2003.

Enard, W.; Khaitovich, P.; Klose, J.; Zollner, S.; Heissig, F.; Giavalisco, P.; Nieselt-Struwe, K.; Muchmore, E.; Varki, A.; Ravid, R.; Doxiadis, G.; Bontrop, R.; Paabo, S. Intra- and interspecific variation in primate gene expression patterns. Science 296: 340–343, 2002.

Esser, K.; Condon, C.; Suga, N.; Kanwal, J. Syntax processing by auditory cortical neurons in the FM-FM area of the mustached bat *Pteronotus parnellii*. Proceedings of the National Academy of Sciences of the United States of America 94: 14019–14024, 1997.

Falzi, G.; Perrone, P.; Vignolo, L. A. Right–left asymmetry in anterior speech region. Archives of Neurology 39(4): 239–240, 1982.

Foundas, A. L.; Eure, K. F.; Luevano, L.; Weinberger, D. R. MRI asymmetries of Broca's area: The pars triangularis and pars opercularis. Brain and Language 64(3): 282–296, 1998.

Foundas, A. L.; Faulhaber, J. R.; Kulynych, J. J.; Browning, C. A.; Weinberger, D. R. Hemispheric and sex-linked differences in Sylvian fissure morphology: A quantitative approach using volumetric magnetic resonance imaging. Neuropsychiatry, Neuropsychology, and Behavioral Neurology 12(1): 1–10, 1999.

Foundas, A. L.; Weisberg, A.; Browning, C. A.; Weinberger, D. R. Morphology of the frontal operculum: A volumetric magnetic resonance imaging study of the pars triangularis. Journal of Neuroimaging 11(2): 153–159, 2001.

Galaburda, A.; Sanides, F. Cytoarchitectonic organization of the human auditory-cortex. Journal of Comparative Neurology 190(3): 597–610, 1980.

Galaburda, A. M.; Aboitiz, F.; Rosen, G. D.; Sherman, G. F. Histological asymmetry in the primary visual-cortex of the rat: Implications for mechanisms of cerebral asymmetry. Cortex 22(1): 151–160, 1986.

Galaburda, A. M.; Sanides, F.; Geschwind, N. Human-Brain: Cytoarchitectonic left–right asymmetries in temporal speech region. Archives of Neurology 35(12): 812–817, 1978.

Galuske, R. A. W.; Schlote, W.; Bratzke, H.; Singer, W. Interhemispheric asymmetries of the modular structure in human temporal cortex. Science 289(5486): 1946–1949, 2000.

Gannon, P. J.; Hof, P. R.; Kheck, N. Brain language area evolution: Human-like pattern of cytoarchitectonic, but not gross anatomic, L>R hemispheric asymmetry of planum temporale homolog in *Macaca fascicularis*. American Journal of Physical Anthropology Supplement 30: 2000.

Gannon, P. J.; Kheck, N.; Hof, P. Language areas of the hominoid brain: A dynamic communicative shift on the upper east side planum. Pp. 216–240 in D. Falk; K. Gibson, eds. Evolutionary Anatomy of the Primate Cerebral Cortex. Cambridge, Cambridge University Press, 2001.

Garcia, R.; Montiel, J.; Villalon, A.; Gatica, M.; Aboitiz, F. AChE-rich magnopyramidal neurons have a left–right size asymmetry in Broca's area. Brain Research 1026: 313–316, 2004.

George, I.; Cousillas, H.; Richard, J.; Hausberger, M. Song perception in the European starling: Hemispheric specialisation and individual variations. Comptes Rendus Biologie 325: 197–204, 2002.

Geschwind, D. Mice, microarrays, and the genetic diversity of the brain. Proceedings of the National Academy of Sciences of the United States of America 97: 10676–10678, 2000.

Grabowski, T. J.; Damasio, H.; Tranel, D.; Ponto, L. L. B.; Hichwa, R. D.; Damasio, A. R. A role for left temporal pole in the retrieval of words for unique entities. Human Brain Mapping 13(4): 199–212, 2001.

Hauser, M.; Andersson, K. Left hemisphere dominance for processing vocalizations in adult, but not infant, rhesus monkeys: Field experiments. Proceedings of the National Academy of Sciences of the United States of America 91: 3946–3948, 1994.

Hayes, T. L.; Lewis, D. A. Anatomical specialization of the anterior motor speech area: Hemispheric-differences in magnopyramidal neurons. Brain and Language 49(3): 289–308, 1995.

Hayes, T. L.; Lewis, D. A. Magnopyramidal neurons in the anterior motor speech region: Dendritic features and interhemispheric comparisons. Archives of Neurology 53(12): 1277–1283, 1996.

Heffner, H.; Heffner, R. Temporal lobe lesions and perception of species-specific vocalizations by macaques. Science 226: 75–76, 1984.

Holloway, R. L.; de Lacoste-Lareymondie, M. C. Brain endocast asymmetry in Pongids and Hominids: Some preliminary findings on the paleontology of cerebral-dominance. American Journal of Physical Anthropology 58(1): 101–110, 1982.

Hopkins, W. D.; Cantalupo, C. Handedness in chimpanzees (*Pan troglodytes*) is associated with asymmetries of the primary motor cortex but not with homologous language areas. Behavioral Neuroscience 118(6): 1176–1183, 2004.

Hopkins, W. D.; Marino, L.; Rilling, J.; MacGregor, L. Planum temporale asymmetries in great apes as revealed by magnetic resonance imaging (MRI). Neuroreport 9: 2913–2918, 1998.

Hutsler, J.; Galuske, R. A. W. Hemispheric asymmetries in cerebral cortical networks. Trends in Neurosciences 26(8): 429–435, 2003.

Hutsler, J. J.; Gazzaniga, M. S. Acetylcholinesterase staining in human auditory and language cortices: Regional variation of structural features. Cerebral Cortex 6(2): 260–270, 1996.

Irle, E.; Lange, C.; Sachsse, U. Reduced size and abnormal asymmetry of parietal cortex in women with borderline personality disorder. Biological Psychiatry 57(2): 173–182, 2005.

Jacobs, B.; Schall, M.; Prather, M.; Kapler, E.; Driscoll, L.; Baca, S.; Jacobs, J.; Ford, K.; Wainwright, M.; Treml, M. Regional dendritic and spine variation in human cerebral cortex: A quantitative Golgi study. Cerebral Cortex 11(6): 558–571, 2001.

Jacobs, B.; Schall, M.; Scheibel, A. B. A quantitative dendritic analysis of Wernicke area in humans. 2. Gender, hemispheric, and environmental-factors. Journal of Comparative Neurology 327(1): 97–111, 1993.

Jacobs, B.; Scheibel, A. B. A quantitative dendritic analysis of Wernicke area in humans. 1. Life-span changes. Journal of Comparative Neurology 327(1): 83–96, 1993.

Jancke, L.; Schlaug, G.; Huang, Y. X.; Steinmetz, H. Asymmetry of the planum parietale. Neuroreport 5(9): 1161–1163, 1994.

Jenner, A. R.; Rosen, G. D.; Galaburda, A. M. Neuronal asymmetries in primary visual cortex of dyslexic and nondyslexic brains. Annals of Neurology 46(2): 189–196, 1999.

Kaas, J. Why is brain size so important: Design problems and solutions as neocortex gets bigger or smaller. Brain and Mind 1: 7–23, 2000.

Kanwal, J.; Matsumura, S.; Ohlemiller, K.; Suga, N. Analysis of acoustic elements and syntax in communication sounds emitted by mustached bats. Journal of the Acoustical Society of America 96: 1229–1254, 1994.

Karni, A.; Meyer, G.; Jezzard, P.; Adams, M. M.; Turner, R.; Ungerleider, L. G. Functional MRI evidence for adult motor cortex plasticity during motor skill learning. Nature 377(6545): 155–158, 1995.

Khaitovich, P.; Muetzel, B.; She, X.; Lachmann, M.; Hellmann, I.; Dietzsch, J.; Steigele, S.; Do, H.; Weiss, G.; Enard, W.; Heissig, F.; Arendt, T.; Nieselt-Strump, K.; Eichler, E. E.; Paabo, S. Regional patterns of gene expression in human and chimpanzee brains. Genome Research 14: 1462–1473, 2004.

Kheck, N.; Hof, P. R.; Deftereos, M.; Lo, T.; Gannon, P. J. Brain language area evolution IV: Chemoarchitectonic interhemispheric asymmetries of the planum temporale (PT) homolog in Old World monkeys. Society of Neuroscience Abstracts 25: 105, 1999.

Kimura, D.; Durnford, M. Normal studies in the function of the right hemisphere in vision. Pp. 25–47 in S. Dimond; J. Beaumont, eds. Hemisphere Function in the Brain. New York, Halstead, 1974.

Knaus, T. A.; Bollich, A. M.; Corey, D. M.; Lemen, L. C.; Foundas, A. L. Variability in perisylvian brain anatomy in healthy adults. Brain and Language 97(2): 219–232, 2006.

Krubitzer, L.; Huffman, K. J. Arealization of the neocortex in mammals: Genetic and epigenetic contributions to the phenotype. Brain, Behavior and Evolution 55(6): 322–335, 2000.

Krubitzer, L.; Kahn, D. M. Nature versus nurture revisited: An old idea with a new twist. Progress in Neurobiology 70(1): 33–52, 2003.

Miki, A.; Liu, G. T.; Englander, S. A.; van Erp, T. G. M.; Bonhomme, G. R.; Aleman, D. O.; Liu, C. S. J.; Haselgrove, J. C. Functional magnetic resonance imaging of eye dominance at 4 tesla. Ophthalmic Research 33(5): 276–282, 2001.

Miki, A.; Liu, G. T.; Raz, J.; van Erp, T. G. M.; Liu, C. S. J.; Leuthardt, E. C.; Haselgrove, J. C. Contralateral monocular dominance in anterior visual cortex confirmed by functional magnetic resonance imaging. American Journal of Ophthalmology 130(6): 821–824, 2000.

Mountcastle, V. B. The columnar organization of the neocortex. Brain 120(Pt 4): 701–722, 1997.

Murphy, G. M. Volumetric asymmetry in the human striate cortex. Experimental Neurology 88(2): 288–302, 1985.

Niznikiewicz, M.; Donnino, R.; McCarley, R. W.; Nestor, P. G.; Iosifescu, D. V.; O'Donnell, B.; Levitt, J.; Shenton, M. E. Abnormal angular gyrus asymmetry in schizophrenia. American Journal of Psychiatry 157(3): 428–437, 2000.

Nudo, R. J.; Milliken, G. W.; Jenkins, W. M.; Merzenich, M. M. Use-dependent alterations of movement representations in primary motor cortex of adult squirrel monkeys. Journal of Neuroscience 16(2): 785–807, 1996.

Ohlemiller, K.; Kanwal, J.; Suga, N. Facilitative responses to species-specific calls in cortical FM-FM neurons of the mustached bat. Neuroreport 7: 1749–1755, 1996.

Ojemann, G. Cortical organization of language. Journal of Neuroscience 11: 2281–2287, 1991.

Pascual-Leone, A.; Wassermann, E. M.; Sadato, N.; Hallett, M. The role of reading activity on the modulation of motor cortical outputs to the reading hand in Braille readers. Annals of Neurology 38(6): 910–915, 1995.

Penhune, V. B.; Zatorre, R. J.; MacDonald, J. D.; Evans, A. C. Interhemispheric anatomical differences in human primary auditory cortex: Probabilistic mapping and volume measurement from magnetic resonance scans. Cerebral Cortex 6(5): 661–672, 1996.

Pesenti, M.; Zago, L.; Crivello, F.; Mellet, E.; Samson, D.; Duroux, B.; Seron, X.; Mazoyer, B.; Tzourio-Mazoyer, N. Mental calculation in a prodigy is sustained by right prefrontal and medial temporal areas. Nature Neuroscience 4(1): 103–107, 2001.

Peters, A.; Kara, D. A. The neuronal composition of area 17 of rat visual-cortex. 4. The organization of pyramidal cells. Journal of Comparative Neurology 260(4): 573–590, 1987.

Peters, A.; Sethares, C. Organization of pyramidal neurons in area-17 of monkey visual-cortex. Journal of Comparative Neurology 306(1): 1–23, 1991.

Petersen, M.; Beecher, M.; Zoloth, S.; Green, S.; Marler, P.; Moody, D.; Stebbins, W. Neural lateralization of vocalizations by Japanese macaques: Communicative significance is more important than acoustic structure. Behavioral Neuroscience 98: 779–790, 1984.

Petersen, M.; Beecher, M.; Zoloth, S.; Moody, D.; Stebbins, W. Neural lateralization of species-specific vocalizations by Japanese macaques (*Macaca fuscata*). Science 202: 324–327, 1978.

Petersen, S.; Fox, P.; Posner, M.; Minton, M.; Raichle, M. Positron emission tomographic studies of the cortical anatomy of single-word processing. Nature 331: 585–589, 1988.

Phillips, K. A.; Sherwood, C. C. Primary motor cortex asymmetry is correlated with handedness in capuchin monkeys (*Cebus apella*). Behavioral Neuroscience 119(6): 1701–1704, 2005.

Poremba, A.; Malloy, M.; Saunders, R.; Carson, R.; Herscovitch, P.; Mishkin, M. Species-specific calls evoke asymmetric activity in the monkey's temporal poles. Nature 427: 448–451, 2004.

Rakic, P. Radial versus tangential migration of neuronal clones in the developing cerebral-cortex. Proceedings of the National Academy of Sciences of the United States of America 92(25): 11323–11327, 1995.

Rao, S. M.; Binder, J. R.; Hammeke, T. A.; Bandettini, P. A.; Bobholz, J. A.; Frost, J. A.; Myklebust, B. M.; Jacobson, R. D.; Hyde, J. S. Somatotopic mapping of the human primary motor cortex with functional magnetic resonance imaging. Neurology 45(5): 919–924, 1995.

Rilling, J. K.; Seligman, R. A. A quantitative morphometric comparative analysis of the primate temporal lobe. Journal of Human Evolution 42(5): 505–533, 2002.

Ringo, J; Doty, R; Demeter, S; Simard, P. Time is of the essence: A conjecture that hemispheric specialization arises from interhemispheric conduction delay. Cerebral Cortex 4: 331–343, 1994.

Rockland, K. S.; Ichinohe, N. Some thoughts on cortical minicolumns. Experimental Brain Research 158(3): 265–277, 2004.

Rosen, G. D. Cellular, morphometric, ontogenetic and connectional substrates of anatomical asymmetry. Neuroscience and Biobehavioral Reviews 20(4): 607–615, 1996.

Rosen, G. D.; Sherman, G. F.; Galaburda, A. Neuronal subtypes and anatomic asymmetry: Changes in neuronal number and cell-packing density. Neuroscience 56: 833–839, 1993.

Rowe, J. B.; Passingham, R. E. Working memory for location and time: Activity in prefrontal area 46 relates to selection rather than maintenance in memory. Neuroimage 14(1): 77–86, 2001.

Scheibel, A. B.; Paul, L. A.; Fried, I.; Forsythe, A. B.; Tomiyasu, U.; Wechsler, A.; Kao, A.; Slotnick, J. Dendritic organization of the anterior speech area. Experimental Neurology 87(1): 109–117, 1985.

Schenker, N. M.; Desgouttes, A.-M.; Semendeferi, K. Neural connectivity and cortical substrates of cognition in hominoids. Journal of Human Evolution 49(5): 547–569, 2005.

Schleicher, A.; Amunts, K.; Geyer, S.; Morosan, P.; Zilles, K. Observer-independent method for microstructural parcellation of cerebral cortex: A quantitative approach to cytoarchitectonics. Neuroimage 9(1): 165–177, 1999.

Seldon, H. L. Structure of human auditory-cortex. 1. Cytoarchitectonics and dendritic distributions. Brain Research 229(2): 277–294, 1981.

Sherwood, C. C.; Broadfield, D. C.; Holloway, R. L.; Gannon, P. J.; Hof, P. R. Variability of Broca's area homologue in African Great Apes: Implications for language evolution. Anatomical Record Part A 271A: 276–285, 2003.

Sherwood, C. C.; Raghanti, M. A.; Wenstrup, J. J. Is humanlike cytoarchitectural asymmetry present in another species with complex social vocalization? A stereologic analysis of mustached bat auditory cortex. Brain Research 1045(1–2): 164–174, 2005.

Stiebler, I.; Neulist, R.; Fichtel, I.; Ehret, G. The auditory cortex of the house mouse: Left–right differences, tonotopic organization and quantitative analysis of frequency representation. Journal of Comparative Physiology A, Sensory, Neural, and Behavioral Physiology 181: 559–571, 1997.

Suner, I.; Rakic, P. Numerical relationship between neurons in the lateral geniculate nucleus and primary visual cortex in macaque monkeys. Visual Neuroscience 13(3): 585–590, 1996.

Tomaiuolo, F.; MacDonald, J. D.; Caramanos, Z.; Posner, G.; Chiavaras, M.; Evans, A. C.; Petrides, M. Morphology, morphometry and probability mapping of the pars opercularis of the inferior frontal gyrus: An *in vivo* MRI analysis. European Journal of Neuroscience 11(9): 3033–3046, 1999.

Uddin, M.; Wildman, D.; Liu, G.; Xu, W.; Johnson, R.; Hof, P. R.; Kapatos, G.; Grossman, L.; Goodman, M. Sister grouping of chimpanzees and humans as revealed by genome-wide phylogenetic analysis of brain gene expression profiles. Proceedings of the National Academy of Sciences of the United States of America 101: 2957–2962, 2004.

Uylings, H. B. M.; Jacobsen, A. M.; Zilles, K.; Amunts, K. Left–right asymmetry in volume and number of neurons in adult Broca's area. Cortex 42(4): 652–658, 2006.

Wada, J. A.; Clarke, R.; Hamm, A. Cerebral hemispheric asymmetry in humans: Cortical speech zones in 100 adult and 100 infant brains. Archives of Neurology 32(4): 239–246, 1975.

Watkins, K. E.; Paus, T.; Lerch, J. P.; Zijdenbos, A.; Collins, D. L.; Neelin, P.; Taylor, J.; Worsley, K. J.; Evans, A. C. Structural asymmetries in the human brain: A voxel-based statistical analysis of 142 MRI scans. Cerebral Cortex 11(9): 868–877, 2001.

Wernicke, C. Das Aphasiche Symptomenkomplex. Breslau, Poland, Cohn and Weigart, 1874.

White, L. E.; Andrews, T. J.; Hulette, C.; Richards, A.; Groelle, M.; Paydarfar, J.; Purves, D. Structure of the human sensorimotor system. 1. Morphology and cytoarchitecture of the central sulcus. Cerebral Cortex 7(1): 18–30, 1997a.

White, L. E.; Andrews, T. J.; Hulette, C.; Richards, A.; Groelle, M.; Paydarfar, J.; Purves, D. Structure of the human sensorimotor system. 2. Lateral symmetry. Cerebral Cortex 7(1): 31–47, 1997b.

Zilles, K.; Dabringhaus, A.; Geyer, S.; Amunts, K.; Qu, M.; Schleicher, A.; Gilissen, E.; Schlaug, G.; Steinmetz, H. Structural asymmetries in the human forebrain and the forebrain of non-human primates and rats. Neuroscience and Biobehavioral Reviews 20(4): 593–605, 1996.

Zilles, K.; Palomero-Gallagher, N.; Grefkes, C.; Scheperjans, F.; Boy, C.; Amunts, K.; Schleicher, A. Architectonics of the human cerebral cortex and transmitter receptor fingerprints: Reconciling functional neuroanatomy and neurochemistry. European Neuropsychopharmacology 12(6): 587–599, 2002.

Zilles, K.; Palomero-Gallagher, N.; Schleicher, A. Transmitter receptors and functional anatomy of the cerebral cortex. Journal of Anatomy 205(6): 417–432, 2004.

The chimpanzee Elwood producing a pant-hoot vocalization following the discovery of food (a coconut). (Photograph by Jared Taglialatela.)

The Evolution of Hemispheric Specialization in Primates
W.D. Hopkins (Editor)

FUNCTIONAL AND STRUCTURAL ASYMMETRIES FOR AUDITORY PERCEPTION AND VOCAL PRODUCTION IN NONHUMAN PRIMATES

Jared P. Taglialatela
Division of Psychobiology, Yerkes National Primate Research Center, Atlanta, GA, USA

ABSTRACT

Language processes are lateralized to one hemisphere or the other in the human brain. Whether similar asymmetries in the perception and production of communicative signals are present in other animals remains a topic of considerable debate. This paper reviews the data on behavioral asymmetries in nonhuman primate auditory perception and vocal production. In addition, data on neuroanatomical asymmetries as well as more recent functional asymmetries are reviewed and discussed in the context of different evolutionary theories of brain asymmetry and language.

Keywords: speech perception, speech production, brain imaging, Wernicke's, planum temporale, primate asymmetry, language homologs

INTRODUCTION

One salient feature of language processing in the human brain is lateralization, or the predominant localization of function to one hemisphere as compared to the other. According to the classical model popularized in the second half of the nineteenth century, two cerebral cortical areas in the left hemisphere are most commonly associated with language functions. One, a receptive region for the processing and integration of auditory sensory information commonly referred to as Wernicke's [1874] area, and the second, a productive region concerned with the encoding of vocal signals into meaningful words and sentences, Broca's area [Broca, 1861]. In other words, Broca's area functions primarily in the planning and execution of speech, whereas Wernicke's area functions to "make sense" of the speech that a listener perceives. Although this modular view of linguistic processing is now considered outdated [e.g., see Poeppel & Hickock, 2004], recent evidence from neuroimaging studies confirm that both men and women show significant left lateralized patterns of activation during the execution of language-related tasks [Binder & Frost, 1998; Binder et al., 1997; Buchanan et al., 2000; Frost et al., 1999].

ISSN: 1936-8526 DOI: 10.1016/S1936-8526(07)05005-1

Although left lateralization is most common in humans, it is not universal. Less than 10 percent of the population show either a right-hemisphere bias or ambiguous lateralization [Deacon, 1997], and recent imaging studies have revealed a graded continuum from left to right lateralization [Knecht et al., 2002]. Right language lateralization, although uncommon, appears to be a natural variant as opposed to a compensatory phenotype [Knecht et al., 2003], and does not appear to interfere in any way with the realization of full linguistic competency [Knecht et al., 2001].

In contrast to the classical model, contemporary theories implicate the right cerebral hemisphere in the processing of a variety of language functions. For example, emotive and expressive vocalizations in humans involve the recruitment of different brain regions [Binder et al., 1997]. A recent functional magnetic resonance imaging (fMRI) study of human subjects compared activation between the left and right inferior frontal lobes during tasks that required the recognition of emotional prosody and the verbal components of speech production [Buchanan et al., 2000]. The investigators found a right-hemisphere lateralization of function for the detection of emotion when compared to the detection of verbal components, and a left-hemisphere lateralization in the reverse condition. In addition, the lateralized activation pattern observed during human language tasks is modality independent [Grossi et al., 1996; Hickok et al., 1998], and is associated not merely with the perception or production of words, but with their meaning [Thiel et al., 1998; Zahn et al., 2000]. Additional evidence points to attention as a modulating component during the processing of auditory linguistic information [Hugdahl et al., 2003b]. Therefore, although not all aspects of human language perception, processing, or production are restricted to the left hemisphere, most do indeed exhibit lateralized activity patterns. This suggests that lateralization plays an important functional role in the mediation of linguistic behaviors in humans.

Humans are not the only species to show asymmetries during the perception of conspecific vocalizations. This chapter will review the literature in nonhuman primate auditory perception and vocal production, but evidence in non-primate species suggests that hemispheric specialization for the perception and processing of communicative signals may not be restricted to the primate order [Boye et al., 2005; Ehret, 1987; George et al., 2005; Okanoya et al., 2001; Palleroni & Hauser, 2003]. For example, adult and subadult California sea lions preferentially orient their right ear to a sound source when presented with conspecific vocalizations [Boye et al., 2005]. In addition, there is considerable evidence of brain lateralization for bird song production [see Schmidt et al., 2004, for a recent review], which involves the bilateral coordination of functionally distinct hemispheres, as well as some data suggesting hemispheric specialization for the perception and processing of songs [George et al., 2005; Okanoya et al., 2001].

AUDITORY PERCEPTION

Indirect Behavioral Assays

Orienting Asymmetries. The study of nonhuman primate vocal and auditory neurophysiology, psychophysics, and behavior continue to provide novel insights into human language, its phylogeny, as well as the evolutionary origins of the human brain. Given the prevalence and significance of lateralization for linguistic behavior in humans, an examination of the structures involved in the mediation of nonhuman primate auditory–vocal systems may reveal if, and to what extent, this lateralization is unique to humans. To this end, researchers have employed a variety of methods in an attempt to assess the presence and magnitude of anatomical, behavioral, and functional asymmetries in nonhuman primates (Table 1). A popular approach has been to utilize psychophysical assays in an attempt to non-invasively address questions of functional lateralization. One method has been to measure orienting asymmetries in response to the presentation of auditory stimuli [Ghazanfar et al., 2001; Gil-da-Costa & Hauser, 2006; Hauser et al., 1998; Hauser & Andersson, 1994]. In this procedure, an auditory stimulus is presented to a subject, and the subject's change in orientation relative to the sound source is recorded (Figure 1). This method is particularly useful for it permits the behavioral evaluation of subjects without the need for training or invasive procedures. The idea behind this approach stems from the fact that spatially biasing one ear towards the source of an auditory stimulus increases the intensity of the signal at that ear as well as the contralateral hemisphere. In fact, speech sound signals presented monaurally to human subjects caused broader regional cerebral blood flow increases in the contralateral Heschl's gyrus when compared to the ipsilateral hemisphere [Hirano et al., 1997]. Researchers employing this method with nonhuman primates interpret orienting biases as evidence of lateralized processing in the contralateral hemisphere. Before discussing the potential pitfalls of this assumption, let us review the data from these studies.

Hauser and Andersson [1994], conducted playback experiments on a relatively large group ($N = 80$) of semi-free-ranging rhesus macaques *(Macaca mulatta)* on the island of Cayo Santiago in order to determine whether the subjects would preferentially show an orienting asymmetry in response to the presentation of conspecific vocalizations. Samples of 12 rhesus monkey call types spanning the vocal repertoire and broadly categorized into the three communicative categories of aggressive, fearful, and affiliative were used as stimuli. The authors report that significantly more adult subjects turned to the right than to the left following the presentation of all three types of conspecific vocalizations. In contrast, a left turning bias was observed following the presentation of a heterospecific vocalization (the alarm call of a sympatric seabird), and no orienting asymmetries were observed in infants less than one year of age following any of the stimuli types. In a second study, Hauser et al. [1998] sought to determine if changes in the acoustic structure of a vocalization beyond the

Table 1. Studies concerning nonhuman primate behavioral asymmetries as well as neuroanatomical and functional lateralization.

Assay	Species	Results	Reference
Orienting asymmetry (OA)	*Macaca mulatta*	Adults show right OA for conspecific vocalizations	Hauser and Andersson [1994]
Orienting asymmetry (OA)	*M. fuscata*	Right OA eliminated or reversed for acoustically modified calls	Hauser et al. [1998]
Orienting asymmetry (OA)	*M. mulatta*	Right OA transferred to left for time-reversed calls	Ghazanfar et al. [2001]
Orienting asymmetry (OA)	*Chlorocebus aethiops*	Left OA for conspecific vocalizations (familiar and unfamiliar)	Gil-da-Costa and Hauser [2006]
Ear advantages (EA)	*M. fuscata*	Right EA for discriminating conspecific calls based on "communicatively relevant" feature, but not for pitch	Petersen et al. [1978]
Ear advantages (EA)	*M. fuscata*	Right EA for discriminating calls based on "communicatively relevant" feature, but not for pitch of the same vocalizations	Beecher et al. [1979]
Ear advantages (EA)	*M. fuscata*	Right EA for discriminating conspecific calls; control subjects attending to same acoustic properties, but no EA	Petersen et al. [1984]

Ear advantages (EA)	*M. mulatta*	Right EA for discriminating among exemplars from two acoustically distinct classes of conspecific vocalizations	Le Prell et al. [2002]
Lesion	*M. fuscata*	Left ablation of the STG temporarily disrupts discrimination of conspecific calls; right ablation does not	Heffner and Heffner [1984]
PET	*M. mulatta*	Conspecific vocalizations evoke significantly greater metabolic activity in the left STG	Poremba et al. [2004]
PET	*M. mulatta*	Conspecific vocalizations evoke activity in the homologues of perisylvian language areas	Gil-da-Costa et al. [2006]
Cadaver specimens	*Pan troglodytes*	Left hemisphere PT larger than right in 94 percent of specimens	Gannon et al. [1998]
Structural MRI	*Gorilla gorilla, P. paniscus, P. troglodytes, P. pygmaeus*	PT is anatomically identifiable in all four great apes and is larger in the left hemisphere	Hopkins et al. [1998]
Structural MRI	*G. gorilla, P. paniscus, P. troglodytes*	Left hemisphere area 44 larger than right	Cantalupo and Hopkins [2001]
Potential recording	*M. fuscata*	Right cerebellar hemispherectomy eliminated s-N, d-P potentials in face motor cortex and Broca's area	Gemba et al. [1995]

Table 1. (Continued)

Assay	Species	Results	Reference
		homologue in the left hemisphere	
Oro-facial asymmetry (OFA)	*Callithrix jacchus*	Larger right hemi-mouth during social call production, larger left for fear	Hook-Costigan and Rogers [1998]
Oro-facial asymmetry (OFA)	*P. troglodytes*	Leftward OFA for hoots and screams	Fernandez-Carriba et al. [2002]
Oro-facial asymmetry (OFA)	*P. troglodytes*	Leftward OFA for hoots and food barks, rightward for "volitional" vocalizations	Reynolds et al. [2005]

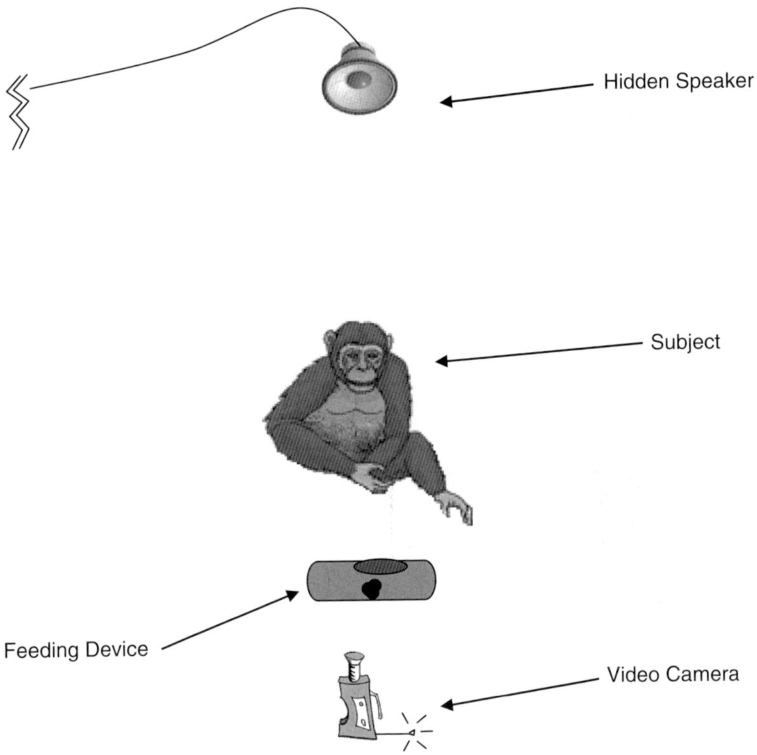

Figure 1. Playback experiment diagram. The figure depicts the equipment and basic setup used for playback experiments.

species-typical variation would eliminate or reverse the observed orienting asymmetry. To this end, the authors again presented semi-free-ranging rhesus monkeys on the island of Cayo Santiago with prerecorded conspecific vocalizations. Three pulsed call types were selected, affilitative grunts, alarm shrill barks, and copulation screams. The vocalizations were then modified by either increasing or decreasing the inter-pulse interval beyond the species norms. Consistent with previous findings, presentation of unaltered vocalizations from all three-call types evoked right turning biases. However, elimination of inter-pulse intervals or increasing them to the population maximum eradicated turning biases for the grunts and barks, but not the screams. Doubling the maximum inter-pulse interval reversed the orienting asymmetry, but again only for grunts and barks. The data from this study indicate that, for at least some call types, the observed right orienting asymmetry for unaltered conspecific vocalizations was eliminated or reversed following the presentation of structurally modified calls.

Ghazanfar et al. [2001] conducted a similar study using the same semi-free-ranging rhesus monkeys on the island of Cayo Santiago. The authors utilized two different call types from the subjects' repertoire: alarm shrill barks, and harmonic arches produced upon discovering high-quality food. Normal and time-reversed vocalizations were presented to the monkeys. Rightward orienting biases were reported following normal calls, and left turning following reversed vocalizations.

Finally, Gil-da-Costa and Hauser [2006], working with vervet monkeys, reported a similar orienting asymmetry following the presentation of conspecific vocalizations, but not in response to heterospecific calls or non-biological sounds. However, the asymmetry was in the opposite direction compared to the studies discussed above. Using an orienting asymmetry paradigm adapted for the laboratory setting, the authors reported a leftward turning bias following the presentation of both familiar and unfamiliar conspecific vocalizations. This somewhat surprising finding warrants careful consideration. Although the implications of this leftward asymmetry will be addressed later in the chapter, the report itself lacks sufficient detail with regard to the experimental method and analyses of results to accept the authors' conclusions at face value. Specifically, the study was completed with a relatively small sample size for a behavioral study of this kind ($N = 5$ for condition I; $N = 4$ for condition II). In addition, the number of trials of each stimulus type each individual received as well as whether these numbers were the same for all individuals is not known. These details are critical for evaluating the results of this study.

Collectively, the results from the studies with rhesus macaques are consistent with the claim that, "The development of hemispheric dominance for processing conspecific calls may depend upon a fully mature brain, as well as sufficient exposure to and understanding of the vocal repertoire," [Hauser, 1996, p 209]. Although this is an intriguing hypothesis, to date there are no data in nonhuman primates correlating orienting biases with either anatomical asymmetries or functional lateralization. Therefore, any association between orienting asymmetries and cerebral lateralization remain unsubstantiated. Furthermore, details regarding individual responses to more than one stimulus type are not reported [Hopkins & Fernandez-Carriba, 2002]. Without information on baseline responding or individual variation, it is somewhat difficult to interpret the results. Notwithstanding, the results do seem robust for this population. However, a better approach may be to present the same subject with repeated experimental and control trials using a number of exemplars from each category.

Ear Advantages. A second psychophysical assay frequently employed with human subjects involves the examination of asymmetries in performance and reaction times following the monaural or dichotic presentation of auditory stimuli [Alho et al., 2003; Hiscock & Chipuer, 1993; Kallman, 1977; Mathiak et al., 2000; Welsh & Elliot, 2001; Wioland et al., 1999; Young & Ellis, 1980]. In contrast to orienting asymmetries, experimental evidence supports the claim that right- and left-ear advantages on such tasks reflect differential activation of the left and right hemispheres, respectively [Ahonniska et al., 1993]. Therefore, this procedure presumably permits an indirect measure of functional asymmetry. Petersen et al. [1978] were the first to employ this method with nonhuman primates. The authors trained five Japanese macaques *(Macaca fuscata)* to discriminate among previously recorded conspecific vocalizations based on one of the two acoustic properties. The Japanese macaques displayed a right-ear advantage for discriminating vocalizations based on the "communicatively relevant" feature, but showed no right- or left-ear advantage for discriminating pitch. Heterospecific control animals did not show an ear advantage for discriminating vocalizations based on either acoustic feature. The authors conclude that Japanese macaques may employ lateralized mechanisms for the processing of communicatively relevant conspecific vocalizations that are analogous to those used by humans during the perception and processing of speech [Petersen et al., 1978]. In addition, Japanese macaques exhibit a right-ear advantage for discriminating the communicatively salient characteristics of conspecific vocalizations, but show no ear advantage for discriminating the non-communicative parameters (i.e., pitch) of the identical calls [Beecher et al., 1979]. These data suggest that there may in fact be an association between the communicative relevance of a sound, and the recruitment of a lateralized system for the perception and processing of that signal.

Petersen et al. [1984] sought to investigate further the relationship between ear advantages and communicative significance. Specifically, the authors sought to determine if previously observed ear advantages for discriminating conspecific vocalizations were due to the acoustic properties of a call or its communicative significance. To this end, four monkey subjects (two Japanese macaques and two comparison macaques) were trained to discriminate among vocalizations presented monaurally to the left and right ears on an automated hold-release task. Consistent with previous findings, the Japanese macaques showed a right-ear advantage for discriminating among the vocalizations, whereas the control subjects did not. The authors then presented the same four subjects with 27 novel vocalizations in a generalization task to determine if all of the subjects were attending to the same acoustic feature – the relative position of the frequency peak within the call. Data indicate that responding of both the comparison monkeys and the Japanese monkey subjects was dependent upon the relative location of the peak in the call. These data suggest that although all of the subjects have the capacity to perform the auditory discrimination and do so while attending to the same acoustic feature, lateralized activity is related to the processing of the communicative significance of the vocalizations [Petersen et al., 1984]. Recent evidence suggests that rhesus macaques may show a similar right-ear performance and latency advantage for discriminating among exemplars of two types of distinct conspecific scream vocalizations [Le Prell et al., 2002]. However, this ear advantage was no longer evident following the second month of testing.

These results present a compelling case for some degree of functional lateralization for the processing of communicative conspecific vocalizations. In sharp contrast to the orienting asymmetry data discussed previously, these psychophysical assays have the advantage of (a) involving active discrimination as opposed to passive listening to stimuli and (b) correlating with direct functional data indicating differential activation of the left and right hemispheres [Ahonniska et al., 1993].

Direct Functional Assays

In addition to these relatively indirect behavioral measures, direct evidence regarding the lateralization of function during the perception of species-specific vocalizations is provided by an additional study with Japanese macaques, *Macaca fuscata* [Heffner & Heffner, 1984]. Researchers evaluated the ability of their subjects to discriminate between two variants of a species-specific call type before and after unilateral or bilateral ablation of the superior temporal gyrus. Whereas the discriminative performance of monkeys who sustained unilateral ablation of the right superior temporal gyrus was unaffected, those subjects with unilateral ablation of the left superior temporal gyrus were temporarily unable to complete the auditory discrimination task. Those individuals that received subsequent bilateral ablations never recovered their discriminative ability. The authors concluded that perception of species-typical vocalizations is mediated in the superior temporal gyrus with the left hemisphere playing a predominant role, supporting the claim that Japanese macaques possess an area analogous to Wernicke's area in humans [Heffner & Heffner, 1984]. Although the claim of homologous structures may be an oversimplification, these data point to a functional role of the left superior temporal gyrus in the discrimination of conspecific vocalizations. It is noteworthy that monkeys that received only an ablation of the left superior temporal gyrus did subsequently regain their discriminative abilities.

Recently, Poremba et al. [2004] utilized positron emission tomography (PET) to determine whether increased metabolic activity is observed in the superior temporal gyrus of rhesus monkeys during passive listening to a variety of auditory stimuli. The authors report that only species-specific vocalizations resulted in significantly greater metabolic activity in the left dorsal temporal pole of the superior temporal gyrus. This is an interesting finding, if not somewhat surprising given the location of the leftward asymmetry – in the temporal pole as opposed to posterior temporal regions. In fact, the data in this report indicate a rightward asymmetry in the posterior regions of the superior temporal gyrus for all 6 sound classes presented to the monkeys. Close examination of the data reveal that normalized radioactivity values in the posterior regions of the superior temporal gyrus range from 100 to 120 percent of the whole-brain radioactivity. In other words, activity in these regions exceeds that of the average radioactivity for the entire brain by as much as 20 percent. In contrast, normalized radioactivity values in the temporal pole sector do not exceed 90 percent of the whole brain activity for any of the sound classes. Although clearly reflecting a leftward asymmetry, the lack of significant activation in this region weakens the relevance of this finding.

In a second study, Gil-da-Costa et al. [2006] similarly utilized PET to visualize cerebral metabolic activity in the rhesus monkey brain during the presentation of conspecific vocalizations. In contrast to the results reported by Poremba et al. [2004], the authors did not find any lateralized effects in response to conspecific coos or screams. However, the authors do report significant activation in the posterior region of the temporal lobe [their "temporoparietal" (Tpt) area] in response to passive listening to conspecific vocalizations when compared to non-biological sounds. Significant activation was also observed in the monkey ventral premotor cortex, but once again, no lateralized activity was observed. No differences between conspecific call types (coos vs. screams) were observed, but both vocalizations evoked greater activity in the monkey temporoparietal area, ventral premotor cortex, and the posterior parietal cortex as compared to non-biological sounds.

Reconciling these two different sets of results presents somewhat of a challenge. On the one hand, right-lateralized activity is observed in posterior regions of the superior temporal gyrus nonspecifically in response to auditory stimuli whereas left-lateralized activity exclusively in response to conspecific vocalizations is localized to the left temporal pole, but far below average whole-brain activity [Poremba et al., 2004]. On the other hand, significant activation selectively in response to conspecific vocalizations is observed in monkey temporoparietal, posterior parietal, and ventral premotor cortex, but no lateralized activation is observed [Gil-da-Costa et al., 2006]. Furthermore, Gil-da-Costa et al. [2006] failed to observe any significant activation in the temporal pole of either hemisphere. Given this considerable disparity, any conclusions regarding hypothesized functional lateralization for processing conspecific vocalizations in macaque monkeys are in question.

VOCAL PRODUCTION

In humans, oro-facial asymmetries have been studied as this physiological measure is thought to reflect neurofunctional lateralization for the production of sounds. Data indicate that the right side of the mouth opens before the left during the production of speech sounds [Graves et al., 1982, 1985; Wolf & Goodale, 1987]. Although there have been attempts to quantify nonhuman primate oro-facial asymmetries, surprisingly few reports have attempted to non-invasively assess asymmetries during vocal production [Fernandez-Carriba et al., 2002; Hauser & Akre, 2001; Hook-Costigan & Rogers, 1998; Reynolds et al., 2005]. Hook-Costigan and Rogers [1998] examined lateralized use of the mouth during the production of two types of vocalizations in a New World primate, the common marmoset. Oro-facial asymmetries were determined by measuring the area of the mouth on either side of a defined midline. The authors report a significant leftward asymmetry during the production of fear vocalizations and a larger right hemi-mouth during the production of social contact calls. This finding parallels similar work with humans [Wolf & Goodale, 1987]. Employing an identical method of quantification, a leftward oro-facial asymmetry is reported for chimpanzees during the production of hoots and screams [Fernandez-Carriba et al., 2002]. A second study on chimpanzees replicated the findings for hoots and found a similar pattern for food barks. In contrast, the so-called "volitional" vocalizations resulted in rightward oro-facial asymmetries [Reynolds et al., 2005]. The authors conclude that the right and left hemispheres are differentially involved in the production of voluntary and involuntary vocal signals in chimpanzees.

Hauser and Akre [2001] examined mouth-opening asymmetries during the production of 8 types of vocalizations. By examining still frames of video, vocalizations were categorized as "asymmetrical" if one side of the mouth opened before the other at the call's onset. The authors report that asymmetrical mouth opening was more common than symmetrical mouth opening overall. However, the data indicate a directional bias (left mouth opening before right) for only three-call types (coos, copulation screams, and harmonic arches). Unfortunately, small sample sizes prevented comprehensive statistical analyses of these results.

Although the oro-facial asymmetry data considered collectively are somewhat intriguing, the significance of these results is questionable. As is the case with the orienting asymmetry assay discussed above, no association with neurological function has been established. Given this, interpreting these results in the context of cerebral lateralization for vocal production is theoretical at best. Of course, advances in neuroimaging techniques should make it possible to compare such oro-facial biases with anatomical and even functional asymmetries in nonhuman primates.

ANATOMICAL ASYMMETRIES

Ever since the early observations of Broca [1861] and Wernicke [1874], there has been considerable interest in neuroanatomical asymmetries in the perisylvian regions of the human brain [Galaburda, 1995]. The majority of attention has focused on the measurement of asymmetries in the posterior region of the temporal lobe demarking the planum temporale (PT) [see Beaton, 1997]. Considered part of Wernicke's area, the PT is located on the surface of the superior temporal gyrus just rostral to Heschl's gyrus. Data indicate that the PT is larger in the left hemisphere than it is in the right [Foundas et al., 1995; Geschwind & Levitsky, 1968].

Historically, there has also been interest in the evolutionary origins of sylvian fissure asymmetry because it presumably reflects expansion of the posterior temporal cortex or Wernicke's area. A number of studies have described variability in the posterior region of the sylvian fissure, particularly as it relates to the occurrence of a bifurcation into a posterior ascending and posterior descending branch. For example, at least four different patterns of bifurcation of the posterior sylvian fissure have been identified in human cadaver brains [Ide et al., 1996; Witelson & Kigar, 1992]. Moreover, variations in the patterns of ascending and descending limbs are associated with the sex and handedness of subjects. Ide et al. [1996] reported that a larger ascending compared to descending limb of the sylvian fissure was more prevalent in males and more frequently seen in the right compared to the left hemisphere. In contrast, females showed more symmetrical ascending and descending limbs of the sylvian fissure but the branches differed in orientation with the ascending branch oriented rostrally, whereas the descending limb was oriented more caudally.

Recently, MRI imaging technology has provided an alternative means of assessing asymmetries in the perisylvian fissure. As observed in cadaver brains, analysis of MRI scans reveal a rightward asymmetry of the posterior wall of the posterior ascending ramus of the sylvian fissure, an area referred to as the planum parietale (PP), in humans [Foundas et al., 2002; Heiervang et al., 2000; Jancke et al., 1994; Steinmetz et al., 1990]. Also consistent with the cadaver results, it has been demonstrated that handedness and gender have an interactive influence on PP asymmetry such that right-handed men and left-handed women show pronounced rightward asymmetries [Jancke et al., 1994]. In addition, a larger PP in the right hemisphere is more common in individuals who utilize their right hand to write when compared to those who use their left hand [Foundas et al., 2002].

Direct evidence of asymmetries in the posterior temporal lobe have recently been documented with the reports of population-level leftward asymmetries in the PT of great apes, the tissue that lies posterior to Heschl's gyrus and whose terminal point is the posterior end of the sylvian fissure [see Figure 2; Cantalupo et al., 2003; Gannon et al., 1998; Hopkins et al., 1998]. Gannon et al. [1998] quantified the PT in 18 chimpanzee brains. The authors report that the PT was significantly larger in the left hemisphere in 94 percent of the specimens examined. Hopkins et al. [1998] replicated this finding using MR images, and report a similar left-hemisphere bias for all four species of great apes [Hopkins et al., 1998]. Interestingly, the authors also report that they were not able to measure a PT in both their Old or New World monkey specimens given that the transverse temporal sulcus (the anterior border of Heschl's gyrus) was not detectable.

Early work with endocasts and cadaver brains reported evidence of asymmetries in the length and height of the sylvian fissure in primates [LeMay, 1985; Yeni-Komshian & Benson, 1976]. The sylvian fissure was reported to be higher in the right compared to the left hemisphere, particularly among great apes, presumably reflecting an asymmetry of the PP [LeMay & Geschwind, 1975]. However, in two species of macaques, no significant differences were found in either the height or the length of the ascending limb of the sylvian fissure

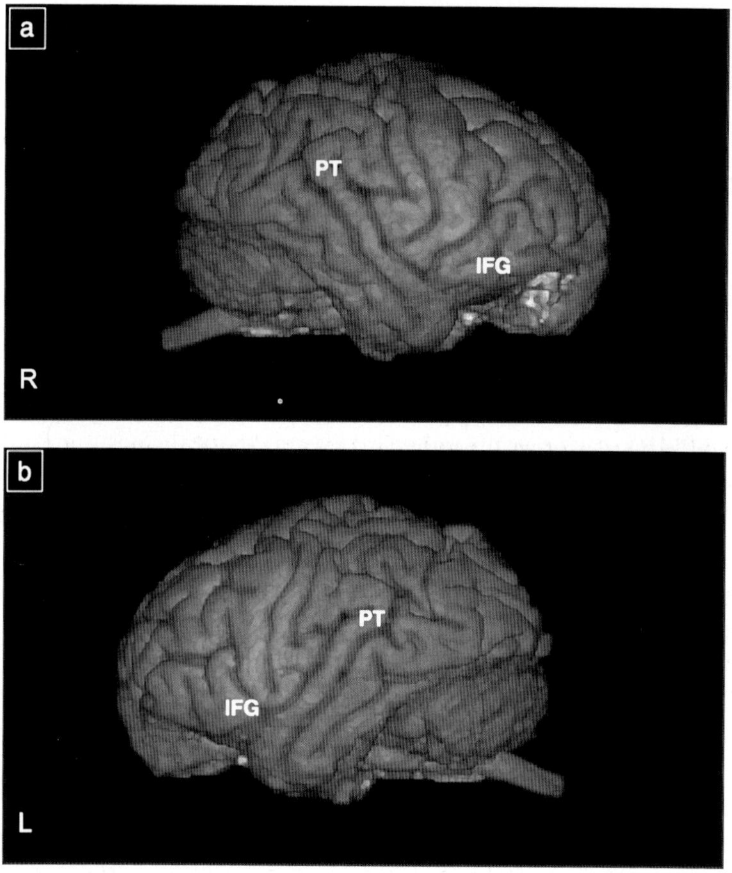

Figure 2. (a) Planum temporale (PT) and (b) Inferior frontal gyrus (IFG) indicated on a surface rendered MRI of a chimpanzee brain. The image of the brain was constructed in Analayze, by Dr. Jared Taglialatela.

[Heilbronner & Holloway, 1988]. With respect to sylvian fissure length, Yeni-Komshian and Benson [1976] reported that human and chimpanzee cadaver specimens showed a leftward asymmetry whereas rhesus monkeys did not. Subsequent MRI studies have revealed similar leftward asymmetries in the sylvian fissure in great apes, replicating the findings from cadaver specimens [Hopkins et al., 2000]. In monkeys, the results have been less consistent with some reports of leftward asymmetries in cadaver specimens [Heilbronner & Holloway, 1988], but no leftward asymmetry when measured from endocasts [Falk et al., 1990].

Notwithstanding, there still remains little investigation of sulci variation in the sylvian fissure in the comparative literature on brain asymmetry, although it has been claimed recently that a rightward asymmetry in the PP is present in two species of great apes, chimpanzees (*Pan*) and orangutans (*Pongo*) [Gannon et al., 2005]. Additionally, examination of a relatively large sample of MR images of chimpanzee brains ($N = 63$) revealed both sex and hemisphere differences in the posterior bifurcation of the sylvian fissure [Taglialatela et al., unpublished data].

In humans, Broca's area has been identified as a critical region for speech production, and it is typically larger in the left hemisphere compared to the right [Amunts et al., 1999; Broca, 1861; Foundas et al., 1998; Galaburda, 1984]. A similar structural asymmetry has been identified in three species of great apes [Cantalupo & Hopkins, 2001]. Furthermore, it has been reported recently that chimpanzees that reliably employ their right hand for manual gestures have larger inferior frontal gyri (presumed site of the Broca's area homologue in the chimpanzee brain) in the left hemisphere than those apes that do not show consistent hand use for gestures [Taglialatela et al., 2006]. Save this single study, no data are available with regard to whether this structural asymmetry in the great apes is related in any way to vocal production or auditory perception.

ASYMMETRIES AND COMMUNICATIVE SIGNIFICANCE

Despite the well-documented hemispheric specializations for human language processing, and the potential mechanisms underlying these systems, the basis for this lateralization is still speculative. Passingham [1981] proposed that the lateralization of function evolved as a result of selection pressure for a system that supports the complex integration of information as well as the organization and control of the precise movements that are required during the comprehension and production of language. It has been argued that the control by a single central program localized to one hemisphere is more efficient than a bilateral system that would need to coordinate activity via the relatively long commissural pathways [Passingham, 1981]. This claim is supported by data from neuroimaging studies that implicate the failure to develop a left-hemisphere dominance for speech as a possible cause of stuttering [Fox et al., 1996]. In addition, Knecht et al. [2002] reported that an individual's degree of language lateralization determined a susceptibility to temporary unilateral brain lesions confirming that these asymmetries are functionally significant.

A compelling model concerning the lateralization of language functions in the human brain proposes that temporal information is processed predominantly in the left hemisphere, whereas spectral information is handled by the right [Zatorre & Belin, 2001]. Recently, however, both hemispheres have been shown to be highly sensitive to temporal information, but on different time scales [Boemio et al., 2005]. Thus, the picture that emerges is one in which speech signals are represented bilaterally at the primary level, and are lateralized in higher-order regions based on the temporal information (relatively long or short) that is extracted by the relevant hemisphere [Poeppel, 2003].

Speech utterances certainly possess information on more than a single time scale (e.g., phonemes vs. prosody), so this model plausibly accounts for both observations of differential lateralization patterns associated with verbal vs. emotional components of language, as well as for differences in lateralized activation patterns as a function of attention [Buchanan et al., 2000; Grimshaw et al., 2003; Hugdahl et al., 2003b]. With regard to hypothesized functional lateralization in nonhuman primates, this model may similarly help to reconcile some of the disparities that exist among the published studies.

First, however, careful consideration of a number of methodological issues is appropriate. Group level systematic structural variation has been reported in the calls of both monkeys and chimpanzees in the wild as well as in captivity [Arcadi, 1996; Crockford et al., 2004; Eiler & Banack, 2004; Fischer et al., 1998; Green, 1975; Marshall et al., 1999; Mitani et al., 1992]. For example, Crockford et al. [2004] reported structural differences in the pant hoot vocalizations of male chimpanzees living in neighboring communities, but not between groups from a distant community. These results could not be accounted for by genetic or habitat differences suggesting that the male chimpanzees may be actively modifying the structure of their calls to facilitate group identification [Crockford et al., 2004]. In a second study, Fischer et al. [1998] reported the results of playback experiments that indicate that observed structural variation was perceptually salient to the monkey subjects. Systematic variation in vocal structure has also been observed in a variety of primate species with regard to social situation and semantic context [Snowdon & Elowson, 1999; Taglialatela et al., 2003]. In addition, female squirrel monkeys produce chuck calls that are structurally different from one another [Smith et al., 1982a], preferentially respond to the vocalizations of familiar group members compared to those produced by unknown individuals [Biben & Symmes, 1991; McCowan & Newman, 2000], exchange calls preferentially with their closely affiliated partners [Biben et al., 1986; Masataka & Biben, 1987; Smith et al., 1982b], and when they do respond to unfamiliar calls are more likely to do so if the structure of the novel vocalization is similar to those produced by familiar group members [Soltis et al., 2002]. Finally, Kojima [2003] reports that, following the presentation of a sample vocalization, a chimpanzee was able to select the photograph corresponding to the individual that produced the sound.

These data indicate that there are a number of different types of potentially meaningful information that can be encoded within a vocalization (e.g., caller identity, group identity, social grouping, social status, etc.). As certain types of learning have been shown to play a role in the vocal development of a variety of nonhuman primate species [Roian Egnor & Hauser, 2004], it is possible that in order for a given individual to successfully extract relevant information from a vocalization, that individual must have prior exposure to these calls as well as their appropriate usage. In other words, not all calls may be functionally equivalent to all members of the same species. Therefore, *meaningful* information contained within such signals could only be extracted by those individuals who have had exposure to the sounds in the relevant vocal repertoire. (Gil-da-Costa & Hauser [2006] did not find an effect of familiarity on orienting asymmetries, but due to the methodological issues raised earlier in this chapter, this study should not be considered the definitive word on this subject.) Similarly, as has been demonstrated with humans, the same information may show different activation patterns depending upon the subject's attention [Hugdahl et al., 2003b; see Beecher et al., 1979]. A given vocalization most likely contains multiple pieces of information. Thus, determining what the subject is attending to is critical.

In previous studies that sought to evaluate the neurological foundations of nonhuman primate vocal communication, consideration has been given to the communicative salience of vocalizations. However, "species-typical" vocalizations have been treated as just that – utterances that have acoustic properties that may be described as produced by members of a single species. If asymmetric processing of auditory information is a function of the potentially *meaningful* information encoded within a call, then one might even expect to observe different results in response to various types of species-typical vocalizations. Whereas some species-typical calls may be considered *communicatively* salient to a given individual, that same vocalizations may be only *acoustically* significant to another [Beecher et al., 1979].

FUTURE DIRECTIONS

Although the study of nonhuman primate vocal-auditory asymmetries has contributed significantly to our understanding of the evolution of language and its biological foundations, the data are limited by two important factors. First, the evidence for neural lateralization is predominantly based on indirect behavioral evaluations that measured orienting asymmetries or ear advantages in response to the presentation of auditory stimuli. These psychophysical assays can provide only an indirect indication of structural or functional lateralization, and have yet to be corroborated with direct neurobiological evidence. However, advances in neuroimaging techniques have made it possible to assess the appropriateness of these methods as an indicator of neural lateralization. In humans, a number of studies have examined the relationship between behavioral asymmetries and anatomical or functional asymmetries by employing both functional neuroimaging techniques as well as the WADA test [see Tervaniemi & Hugdahl, 2003, for a review]. In addition, Hugdahl et al. [2003a] reported a significant relationship between dichotic listening asymmetry and the area of the PT in children with dyslexia. Similar studies have yet to be reported in nonhuman primates, but could provide valuable information regarding the evolution of neuroanatomical and functional asymmetries as well as offer a validation or refutation of indirect behavioral assays such as orienting asymmetries.

Secondly, the majority of direct neurophysiological evidence concerning lateralization in nonhuman primates is based on lesion studies [see Gil-da-Costa et al., 2006; Poremba et al., 2004, for recent exceptions]. Prior to the development of neuroimaging technology, human language was thought to be limited to a few major processing centers. It is now apparent that the traditional areas are not the only regions involved in the mediation of language in the human brain [Binder et al., 1997]. Functional neuroimaging of nonhuman primates during vocalization perception and production will likely provide similar insight and detail regarding the anatomical structures and systems that mediate these behaviors as well as the phylogenetic origins of the neural lateralization associated with human spoken language. In contrast to the early view that lateralization for communication is unique to humans, the data reviewed in this chapter do suggest that functional and anatomical asymmetries may have played a facilitating role in the evolution of spoken language – perhaps providing the brain with the increased computational capacity necessary for linguistic communication.

REFERENCES

Ahonniska, J.; Cantell, M.; Tolvanen, A.; Lyytinen, H. Speech perception and brain laterality: The effect of ear advantage on auditory event-related potentials. Brain and Language 45(2): 127–146, 1993.

Alho, K.; Vorobyev, V. A.; Medvedev, S. V.; Pakhomov, S. V.; Roudas, M. S.; Tervaniemi, M.; van Zuijen, T.; Naatanen, R. Hemispheric lateralization of cerebral blood-flow changes during selective listening to dichotically presented continuous speech. Cognitive Brain Research 17(2): 201–211, 2003.

Amunts, K.; Schleicher, A.; Bürgel, U.; Mohlberg, H.; Uylings, H. B.; Zilles, K. Broca's region revisited: Cytoarchitecture and intersubject variability. The Journal of Comparative Neurology 412(2): 319–341, 1999.

Arcadi, A. C. Phrase structure of wild chimpanzee pant hoots: Patterns of production and interpopulation variability. American Journal of Primatology 39(3): 159–178, 1996.

Beaton, A. A. The relation of planum temporale asymmetry and morphology of the corpus callosum to handedness, gender and dyslexia: A review of the evidence. Brain and Language 60: 255–322, 1997.

Beecher, M. D.; Petersen, M. R.; Zoloth, S. R.; Moody, D. B.; Stebbins, W. C. Perception of conspecific vocalizations by Japanese macaques. Evidence for selective attention and neural lateralization. Brain, Behavior, and Evolution 16(5–6): 443–460, 1979.

Biben, M.; Symmes, D. Playback studies of affiliative vocalizing in captive squirrel monkeys: Familiarity as a cue to response. Behaviour 117(1–2): 1–19, 1991.

Biben, M.; Symmes, D.; Masataka, N. Temporal and structural analysis of affiliative vocal exchanges in squirrel monkeys (*Saimiri sciureus*). Behaviour 98: 259–273, 1986.

Binder, J. R.; Frost, J. A. Functional MRI studies of language processes in the brain. Neuroscience News 1: 15–23, 1998.

Binder, J. R.; Frost, J. A.; Hammeke, T. A.; Cox, R. W.; Rao, S. M.; Prieto, T. Human brain language areas identified by functional magnetic resonance imaging. The Journal of Neuroscience 17(1): 353–362, 1997.

Boemio, A.; Fromm, S.; Braun, A.; Poeppel, D. Hierarchical and asymmetric temporal sensitivity in human auditory cortices. Nature Neuroscience 8: 389–395, 2005.

Boye, M.; Gunturkun, O.; Vauclair, J. Right ear advantage for conspecific calls in adults and subadults, but not infants, California sea lions (*Zalophus californius*): Hemispheric specialization for communication? European Journal of Neuroscience 21: 1727–1732, 2005.

Broca, P. Remarques sur le siege de la faculte du langage articule; suivies d'une observation d'aphemie. Bulletin de la Société Anatomique 6: 330–357, 1861.

Buchanan, T. W.; Lutz, K.; Mirzazade, S.; Specht, K.; Shah, N. J.; Zilles, K.; Jancke, L. Recognition of emotional prosody and verbal components of spoken language: An fMRI study. Cognitive Brain Research 9: 227–238, 2000.

Cantalupo, C.; Hopkins, W. D. Asymmetric Broca's area in great apes. Nature 414: 505, 2001.

Cantalupo, C.; Pilcher, D.; Hopkins, W. D. Are planum temporale and sylvian fissure asymmetries directly related? A MRI study in great apes. Neuropsychologia 41: 1975–1981, 2003.

Crockford, C.; Herbinger, I.; Vigilant, L.; Boesch, C. Wild chimpanzees produce group-specific calls: A case for vocal learning? Ethology 110: 221–243, 2004.

Deacon, T. W. The Symbolic Species: The Coevolution of Language and the Brain. New York, W. W. Norton & Company, 1997.

Ehret, G. Left hemisphere advantage in the mouse brain for recognizing ultrasonic communication calls. Nature 325: 249–251, 1987.

Eiler, K. C.; Banack, S. A. Variability in the alarm call of golden-mantled ground squirrels (*Spermophilus lateralis* and *S. saturatus*). Journal of Mammalogy 85(1): 43–50, 2004.

Falk, D.; Hildebolt, C.; Cheverud, J.; Vannier, M.; Helmkamp, R. C.; Konigsberg, L. Cortical asymmetries in the frontal lobe of rhesus monkeys (*Macaca mulatta*). Brain Research 512: 40–45, 1990.

Fernandez-Carriba, S.; Loeches, A.; Morcillo, A.; Hopkins, W. D. Asymmetry in facial expression of emotions by chimpanzees. Neuropsychologia 40(9): 1523, 2002.

Fischer, J.; Hammerschmidt, K.; Todt, D. Local variation in Barbary macaque shrill barks. Animal Behaviour 56: 623–629, 1998.

Foundas, A.; Leonard, C.; Heilman, K. Morphological cerebral asymmetries and handedness: The pars triangularis and planum temporale. Archives of Neurology 52: 501–508, 1995.

Foundas, A. L.; Eure, K. F.; Luevano, L. F.; Weinberger, D. R. MRI asymmetries of Broca's area: The pars triangularis and pars opercularis. Brain and Language 64(3): 282, 1998.

Foundas, A. L.; Leonard, C. M.; Hanna-Pladdy, B. Variability in the anatomy of the planum temporale and posterior ascending ramus: Do right- and left handers differ? Brain and Language 83: 403–424, 2002.

Fox, P. T.; Ingham, R. J.; Ingham, J. C.; Hirsch, T. B.; Downs, J. H.; Martin, C.; Jerabek, P.; Glass, T.; Lancaster, J. L. A PET study of the neural systems of stuttering. Nature 382: 158–162, 1996.

Frost, J. A.; Binder, J. R.; Springer, J. A.; Hammeke, T. A.; Bellgowan, P. S. F.; Rao, S. M.; Cox, R. W. Language processing is strongly left lateralized in both sexes. Evidence from functional MRI. Brain 122(2): 199–208, 1999.

Galaburda, A. M. Anatomical asymmetries. Pp. 11–25 in N. Geschwind; A. M. Galaburda, eds. Cerebral Dominance: The Biological Foundations. Cambridge, MA, Harvard University Press, 1984.

Galaburda, A. M. Anatomic basis of cerebral dominance. Pp. 1–18 in R. J. Davidson; K. Hugdahl, eds. Brain Asymmetry. Cambridge, MA, MIT Press, 1995.

Gannon, P. J.; Holloway, R. L.; Broadfield, D. C.; Braun, A. R. Asymmetry of chimpanzee planum temporale: Humanlike pattern of Wernicke's language area homolog. Science 279: 220–222, 1998.

Gannon, P. J.; Kheck, N. M.; Braun, A. R.; Holloway, R. L. Planum parietale of chimpanzees and orangutans: A comparative resonance of human-like planum temporale asymmetry. The Anatomical Record 287: 1128–1141, 2005.

Gemba, H.; Miki, N.; Sasaki, K. Cortical field potentials preceding vocalization and influences of cerebellar hemispherectomy upon them in monkeys. Brain Research 697: 143–151, 1995.

George, I.; Cousillas, H.; Richard, J. P.; Hausberger, M. State-dependent hemispheric specialization in the songbird brain. The Journal of Comparative Neurology 488: 48–60, 2005.

Geschwind, N.; Levitsky, W. Human brain: Left-right asymmetries in temporal speech region. Science 161: 837186–837187, 1968.

Ghazanfar, A. A.; Smith-Rohrberg, D.; Hauser, M. D. The role of temporal cues in rhesus monkey vocal recognition: Orienting asymmetries to reversed calls. Brain, Behavior, and Evolution 58: 163–172, 2001.

Gil-da-Costa, R.; Hauser, M. D. Vervet monkeys and humans show brain asymmetries for processing conspecific vocalizations, but with opposite

patterns of laterality. Proceedings of the Royal Society of London: B 273: 2313–2318, 2006.
Gil-da-Costa, R.; Martin, A.; Lopes, M. A.; Munoz, M.; Fritz, J.; Braun, A. R. Species-specific calls activate homologs of Broca's and Wernicke's areas in the macaque. Nature Neuroscience 9: 1065–1070, 2006.
Graves, R.; Goodglass, H.; Landis, T. Mouth asymmetry during spontaneous speech. Neuropsychologia 20(4): 371–381, 1982.
Graves, R.; Landis, T.; Simpson, C. On the interpretation of mouth asymmetry. Neuropsychologia 23(1): 121–122, 1985.
Green, S. Dialects in Japanese monkeys: Vocal learning and cultural transmission of locale-specific vocal behavior? Zeitschrift fur Tierpsychologie 38(304): 304–314, 1975.
Grimshaw, G. M.; Kwasny, K. M.; Covell, E.; Johnson, R. A. The dynamic nature of language lateralization: Effects of lexical and prosodic factors. Neuropsychologia 41(8): 1008–1019, 2003.
Grossi, G.; Semenza, C.; Corazza, S.; Volterra, V. Hemispheric specialization for sign language. Neuropsychologia 34(7): 737–740, 1996.
Hauser, M. D. The Evolution of Communication. Cambridge, MA, MIT Press, 1996.
Hauser, M. D.; Agnetta, B.; Perez, C. Orienting asymmetries in rhesus monkeys: The effect of time-domain changes on acoustic perception. Animal Behaviour 56: 41–47, 1998.
Hauser, M. D.; Akre, K. Asymmetries in the timing of facial and vocal expressions by rhesus monkeys: Implications for hemispheric specialization. Animal Behaviour 61: 391–400, 2001.
Hauser, M. D.; Andersson, K. Left hemisphere dominance for processing vocalizations in adult, but not infant, rhesus monkeys: Field experiments. Proceedings of the National Academy of Sciences 91: 3946–3948, 1994.
Heffner, H. E.; Heffner, R. S. Temporal lobe lesions and perception of species-specific vocalizations by macaques. Science 226: 75–76, 1984.
Heiervang, E.; Hugdahl, K.; Steinmetz, H.; Smievoll, A. I.; Stevenson, J.; Lund, A.; Ersland, L.; Lundervold, A. Planum temporale, planum parietale and dichotic listening in dyslexia. Neuropsychologia 38: 1704–1713, 2000.
Heilbronner, P. L.; Holloway, R. L. Anatomical brain asymmetries in New World and Old World monkeys. Stages of temporal lobe development in primate evolution. American Journal of Physical Anthropology 76: 39–48, 1988.
Hickok, G.; Bellugi, U.; Klima, E. S. The neural organization of language: Evidence from sign language aphasia. Trends in Cognitive Sciences 2(4): 129–136, 1998.
Hirano, S.; Naito, Y.; Okazawa, H.; Kojima, H.; Honjo, I.; Ishizu, K.; Yenokura, Y.; Nagahama, Y.; Fukuyama, H.; Konishi, J. Cortical activation by monaural speech sound stimulation demonstrated by

positron emission tomography. Experimental Brain Research 113(1): 75–80, 1997.

Hiscock, M.; Chipuer, H. Children's ability to shift attention from one ear to the other: Divergent results for dichotic and monaural stimuli. Neuropsychologia 31(12): 1339–1350, 1993.

Hook-Costigan, M. A.; Rogers, L. J. Lateralized use of the mouth in production of vocalizations by marmosets. Neuropsychologia 36(12): 1265–1273, 1998.

Hopkins, W. D.; Fernandez-Carriba, S. Laterality of communicative behaviors in nonhuman primates: A critical analysis. Pp. 445–479 in L. J. Rogers; R. J. Andrew, eds. Comparative Vertebrate Lateralization. Cambridge, Cambridge University Press, 2002.

Hopkins, W. D.; Marino, L.; Rilling, J. K.; MacGregor, L. A. Planum temporale asymmetries in great apes as revealed by magnetic resonance imaging (MRI). Neuroreport 9: 2913–2918, 1998.

Hopkins, W. D.; Pilcher, D. L.; MacGregor, L. Sylvian fissure length asymmetries in primates revisited: A comparative MRI study. Brain, Behavior and Evolution 56: 293–299, 2000.

Hugdahl, K.; Heiervang, E.; Ersland, L.; Lundervold, A.; Steinmetz, H.; Smievoll, A. I. Significant relation between MR measures of planum temporale area and dichotic processing of syllables in dyslexic children. Neuropsychologia 41: 666–675, 2003a.

Hugdahl, K.; Thomsen, T.; Ersland, L.; Rimol, L. M.; Niemi, J. The effects of attention on speech perception: An fMRI study. Brain and Language 85: 37–48, 2003b.

Ide, A.; Rodriguez, E.; Zaidel, E.; Aboitiz, F. Bifurcation patterns in the human sylvian fissure: Hemispheric and sex differences. Cerebral Cortex 6(5): 717–725, 1996.

Jancke, L.; Schlaug, G.; Huang, Y.; Steinmetz, H. Asymmetry of the planum parietale. Neuroreport 5: 1161–1163, 1994.

Kallman, H. J. Ear asymmetries with monaurally-presented sounds. Neuropsychologia 15: 833–835, 1977.

Knecht, S.; Drager, B.; Floel, A.; Lohmann, H.; Breitenstein, C.; Deppe, M.; Henningsen, H.; Ringelstein, E. B. Behavioural relevance of atypical language lateralization in healthy subjects. Brain 124: 1657–1665, 2001.

Knecht, S.; Floel, A.; Drager, B.; Breitenstein, C.; Sommer, J.; Henningsen, H.; Ringelstein, E. B.; Pascual-Leone, A. Degree of language lateralization determines susceptibility to unilateral brain lesions. Nature Neuroscience 5(7): 695–699, 2002.

Knecht, S.; Jansen, A.; Frank, A.; van Randenborgh, J.; Sommer, J.; Kanowski, M.; Heinze, H. J. How atypical is atypical language dominance? Neuroimage 18: 917–927, 2003.

Kojima, S. A Search for the Origins of Human Speech: Auditory and Vocal Functions of the Chimpanzee. Kyoto, Kyoto University Press, 2003.

LeMay, M. Asymmetries of the brains and skulls of nonhuman primates. Pp. 223–245 in S. D. Glick, ed. Cerebral Lateralization in Nonhuman Species. New York, Academic Press, 1985.

LeMay, M.; Geschwind, N. Hemispheric differences in the brains of great apes. Brain, Behavior, and Evolution 11(1): 48–52, 1975.

Le Prell, C. G.; Hauser, M. D.; Moody, D. B. Discrete or graded variation within rhesus monkey screams? Psychophysical experiments on classification. Animal Behaviour 63: 47–62, 2002.

Marshall, A. J.; Wrangham, R. W.; Arcadi, A. C. Does learning affect the structure of vocalizations in chimpanzees? Animal Behaviour 58: 825–830, 1999.

Masataka, N.; Biben, M. Temporal rules regulating affiliative vocal exchanges of squirrel monkeys. Behaviour 101: 311–319, 1987.

Mathiak, K.; Hertrich, I.; Lutzenberger, W.; Ackermann, H. Encoding of temporal speech features (formant transients) during binaural and dichotic stimulus application: A whole-head magnetencephalography study. Cognitive Brain Research 10: 125–131, 2000.

McCowan, B.; Newman, J. D. The role of learning in chuck call recognition by young squirrel monkeys (*Saimiri sciureus*). Behaviour 137: 279–300, 2000.

Mitani, J. C.; Hasegawa, T.; Gros-Luis, J.; Marler, P.; Byrne, R. Dialects in wild chimpanzees? American Journal of Primatology 27: 233–243, 1992.

Okanoya, K.; Ikebuchi, M.; Uno, H.; Watanabe, S. Left-side dominance for song discrimination in Bengalese finches (*Lonchura striata* var. *domestica*). Animal Cognition 4: 241–245, 2001.

Palleroni, A.; Hauser, M. Experience-dependent plasticity for auditory processing in a raptor. Science 299: 1195, 2003.

Passingham, R. E. Broca's area and the origins of human vocal skill. Philosopical Transactions of the Royal Society of London B 292: 167–175, 1981.

Petersen, M. R.; Beecher, M. D.; Zoloth, S. R.; Green, S.; Marler, P. R.; Moody, D. B.; Stebbins, W. C. Neural lateralization of vocalizations by Japanese macaques: Communicative significance is more important than acoustic structure. Behavioural Neuroscience 98(5): 779–790, 1984.

Petersen, M. R.; Beecher, M. D.; Zoloth, S. R.; Moody, D. B.; Stebbins, W. C. Neural lateralization of species-specific vocalizations by Japanese macaques (*Macaca fuscata*). Science 202(4365): 324–327, 1978.

Poeppel, D. The analysis of speech in different temporal integration windows: Cerebral lateralization as 'asymmetric sampling in time'. Speech Communication 41(1): 245–255, 2003.

Poeppel, D.; Hickock, G. Towards a new functional anatomy of language. Cognition 92: 1–12, 2004.

Poremba, A.; Malloy, M.; Saunders, R. C.; Carson, R. E.; Herscovitch, P.; Mishkin, M. Species-specific calls evoke asymmetric activity in the monkey's temporal poles. Nature 427: 448–451, 2004.

Reynolds, E.; Freeman, H.; Russell, J. L.; Hopkins, W. D. Oro-facial asymmetry of voluntary and involuntary vocal signals in chimpanzees (*Pan troglodytes*). American Journal of Primatology 66(Suppl. 1): 99, 2005.

Roian Egnor, S. E.; Hauser, M. D. A paradox in the evolution of primate vocal learning. Trends in Neurosciences 27(11): 649–654, 2004.

Schmidt, M. F.; Ashmore, R. C.; Vu, E. T. Bilateral control and interhemispheric coordination in the avian song motor system. Annals of the New York Academy of Sciences 1016: 171–186, 2004.

Smith, H. J.; Newman, J. D.; Hoffman, H. J.; Fetterly, K. Statistical discrimination among vocalizations of individual squirrel monkeys (*Saimiri sciureus*). Folia Primatologica 37: 267–279, 1982a.

Smith, H. J.; Newman, J. D.; Symmes, D. Vocal concomitants of affiliative behavior in squirrel monkeys. Pp. 30–49 in C. T. Snowdon; C. H. Brown; M. R. Petersen, eds. Primate Communication. Cambridge, Cambridge University Press, 1982b.

Snowdon, C. T.; Elowson, A. M. Pygmy marmosets modify call structure when paired. Ethology 105(10): 893, 1999.

Soltis, J.; Bernhards, D.; Donkin, H.; Newman, J. D. Squirrel monkey chuck call: Vocal response to playback chucks based on acoustic structure and affiliative relationship with the caller. American Journal of Primatology 57(3): 119–130, 2002.

Steinmetz, H.; Rademacher, J.; Jancke, L.; Huang, Y.; Thron, A.; Zilles, K. Total surface of temporoparietal intrasylvian cortex: Diverging left-right asymmetries. Brain and Language 39: 357–372, 1990.

Taglialatela, J. P.; Cantalupo, C.; Hopkins, W. D. Gesture handedness predicts asymmetry in the chimpanzee inferior frontal gyrus. Neuroreport 17(9): 923–927, 2006.

Taglialatela, J. P.; Savage-Rumbaugh, E. S.; Baker, L. A. Vocal production by a language-competent *Pan paniscus*. International Journal of Primatology 24(1): 1–17, 2003.

Tervaniemi, M.; Hugdahl, K. Lateralization of auditory-cortex functions. Brain Research Reviews 43: 231–246, 2003.

Thiel, A.; Herholz, K.; von Stockhausen, H. M.; van Leyen-Pilgram, K.; Peitrzyk, U.; Kessler, J.; Wienhard, K.; Klug, N.; Heiss, W.-D. Localization of language-related cortex with 15O-labeled water: PET in patients with gliomas. Neuroimage 7: 284–295, 1998.

Welsh, T. N.; Elliot, D. Gender differences in a dichotic listening and movement task: Laterization or strategy? Neuropsychologia 39: 25–35, 2001.

Wernicke, C. Der aphasische Symptomenkomplex. Breslau, Cohn & Weigert, 1874.

Wioland, N.; Rudolf, G.; Metz-Lutz, M. N.; Mutschler, V.; Marescaux, C. Cerebral correlates of hemispheric lateralization during a pitch discrimination task: An ERP study in dichotic situation. Clinical Neurophysiology 110: 516–523, 1999.

Witelson, S. F.; Kigar, D. L. Sylvian fissure morphology and asymmetry in men and women: Bilateral differences in relation to handedness in men. Journal of Comparative Neurology 323(3): 326–340, 1992.

Wolf, M. E.; Goodale, M. A. Oral asymmetries during verbal and non-verbal movements of the mouth. Neuropsychologia 25(2): 375, 1987.

Yeni-Komshian, G.; Benson, D. Anatomical study of cerebral asymmetry in the temporal lobe of humans, chimpanzees and monkeys. Science 192: 387–389, 1976.

Young, A. W.; Ellis, H. D. Ear asymmetry for the perception of monaurally presented words accompanied by binaural white noise. Neuropsychologia 18: 107–110, 1980.

Zahn, R.; Huber, W.; Drews, E.; Erberich, S.; Krings, T.; Willmes, K.; Schwarz, M. Hemispheric lateralization at different levels of human auditory word processing: A functional magnetic resonance imaging study. Neuroscience Letters 287: 195–198, 2000.

Zatorre, R. J.; Belin, P. Spectral and temporal processing in human auditory cortex. Cerebral Cortex 11: 946–953, 2001.

Shown in the bottom panel are axial views of magnetic resonance images of chimpanzee, gorilla, and orangutan brains. The yellow colored area represents the region of the precentral gyrus that corresponds to where the hand is located in the motor strip of each genus. This brain area is referred to as the motor-hand area or KNOB because of its omicron or omega shape. In the top panel are photographs of each species engaged in bimanual feeding, a behavioral measures that correlates with asymmetries in the motor hand area. (Photographs by William Hopkins.)

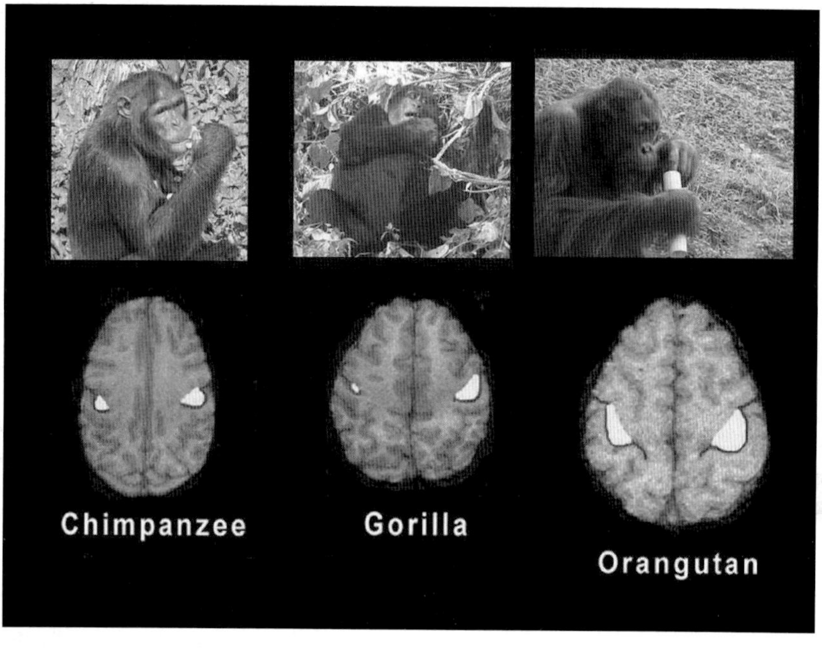

The Evolution of Hemispheric Specialization in Primates
W.D. Hopkins (Editor)

HANDEDNESS AND NEUROANATOMICAL ASYMMETRIES IN CAPTIVE CHIMPANZEES: A SUMMARY OF 15 YEARS OF RESEARCH

William D. Hopkins, Jamie L. Russell, Susan Lambeth and Steven J. Schapiro
Division of Psychobiology, Yerkes National Primate Research Center, Atlanta, Georgia, USA (WDH, JLR); Department of Psychology, Agnes Scott College, Decatur, Georgia, USA (WDH); Department of Veterinary Sciences, The University of Texas, M. D. Anderson Cancer Center, Bastrop, Texas, USA (SL, SJS); Department of Psychology, Southwestern University, Georgetown, Texas, USA (SJS)

ABSTRACT

For the past 15 years, we have conducted a series of studies examining the role of genetic and non-genetic factors on the development of handedness, motor skill asymmetries and other behavioral asymmetries in captive chimpanzees. This chapter summarizes the collective results of these studies. In addition, recent findings on neuroanatomical asymmetries are presented as well as associations between behavioral and brain asymmetries. Collectively, we argue for continuity in the evolution of behavioral and brain asymmetries between humans and chimpanzees.

Keywords: chimpanzees, cognition, brain asymmetry, manual skill, tool use

Hemispheric specialization refers to the differential abilities of the left and right cerebral hemispheres for perceptual, motor and cognitive functions [Hellige, 1993]. One of the most pronounced behavioral asymmetries in humans is handedness. Right-handedness is a universal trait in humans, although the proportion of self-reported right-handed individuals does vary between cultures to some extent [Annett, 2002; Porac & Coren, 1981; Perelle & Erhman, 1994]. It has been suggested that right-handedness reflects a left hemisphere specialization for motor skill and is thought to be an indirect marker of lateralization for language functions. For example, 96% of right-handed individuals are left-hemisphere dominant for language, whereas only about 70% of left-handed individuals are left-hemisphere dominant for language [Knecht et al., 2000; Rasmussen & Milner, 1977]. The association between hand preference and language dominance has led to a number of evolutionary and biological theories proposing that hemispheric specialization and language (or perhaps other higher cognitive functions) co-evolved and were unique characteristics of the human brain [see Corballis, 1992, 2002, for review]. The historical lack of

ISSN: 1936-8526 DOI: 10.1016/S1936-8526(07)05006-3

evidence of functional asymmetries in nonhuman species, with the exception of birds, was taken as evidence in support of the uniqueness of hemispheric specialization to humans [Bradshaw & Rogers, 1993]. However, based on research carried out in the past 15–20 years, there is now good evidence of population-level hemispheric specialization in a host of nonhuman species for cognitive, perceptual and motor functions [Bradshaw & Rogers, 1993; Rogers & Andrew, 2002; Ward & Hopkins, 1993; Vallortigara & Rogers, 2005]. There is also good evidence of population-level neuroanatomical asymmetries, although it is unclear how they are associated with functional asymmetries associated with hand use and other lateralized behaviors [LeMay, 1982; Hopkins & Rilling, 2000; Hopkins & Cantalupo, 2004]. Taken together, the most recent behavioral and neuroanatomical data clearly challenge the long held belief that hemispheric specialization is a uniquely human attribute and, at a minimum, indicate that language is not a necessary condition for the expression of hemispheric specialization.

Despite the compelling evidence of population-level behavioral asymmetries in lower vertebrates, the question of whether population-level handedness exists in nonhuman primates remains a topic of considerable theoretical and empirical debate [e.g., Crow, 1998, 2004; Palmer, 2002; Papademetriou et al., 2005]. Prior to 1987, the general scientific view was that population-level handedness was unique to humans [Warren, 1980]. This conclusion was largely based on the hand preference data for simple reaching provided from Old World monkeys (macaques) [see Lehman, 1993] and paw preference data in rodents. With respect to apes, there was little data on handedness in both captive and wild settings. Indeed, prior to 1993, the largest single study of handedness was by Finch [1941] who tested handedness in 30 captive chimpanzees on four measures of simple reaching [reviewed in Hopkins, 1996]. Finch [1941] found no evidence of population-level handedness in the 30 chimpanzees he tested but concluded "admittedly, increasing the number of test-situations should result in a more adequate determination of chimpanzee handedness; however, the present work does not pretend to explore more than a rather narrowly limited aspect of chimpanzee lateral organization" (pp. 118–119), a notion that seemed to be lost for the next 50 years. Thus, in our opinion, the conclusions regarding the lack of population-level handedness in apes and other primates was largely generated based on (a) very little data in apes and (b) a rather confined set of behavioral measures of hand preference in many monkey and ape species.

One purpose of this chapter is to provide a summary of data on the assessment of handedness in captive chimpanzees that has taken place over the past 15 years, Hopkins and colleagues have revealed evidence of population-level handedness for a variety of measures and this work has been criticized on both methodological and theoretical grounds. For instance, it has been suggested that the evidence of population-level handedness in captive chimpanzees might be due to (a) observer bias [Palmer, 2002], (b) unequal observations across individuals on the basis of which handedness was determined

[Palmer, 2002], (c) a lack of independence of observations of hand use [McGrew & Marchant, 1997; Palmer, 2003], and (d) biases in the early environments of the captive apes that might influence their hand use [Crow, 2004; McGrew & Marchant, 1997]. By summarizing the results over the past 15 years and integrating the findings from different laboratories housing chimpanzees that have participated in this research, we hope to present a coherent argument in response to the many criticisms and skepticism that accompanied the initial work.

A second aim of this chapter is to provide preliminary findings on consistency on handedness across multiple measures of hand use. Some have suggested that humans show "true handedness", that is, consistent hand use across multiple measures whereas nonhuman primate handedness is characterized as "task-specific" [McGrew et al., 1999]. The issue of "true handedness" compared to "task-specific" is by no means uniformly accepted in the assessment of handedness in humans. There is considerable argument over whether human handedness lies on a continuum from strongly left- to strongly right-handed individuals or whether handedness is a discrete trait with truly left-handed, right-handed and ambidextrous subjects [see Corey et al., 2001; Healey et al., 1986, for various views]. Notwithstanding, to facilitate comparisons between human and nonhuman primate handedness and to more fully explore the dimension of handedness in chimpanzees, we provide some preliminary findings on consistency of hand use in our sample of apes.

Before presenting the results, some information is provided on our chimpanzee sample as well as the housing and testing conditions of the subjects. Additionally, we provide cursory descriptions of the main tasks for which we have assessed handedness in the chimpanzees. The list of behaviors for which handedness has been collected is not exhaustive but rather focuses on those tasks for which at least 100 individuals have been tested.

GENERAL METHODS

Subjects

Initially, the work was carried out at the Yerkes National Primate Research Center (YERKES) but subsequent studies have been carried out at The University of Texas M. D. Anderson Cancer Center (UTMDACC) as well as the Alamogordo Primate Facility (APF).

Data collection at the YERKES began in 1992 and continues to date, while data collection began at UTMDACC in 2001, and at APF in 2003. For most of the measures, at least two replications of tests are done on each animal to provide a means of assessing reliability in hand use as well as increase the number of individual responses obtained for each subject. Because of death by natural causes, retirement of some animals, level of motivation to participate in a task or other mitigating factors, the number of subjects that have been tested on different tasks has varied over time. For example, bimanual feeding and the

TUBE task (see below) were some of the initial tasks we used in the YERKES chimpanzees and subsequent tests were conducted with chimpanzees from UTMDACC and APF. In contrast, the measures of motor skill and studies on grip morphology were developed and employed more recently. Thus, some subjects that were in the initial studies on feeding and the TUBE task were not tested on these later tasks.

General Testing Environments

For the most part, the chimpanzees have been tested in the outdoor portions of their home cages. Housing conditions varied considerably across subjects from singly housed individuals to groups of subjects with as many as 22 chimpanzees. Wherever possible, testing (in larger social groups) has been done using focal animal techniques rather than separating subjects from their natal groups for the purposes of data collection. Because our goal has been to obtain data from as many subjects as possible under the least stressful circumstances, we have opted to collect data under these conditions. Separating the chimpanzees from their groups sometimes caused stress and the animals were less motivated to engage in the task of interest precipitating the decision to use focal sampling techniques. No specific effort has been made to counterbalance the tasks across subjects. Rather, data collection has often been opportunistic and influenced by the development of increasingly sophisticated tasks or motivated by theoretical questions.

One of the main aims of the initial and ongoing research has been to characterize handedness in the chimpanzees for tasks that are less vulnerable to situational or positional factors, such as simple reaching, which is one of the most common measures of hand use. To this end, we have tried to develop measures that assess handedness on tasks that require coordinated actions of the hands such as bimanual feeding and the TUBE task. For unimanual tasks, particularly those that occur naturally such as throwing, we have tried to take into consideration situational and positional factors such as the type of throwing movement (underhand, overhand) or posture (upright, tripedal) to assess their potential influence on hand preferences. Lastly, for structured unimanual tasks, such as simple reaching and manual gestures, we have tried to obtain independent and discrete observations of hand use. Moreover, we have attempted to consider morphological factors such as grip type (precision, nonprecision) or gesture type (whole hand, food beg) as potential factors that might influence hand use.

MEASURES OF HAND PREFERENCE

Coordinated Bimanual Tube Task (TUBE)

The TUBE task entails the use of poly-vinyl-chloride (PVC) tubes (24–31 cm long, 2.5 cm wide) with peanut butter smeared on the inside edge, approximately 2–4 cm in depth. The tubes are given to the subjects in their home cage by pushing them through the cage mesh. The digit and hand used to remove the peanut butter are recorded as either the left or right each time the subjects inserted their finger, removed peanut butter from the tube and placed their finger in their mouth. Observations continue until the subjects stop showing interest in the tube (usually when they have eaten all the peanut butter), dropped it for at least 10 s, or pushed the tube back out of their home cage through the cage mesh.

Fixed-Tube (FIXED-TUBE)

The motor demands of this task are similar to the coordinated bimanual TUBE task, but the positional and postural demands of the task are slightly different from the TUBE task. We attached T-junction to one end of the sections of PVC pipe so that one end of the PVC could be inserted into the cage but could not be pulled into it. The end of the PVC pipe inserted into the cage had peanut butter smeared on the inside edge, as in the TUBE task. We placed the tubes 1 m above the cage floor according to height of the subject. In this task, the subjects were required to adopt a bipedal posture to extract the peanut butter. Typically, subjects propped themselves with one hand on the cage or held the tube and extracted the peanut butter with the opposite hand. We tested subjects with the PVC to the immediate left or right in order to balance out the potential influence of stimulus position on hand use. We obtained a minimum of 50 responses from each subject.

Bimanual Feeding Task (FEED)

The chimpanzees housed at each facility receive fruits and vegetables as part of their daily diet. Each subject usually receives 2 oranges, 1 banana, some celery stalks and/or carrots. The chimpanzees will cache or hold the extra food while simultaneously feeding with the opposite hand. The initial hand used to eat during these contexts is noted as left or right. The hand used in feeding had to occur for a minimum duration of 3 s and the nonfeeding hand had to be holding the remaining portions of food and not used in postural support or holding the cage. A minimum of 10 responses were obtained from each subject.

Simple Reaching

On each trial, a raisin was thrown into the subject's home cage. The raisin was thrown by the experimenter to a location at least 3 m from the focal subject so that the chimpanzees had to locomote to the position of the raisin, pick up the raisin and bring it to their mouth for consumption. When the chimpanzee acquired the raisin, the experimenter recorded the hand used as left or right. One, and only one reaching response, was recorded each trial to assure independence of data points [see McGrew & Marchant, 1997; Hopkins, 1999, for contrasting views]. Thus, raisins were not randomly scattered in home cages but rather an individual raisin was thrown into cages and subjects retrieved the raisin before another was thrown into the cage. Subjects were required to locomote at least three strides between reaching responses to maintain postural readjustment between trials. A minimum of 50 responses were collected from each subject with the range in responses between 50 and 84 for the sample.

Throwing

For anyone that works with captive chimpanzees and other great apes, it is well known that they are renowned for their propensity to throw feces and other materials from their cages at humans, particularly unfamiliar individuals. Beginning in the early 1990s, my colleagues and I began to make note of the hand used and the posture of the chimpanzees when they threw at us, strangers and at other individuals living with or adjacent to the subjects [see Hopkins et al., 1993]. Even though the initial work was published in 1993, we have continued to opportunistically collect throwing data at YERKES and, in 2002, we began to collect similar data on the chimpanzees from the UTMDACC [see Hopkins et al., 2005a]. We have observed throwing in 135 individuals and have obtained at least 6 observations of hand use for throwing in 89 chimpanzees.

Tool Use (Simulated Termite Fishing)

Chimpanzees are well known for their tool-using abilities in the wild and this is the most frequently studied measure of handedness in wild chimpanzees [Biro et al., 2003; Boesch, 1991; Marchant & McGrew, 1991; McGrew & Marchant, 1996; Lonsdorf & Hopkins, 2005]. In an attempt to mimic the data collected in wild chimpanzees, we have examined hand use for a task that mimics the sensory and motor demands of termite fishing. For the unimanual version of this task, each PVC tube in the apparatus was first filled with honey or syrup to about one-third of the whole length of the tube, which made it impossible for the subjects to reach the food directly with their fingers. After placing the device on the cage, sticks or bamboo skewers were supplied to the subjects by handing them directly or dropping them into the cages from observation towers positioned around the enclosures. At UTMDACC, the unimanual tool use task is a standard enrichment device and many of the subjects already know how to use the device. At YERKES, the device was relatively new to the chimpanzees. Therefore, some subjects were tested repeatedly in order to facilitate the learning of the task prior to data acquisition. Specifically, subjects had to successfully use the tool use device on 100 trials before they were considered to be facile in their tool use. Once the subjects met this criterion, hand preference data were collected for use in this study. Despite repeated exposure to the device, some subjects failed to learn or failed to show any interest in the tool use task.

For the bimanual version of the task, for the YERKES device, peanut butter was smeared close to the T-junction end of the PVC tube. This was so far from the endpoint placed into the cage that the subjects could not gain access to the food with their finger. The tube was inserted through the mesh wall of the subject's cage, at a height that led the subject to assume a seated or crouched posture for accessing the peanut butter inside the tube by means of the bamboo skewer. Because the PVC tube was not firmly attached to the cage, but could move and rotate freely around the opening in the cage mesh, subjects had to stabilize the tube with one hand and insert the bamboo skewer with the other to retrieve the peanut butter, thereby engaging in a coordinated bimanual action. For the UTMDACC chimpanzees, as previously noted, the peanut butter was smeared in the inner most, central portion of the PVC pipe, precluding the subjects from efficiently obtaining the food with their fingers. The PVC pipe was placed into the cage and the subjects had to hold the tube with one hand and use an available tool (stick, grass, etc.) with the opposite hand to remove the peanut butter. For both the unimanual and bimanual tasks, the experimenter recorded the hand used by the subject when inserting the stick and the type of manual grip adopted in holding the stick.

Coconut Opening

This measure was designed also to simulate the motor demands of the use of an anvil to open hard foods in wild chimpanzees [McGrew et al., 1999]. Bouts and frequencies of hand use were recorded for subjects as they struck a coconut against a hard surface in order to crack it and consume it. Coconut cracking behavior was defined as any attempt to open the coconut in which the subject held the coconut in one hand and struck it against a surface in the enclosure. In some instances, animals would throw the coconut against the wall in order to crack it, hit the coconut with their hand, drop the coconut from a height onto a hard surface, or bit the coconut. Although these attempts were occasionally successful at opening the coconut, these behaviors were not included in the analysis of handedness. The most commonly observed coconut cracking technique consisted of chimpanzees holding a coconut in one hand with the palm facing down while using an overhand downward and forward motion to strike the coconut on a hard surface in front of them (either at ground level or on the concrete wall). A few subjects employed a different technique in which they held the coconut in one hand with the palm facing up while standing tripedally and striking the coconut against a surface in front of them using an underhanded forward swinging motion.

A bout was defined as a continuous series of coconut strikes. A change in bout was recorded under three conditions: (1) the subject transferred the coconut from one hand to another, (2) the subject put the coconut down before picking it up to resume cracking, or (3) the subject stopped striking the coconut for a period of at least 3 s. Each individual strike of the coconut within a bout was recorded as the frequency score. For example, if a subject made five continuous strikes with his left hand, then switched to his right hand and made 10 continuous strikes before successfully opening and consuming the coconut, he would have one left bout, one right bout, a left frequency score of five, and a right frequency score of ten. Each animal housed at YERKES was given access to a coconut on three separate occasions and observed for a minimum of 30 min for each coconut. Each animal housed at UTMDACC was given access to a coconut on a single occasion and observed for at least 30 min. During the observation period, the experimenter recorded bouts and frequencies of hand use for coconut cracking and at the end of the observation period, the experimenter recorded whether or not the subject had successfully opened the coconut.

Grooming

For this behavior, data were collected from video tapes of apes grooming as well as real-time observation [Hopkins et al., in press]. There were several criteria employed when selecting which animals to observe both for filming and for real-time observations. First, the experimenter chose focal subjects opportunistically; whichever chimpanzees happened to be grooming at the time of observation became possible subjects. Second, if multiple animals were grooming, the experimenter chose the animal that was located in the most visible position. Third, focal subjects were chosen by the number of data points previously recorded for each animal. For example, if subject A had 25 previously observed bouts of grooming while subject B had only five and both subjects were equally visible to the experimenter, then subject B was chosen as the focal subject in an effort to maintain equal numbers of observations for each subject. Obviously, some subjects groomed much more than others and, therefore, some subjects have more data points than others. However, a consistent effort was made to focus on subjects that had the fewest observations whenever possible in order to increase our overall sample size.

Both real-time and filming observations were done in 5-min increments. Once a focal subject had been chosen, the experimenter would observe and record behaviors of that animal for 5 min. If the focal subject's grooming episode stopped within the 5-min period (defined by a 30-s period of time in which neither the focal subject nor his or her grooming partner showed any grooming activity), then the experimenter would stop observations and select another focal subject actively engaged in a grooming episode. Occasionally, only one chimpanzee would be grooming at a time. In this case, if after 5 min of observation no other chimpanzees were seen to be grooming, the experimenter would begin a new 5-min observation period with the same focal subject.

Because grooming is such a fluid, dynamic behavior, coding the behavior real-time and defining the onset and offset of different dimensions of this behavior are quite formidable. Therefore, to simplify the data collection, time sampling was chosen as the most reliable data collection method. For both video and real-time data collection, observations of hand use for grooming were recorded using 15-s time sampling intervals. A stopwatch was set to beep every 15 s allowing the experimenter to vigilantly watch grooming behavior and record hand(s) used and whether or not the mouth was involved in the grooming action during each sampling interval. If no manual grooming occurred at the 15-s interval, "no grooming" was recorded.

Hand use was characterized as unimanual or bimanual. Grooming was scored as being unimanual if only one hand was involved in the grooming process and the other hand was in a resting position or being used for some other purpose such as postural support. Unimanual grooming was simply categorized as either left or right depending on the hand being used in the grooming action. Bimanual grooming consisted of three classifications: left-dominant/right-subordinate (L-BIM), right-dominant/left-subordinate (R-BIM), or yoked (YOKED). Bimanual grooming was scored as L-BIM if the right hand was used to hold or stabilize an area of skin for grooming, while the left hand performed the action of plucking, picking, etc. In the case of R-BIM, the left hand was used to stabilize the skin while the right hand performed the grooming action. In this way, L-BIM represented a coordinated bimanual action in which the action of the right hand was subordinate to the action of the left hand while R-BIM represented the same type of coordinated bimanual action in which the left hand was subordinate to the right hand. Grooming was scored as YOKED if both hands played an active role in the grooming action with neither hand being subordinate to the other.

PERFORMANCE MEASURES OF HAND SKILL

Grasping Errors

Raw peanuts were removed from their shells and split into half. The peanuts were then thrown into the subjects' home cage. Peanuts were thrown into the cage until exactly 20 responses with each hand had been obtained from the subjects. Because the emphasis of this particular experiment was on obtaining 20 responses with each hand from each subject, the subjects were not required to locomote to the peanut in order for the response to be recorded. Instead, the peanut halves were dropped next to one of the subject's hands, thus encouraging him to use one hand over the other. The experimenter watched as the subject attempted to pick up the peanut and noted the sequence of grip types attempted in grasping the food item. An error was recorded when the subjects attempted to grasp but dropped the peanut or failed to grasp the peanut from the floor upon their attempt. If at any point the subjects switched hands or used their mouth to obtain the peanut, the trial was not included. Data were collected in two test sessions with the first 20 responses (10 from each hand) being recorded over a one-week period. The second test session took place three weeks later when the second set of data was collected. During testing, the peanuts were thrown to either side of the subject's body in a random order to facilitate obtaining responses with both hands.

Gesture Latency

For this task, the aim was to assess the latency to produce a left or right hand gesture from the chimpanzees. In this task, the experimenters positioned themselves either to the extreme right or extreme left of the subject and offered a piece of banana to the chimpanzee. The amount of time it took to produce a gesture with the hand ipsilateral to the position of the experimenter was recorded on each trial. Each subject received two trials from each position and the mean latency for the left and right hands was calculated and used as the dependent measure.

Bent Wire Task

The YERKES chimpanzees were also tested on a motor task called the Life Saver task. This task looked at the amount of time it took chimpanzees to remove a LifeSaver candy from a bent wire using their right compared to their left hand. In order to do the task, a piece of wood with a bent wire sticking out of it was held, with the hook facing down, through the mesh in the chimpanzees enclosure. A LifeSaver was placed on the wire and the amount of time it took for the chimpanzee to remove the lifesaver using only one hand was recorded. In addition, we recorded if the chimpanzees dropped the LifeSaver when it came off the wire or if they caught it. An error was recorded when the LifeSaver was dropped. Five trials were completed for each hand.

CHARACTERIZING HAND PREFERENCE

For each subject and measure, a *z*-score based on the total frequency of right- and left-hand responses was determined. Based on the *z*-scores with $p<.05$, subjects were classified as being either left-handed ($z \leq -1.96$), right-handed ($z \geq 1.96$) or ambiguously handed ($z > -1.96$ and $z < 1.96$). In addition, for each measure and subject, a handedness index (*HI*) was derived by subtracting the number of left- from the number of right-hand responses and dividing by the total number of responses (R−L/R+L). The *HI* values ranged from −1.0 to +1.0 with the sign indicating direction of hand use and the absolute value reflecting the strength in hand preference. Positive *HI* values reflect right hand biases and negative values reflect left hand biases.

RESULTS

Distribution of Hand Preferences and Performance Asymmetries for Each Measure

Shown in Table I are the distributions of hand preference for each of the behavioral measures. Included in Table I also are the mean *HI* scores and R:L odds for each task. Significant population-level rightward asymmetries are found for manual gestures [$t(226) = 6.14$, $p<.001$] throwing [$t(88) = 3.81$, $p<.01$], tube [$t(470) = 5.48$, $p<.01$], fixed tube [$t(125) = 3.38$, $p<.01$], reaching [$t(278) = 2.62$, $p<.01$], and bimanual feeding [$t(202) = 2.45$, $p<.01$].

Table I. Distribution of hand preferences for different behavioral tasks.

Measure	#L	#NP	#R	R:L Odds	Mean *HI*	s.e.
Manual gesture	36	58	133	3.69	.312	.049
Throwing	23	16	50	2.17	.301	.075
Tube	129	105	237	1.84	.137	.025
Fixed tube	36	26	64	1.78	.142	.042
Reaching	58	125	96	1.66	.068	.026
Bimanual feeding	52	84	67	1.29	.076	.031
Grooming	59	76	75	1.27	.034	.025
Simulated termite fishing	66	50	76	1.15	.004	.040
Coconut	65	45	62	0.95	.015	.053

Italicized measures reveal population-level right-handedness as assessed by one-sample t-test.
NP = No Preference

No population-level handedness was found for the simulated termite fishing task, opening coconuts or grooming. With respect to the skill tasks, for the grasping error task, the chimpanzees make fewer errors with the right compared to the left hand [$t(265) = 2.67$, $p < .009$]. For gesture latency, the chimpanzees responded more quickly with the right compared to the left hand [$t(122) = 4.01$, $p < .01$]. No significant difference in motor performance between the left and right hand was found on the bent wire task.

The Effect of Rearing on Hand Preference and Performance Asymmetries

One of the main criticisms of the findings in captive chimpanzees has been the argument that rearing chimpanzees in a captive, right-handed human environment has induced the observed population-level asymmetries for some of the measures of interest. To further test this hypothesis, we compared the *HI* scores for mother-reared, nursery-reared and wild-caught animals using an analysis of variance. Shown in Table II are the mean *HI* scores for each rearing group and measure. There was no significant effect of rearing on any measures of hand preference and, if anything, the influence of rearing has a random influence among the different measures.

Sex Differences in Hand Preference

It is often claimed in humans that females are slightly more right-handed than males [Kimura, 1999; Springer & Deutsch, 1993], although meta-analyses indicate that the effect is relatively small. Evidence of sex differences in handedness in primates is relatively weak [see Fagot & Vauclair, 1993], but one of the more intriguing findings is the recent report of a fairly strong sex difference in bimanual feeding in wild chimpanzees by Corp and Byrne [2004]. Corp and Byrne [2004] reported that females showed population-level right-handedness

Table II. Mean handedness indices for each behavioral measure and rearing group.

Measure	Rearing Group		
	Mother-reared	Human-reared	Wild-caught
Gesture	.306	.315	.352
Reaching	.054	.120	.031
Bimanual feeding	.091	.081	.128
TUBE	.161	.159	.050
Fixed tube	.246	.090	.009
Throwing	.293	.309	.500
Coconut	.004	.078	−.031
Simulated			
Termite fishing (U)	.039	−.037	−.023
Termite fishing (B)	.093	.206	−.001
Grooming (U)	−.039	.010	−.001
Grooming (B)	.137	.112	.261
Grasping error	.094	.059	.049
Gesture latency	.138	.241	.186

U = Unimanual
B = Bimanual

for bimanual feeding whereas males showed population-level left-handedness. Shown in Figure 1 are the mean *HI* scores for females and males for each of the behavioral measures. For any given measure of hand preference, no statistically significant sex differences were found for hand preference, although females have higher *HI* values on more measures than do the males.

Unimanual Contrasted to Bimanual Hand Use

For two of our measures, grooming and simulated termite fishing, we have collected data under conditions where the subjects could solve the task using a unimanual contrasted with bimanual strategy. In light of the fact that evidence of population-level handedness has been found for bimanual tasks in both captive and wild apes [Byrne & Byrne, 1991; Corp & Byrne, 2004], this distinction seems important in the assessment of handedness. Shown in Figure 2 are the mean *HI* scores for unimanual and bimanual hand use for grooming and the simulated termite fishing task. For both tasks, the chimpanzees show a significant rightward shift in hand preference for bimanual compared to unimanual hand use. Indeed, population-level right-handedness is evident for both measures when considering bimanual, but not unimanual measures of hand use.

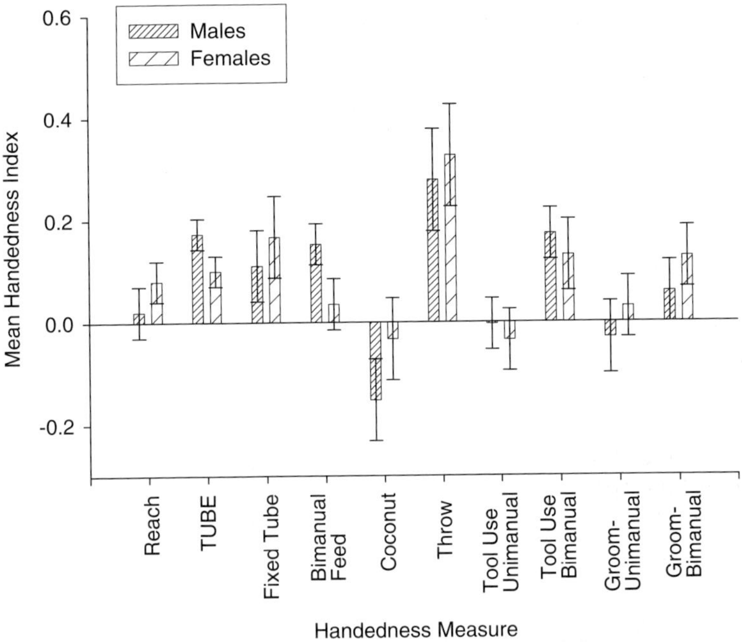

Figure 1. Mean handedness index (*HI*) scores (±s.e.) for males and females on several measures of handedness (see text for descriptions of the measures). Positive *HI* values reflect right hand preferences and negative values reflect left hand preferences.

In addition to these two measures, population-level right-handedness for bimanual feeding has been reported in the YERKES chimpanzees [Hopkins, 1994] and some have been critical of these findings [Palmer, 2002]. Specifically, Palmer [2002] has argued that individual differences in the number of observations used to derive hand preferences for this measure skewed the distribution toward right-handedness. In response, Hopkins and Cantalupo [2003] showed that obtaining the same number of observations had no significant effect on handedness at least for the TUBE task. As a means of further assessing asymmetries in unimanual and bimanual feeding and to address the issue raised by Palmer [2002], we carried out a study in which we measured how much time chimpanzees spent feeding unimanually and bimanually with the left and right hands rather than recording their preferred hand use. In this study, each chimpanzee was tested on three separate occasions with three different types of foods including celery, cabbage, and grapefruit. The foods were provided to the animals and we recorded the amount of time that the chimpanzees engaged in unimanual left, unimanual right, bimanual left, and bimanual right feeding for each food item. An *HI* based on the average amount of time the subjects fed with the left and right hand was calculated for unimanual and bimanual feeding

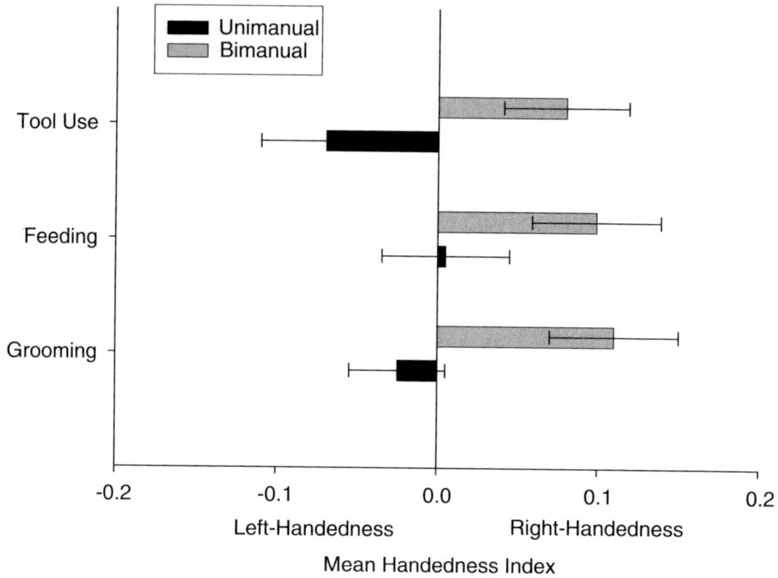

Figure 2. Mean handedness index values (±s.e.) for unimanual and bimanual hand use in tool use, feeding and grooming. Positive *HI* values reflect right hand preferences and negative values reflect left hand preferences.

(see Figure 3). The chimpanzees were significantly more right-handed for bimanual compared to unimanual feeding [$t(89) = 2.66$, $p < .01$]. Moreover, this pattern of results was evident for all three food types. In 90 chimpanzees, hand preference data from the Hopkins [1994] paper as well as *HI* scores for unimanual and bimanual feeding durations were available. To evaluate whether the preferences for bimanual feeding duration reflected inherent preferences, we correlated the *HI* preference scores from the Hopkins [1994] paper with the unimanual and bimanual *HI* duration scores. The *HI* preference scores significantly positively correlated with the bimanual *HI* duration scores ($r = .391$, $df = 88$, $p < .01$) but not the unimanual *HI* duration scores ($r = .048$, $df = 88$, n.s.). Thus, hand preferences in bimanual feeding positively correlate with the amount of lateralized time the chimpanzees spend bimanually, but not unimanually, feeding. Moreover, hand preferences for bimanual feeding are stable over a relatively long period of time (at least 10 years).

Grip Morphology and Hand Use for Simple Reaching

One of the more interesting results to emerge in recent years is the evidence of an association between hand preference and grip morphology in chimpanzees. There are considerable individual differences in grasping techniques used

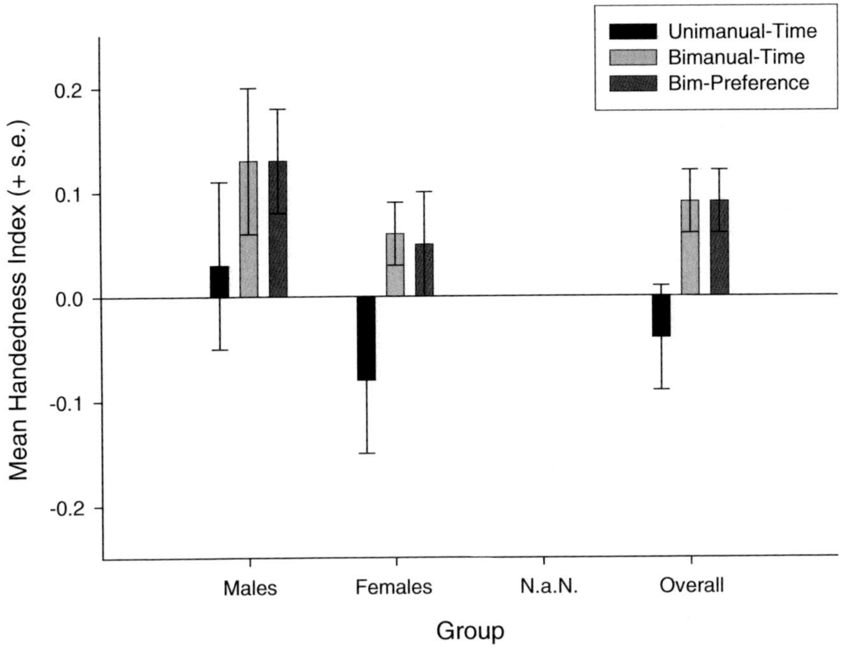

Figure 3. Mean HI values for the duration and preference for bimanual feeding at the YERKES chimpanzees.

by chimpanzees (and other great apes). Primarily, the grasping techniques can be broken down into three types: the thumb-index, middle-index and single digit response. Typically, subjects using a thumb-index grip grasp the object between the thumb and lateral portion of the index fingers (although sometimes they use a tip-to-tip pad response). The middle-index grip is characterized as the subjects grasping an object between the lateral or tip portion of the middle and index fingers. Lastly, the single digit response, which is rare, occurs when the subject presses a single digit down on the object against a hard surface (like the floor) and the objects (usually food) stick to the finger.

At YERKES, we have examined hand use in relation to grip morphology in several studies [Hopkins et al., 2002]. Using four separate foods including quartered peanuts, M&M's®, stick pretzels, and raisins, we have recorded the number of thumb-index, middle-index and single digit responses when the subjects were required to produce an equal number of grasping responses for the left and right hands ($N = 25$ for each food item). Shown in Figure 4 are the mean differences in the number of thumb-index and middle-index responses for the left and right hands for each of the four food items. There were significantly more right compared to left thumb-index responses for peanuts [$t(144) = 4.04$, $p < .001$], M&M's® [$t(139) = 2.53$, $p < .01$], pretzels [$t(135) = 2.56$, $p < .01$], and

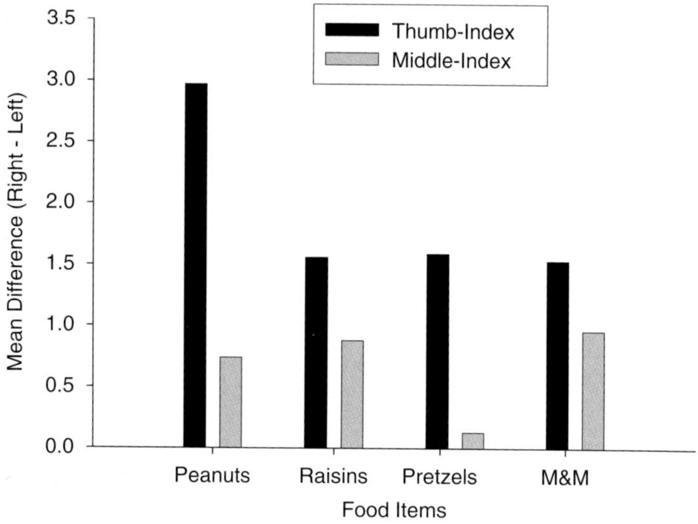

Figure 4. Mean difference between the right and left hands in the number of thumb-index and middle-index grasping responses for four different foods. Positive values reflect a greater number of responses by the right hand.

raisins [$t(142) = 2.25$, $p < .05$]. In contrast, no difference was found for the number of middle-index responses produced by the right and left hands. Thus, chimpanzees produce more thumb-index responses when using their right compared to left hand and these results are consistent with the data from two other studies conducted in relatively large numbers of chimpanzees [see Tonooka & Matsuzawa, 1995; Hopkins et al., 2005b].

Motor Skill and Grasping Errors

To assess possible asymmetries in motor skill, we have measured the number of errors chimpanzees make when grasping small food items. Initially, chimpanzees were tested using quartered peanuts [Hopkins et al., 2002], but subsequent tests were carried out with other food items including M&M's®, TartNTiny and pretzels [Hopkins & Russell, 2004]. In the initial tests using quartered peanuts, each chimpanzee was required to grasp 20 food items with the left and right hands and the experimenter recorded how many errors they made when attempting to grasp the food. As noted above, in the course of this initial study it was observed that many of the chimpanzees adopted different grasping techniques and we believed that this might influence our results. Thus, in a follow-up study, we retested many of the chimpanzees for their grasping errors in two experiments. In Experiment 1, the subjects were required to adopt a thumb-index response for all responses of the left and right hand. In Experiment 2, we characterized each subject as preferring to use a thumb-index or middle-index grasping technique, then tested each subject and recorded responses only when they used their preferred grip technique. We have largely found consistent evidence of a right hand advantage in grasping for the chimpanzees. The mean *HI* scores for each study are shown in Figure 5.

The mouth can also be used as a manipulative agent in chimpanzees. Indeed, they often use the mouth for motor functions such as grooming and to "grasp" food that is difficult to obtain using the hand. In our studies on grasping errors, we have also noted the frequency of occurrence in the use of the mouth to pick up the food. Assuming that chimpanzees are worse at grasping with the left compared to the right hand, it might be hypothesized that increased use of the mouth would be associated with increasing errors in grasping. We have found a significantly positive association between the frequency of the use of the mouth and total errors in grasping ($r = .261$, $df = 11$, $p < .005$) and this association is specific to the number of errors made by the left ($r = .282$, $df = 111$, $p < .001$) but not the right hand ($r = .091$, $df = 111$, n.s.). Thus, chimpanzees that make more errors with the left hand tend to also use their mouth more for grasping the food items.

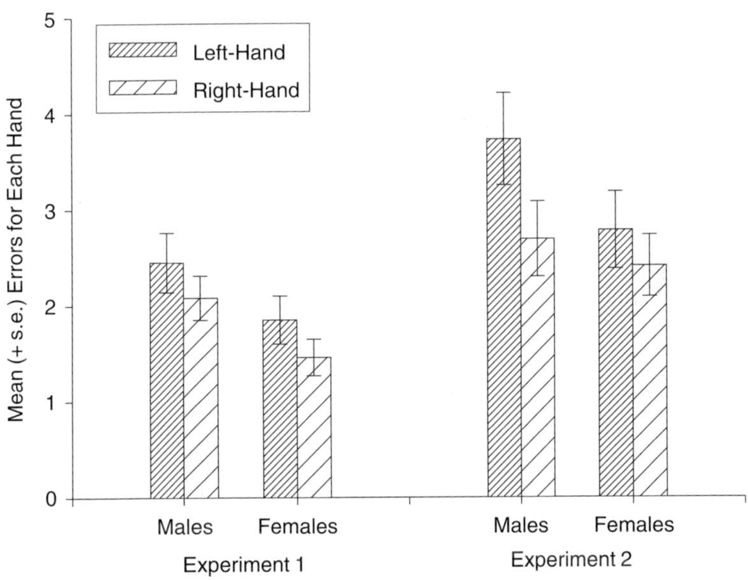

Figure 5. Mean number of errors made by the left and right hands during two different grasping tasks (see Hopkins & Russell, 2004)

Table III. Handedness task loadings for each factor structure.

Task	Factor 1	Factor 2
Simulated termite fishing	**.658**	.222
Reaching	**.738**	.132
Tube	**.693**	−.358
Manual gestures	.217	**.694**
Grasping errors	.063	**.658**

Bold values indicate significant item loading on each factor.

Chimpanzees' Handedness: Unidimensional or Multidimensional Trait?

The acquisition of handedness data in a relatively large sample of apes for multiple measures allows for the assessment of whether handedness in chimpanzees is unidimensional (reflecting a single, underlying trait) or is multidimensional (task specific, with potentially different motor components underlying different aspects of hand use) [see Corey et al., 2001; Healey et al., 1986, for contrasting views]. When combining the data from YERKES and UTMDACC, we have complete data on five measures of hand use in 218 chimpanzees including the TUBE task, simple reaching, manual gestures, termite fishing and grasping skill. The *HI* scores for each task were subjected to a principal components factor analysis using varimax rotation to account for the most variance in the scores. Adopting the criteria of eigenvalues exceeding 1.0 to be considered a reliable factor, this analysis revealed two handedness factors that accounted for 53% of the variance. Shown in Table III are the item loadings for each factor. The TUBE, reaching and termite fishing tasks loaded on Factor 1, while manual gestures and the motor skill task loaded on Factor 2. These results are consistent with previous findings using factor analysis of handedness in chimpanzees [Wesley et al., 2002] and support the interpretation that handedness is a multidimensional construct of handedness rather than a unidimensional, discrete trait, as has been suggested by some [McGrew & Marchant, 1997].

Consistency in Hand Use Across Preference Tasks

Because we have multiple measures of hand preference on 218 chimpanzees, we can begin to consider consistency in hand use across multiple measures in a way that might approximate the approaches employed with humans. If each handedness task is treated as an item, we classified subjects as preferring the right or left hand based on the sign of their *HI* score (positive = right, negative = left). We then assigned values of 0 and 1 to hand preference classification for each task. Summing the classification scores across measures results in 5 possible outcomes ranging from 0 (exclusive left hand use) to 4 (exclusive right hand use). Subjects with a score of 2 are true ambidextrous subjects

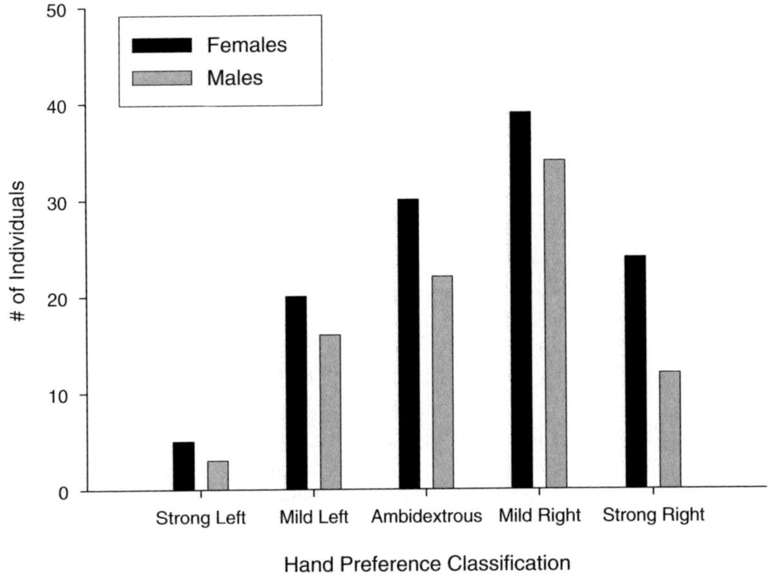

Figure 6. Number of male and female chimpanzees classified as strongly left, mildly left, ambidextrous, mildly right and strongly right when summed across multiple measures of hand use (see text for description of behaviors).

(preferring the left hand on two tasks and the right on two tasks). Subjects with scores of 1 and 3 were characterized as mildly left (left preferent on three-fourth tasks) and mildly right (right preferent on three-fourth tasks). When this is applied to the four hand preference measures (the motor skill task was excluded) used in the factor analysis, the distribution differs significantly from chance [$X^2(4, N = 218) = 55.12, p < .001$]. As can be seen, there are significantly more exclusively right compared to exclusively left handed subjects [$X^2(1, N = 50) = 18.00, p < .001$]. Similarly, there are significantly more mildly right compared to mildly left handed subjects [$X^2(1, N = 112) = 14.29, p < .001$]. The distribution of subjects classified in each group as a function of sex is shown in Figure 6.

Findings in Wild Compared to Captive Chimpanzees

One of the fundamental problems in interpreting the findings on handedness in chimpanzees and other great apes is the alleged discrepancy in results between settings. Indeed, some have suggested that the absence of population-level handedness in wild chimpanzees raises serious questions regarding the validity of evolutionary models of human handedness based entirely on findings from captive apes. While some have suggested that the differences in findings between wild and captive chimpanzees are significant, there are substantial

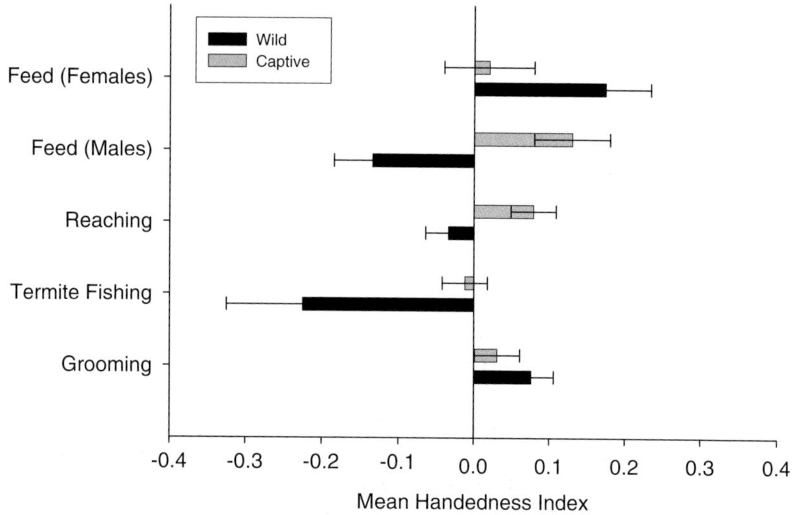

Figure 7. Mean handedness index (*HI*) scores (±s.e.) for wild and captive chimpanzees for bimanual feeding (Feed), reaching, termite fishing and grooming (see text for descriptions of the measures). Positive *HI* values reflect right hand preferences and negative values reflect left hand preferences.

differences between studies in captive and wild apes that might explain differences between the two settings. Some of these differences include the (a) types of measures employed, (b) methods of data collection and quantification of handedness, and (c) statistical power [Hopkins, 1999, 2006; Hopkins & Cantalupo, 2005]. It would be beneficial to have a greater number of studies in which common measures of hand use have been assessed in captive and wild apes. With respect to chimpanzees, there have been some comparable measures of handedness obtained in captive and wild chimpanzees including (a) reaching, (b) grooming, (c) bimanual feeding and (d) tool use, specifically termite fishing. Shown in Figure 7 are the mean *HI* scores for wild and captive chimpanzees for each of the behaviors described above. Wild chimpanzees are significantly more left-handed for termite fishing than captive chimpanzees and significantly more right-handed for grooming. No significant differences are found for bimanual feeding and reaching. For bimanual feeding, wild chimpanzees show a sex difference and a comparison of these findings to those seen in captivity were not consistent.

Comparison to Findings in Humans

One of the serious challenges facing researchers examining handedness in nonhuman primates is how to best compare the results to findings on human handedness. Human handedness is typically assessed using questionnaires and individuals are classified as right-, left- or ambidextrous based on the number of response items on which they report preferring to use the right hand or left hand. Depending on the criteria employed by the investigator, subjects can be classified into the different handedness groups. The most obvious classification are those subjects who report exclusive right or left hand use for all the items comprising the questionnaire but the difficulty is how to resolve the subjects that fall somewhere between exclusive right or left hand use. Most researchers working with human subjects adopt some cut point for classifying the subjects as right- or left-handed but this is a fairly arbitrary distinction and varies considerably between studies. The fact that human handedness (a) relies on verbal report, (b) is heavily culturally determined, and (c) is qualitative in assessment, makes comparisons to nonhuman handedness extremely difficult.

For the purposes of comparison to our findings, we have selected data published by Perelle and Ehrman [1983, 1994] because they used a similar scale to ours. Specifically, Perelle and Ehrman [1983, 1994] surveyed 2,500 participants and requested that the individuals self-report their handedness as (a) strongly right, (b) mildly right, (c) ambidextrous, (d) mildly left and (e) strongly left handed. Note that although Perelle and Ehrman [1983] did obtain specific hand preference data for 14 certain actions, responses to these items were not used to derive handedness classification; rather subjects were simply asked to rate their overall handedness into one of the five previously described categories.

Shown in Table IV is the distribution of strongly left, mildly left, ambidextrous, mildly right and strongly right handed subjects from Perelle and Ehrman [1983, 1994] and for the chimpanzee data presented here. The odds ratios for right- to left-handed individuals are also presented in Table IV. As can be seen, the ratio of right- to left-handed individuals is substantially higher in humans compared to chimpanzees, at least in the international data presented in their 1994 paper. These findings are consistent with the view that, although both species are right-handed, humans are significantly more right-handed than chimpanzees. Why the distributions reported by Perelle and Ehrmann [1983, 1994] differ between the 1983 and 1994 publications is unclear.

Table IV. Distribution of human and chimpanzee handedness.

	Hand preference classification					
	SL	**ML**	**A**	**MR**	**SR**	**R:L Odds**
Perelle and Ehrman (1983)	313	168	120	499	1304	4.17
Perelle and Ehrman (1994)	653	498	410	2281	7220	11.05
Chimpanzees (this study)	10	36	56	76	40	4.00

SL = strongly left, ML = mildly left, A = ambidextrous, MR = mildly right, SR = strongly right.

Mechanisms Influencing Chimpanzee Handedness

In addition to the basic information on hand preferences in the chimpanzees, we have also sought to examine the influence of different biological and nonbiological factors on handedness. Specifically, we have focused on assessing whether handedness runs in families of chimpanzees and whether these associations are modified by life history variables including rearing history and birth order of the subjects. Significant rates of concordance between mothers and offspring as well as between maternal and full siblings have been reported in our chimpanzee sample [Hopkins, 1999; Hopkins et al., 2006]. Birth order has also been shown to have a significant influence on handedness in the YERKES chimpanzees Hopkins & Dahl, 2000]. First-born chimpanzees were found to be significantly more left-handed than nonfirst-born individuals. More importantly, birth order appears to modulate the relationship between offspring and maternal hand preferences as well as between siblings. Adjusting concordance rates in handedness for birth order effects explains a significant proportion of variance in handedness between maternal half-siblings, particularly among related individuals that were reared apart [Hopkins et al., 2001].

Neuroanatomical Asymmetries

Since 1998, our laboratory has been collecting structural magnetic resonance images in chimpanzees residing at YERKES. The aim of this project has been to document (a) whether chimpanzees show population-level asymmetries in specific brain regions of interest and (b) to assess whether any neuroanatomical asymmetries correlate with the handedness measures. One of the major advantages of MRI is that it allows in vivo assessment of brain asymmetries that can subsequently be correlated with various behaviors, in our case handedness. To date, we have scanned 105 apes including 5 gorillas, 6 bonobos, 8 orangutans and 86 chimpanzees. The bulk of the scans have been collected in living apes using T1-weighted protocols but we have also developed T2 imaging protocols for scanning postmortem brains, when apes die from natural causes. In terms of neuroanatomical asymmetries for specific regions of interest, we have focused on three main regions including the inferior frontal gyrus (IFG), planum temporale (PT) and the motor-hand area of the precentral central gyrus, sometimes referred to as the KNOB [see Cantalupo & Hopkins, 2001; Cantalupo et al., 2003; Hopkins & Cantalupo, 2004; Hopkins & Pilcher, 2001, for descriptions]. Analyses have also been conducted on the entire depth of the central sulcus, post-central sulcus, inferior parietal lobe, anterior cingulate gyrus, hippocampus and amygdala [Dadda et al., 2006; Freeman et al., 2004]. For each region of interest, asymmetry quotients have been derived following the formula [$AQ = (R-L)/((R+L) \times .5)$], with positive values reflecting a right hemisphere bias and negative values reflecting a left hemisphere bias. Shown in Figure 8 are the mean AQ values for each of these brain regions. Chimpanzees show a leftward asymmetry for the PT and IFG and right hemisphere bias for

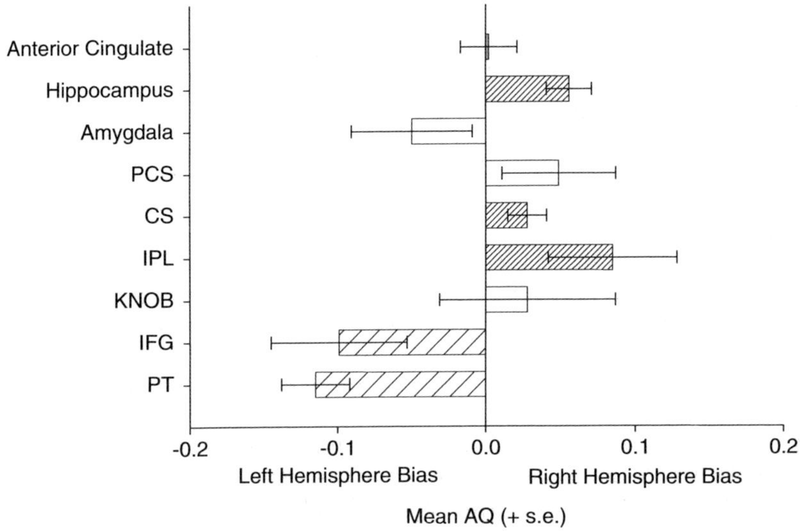

Figure 8. Mean asymmetry quotients (AQ) ($\pm$s.e.) for several brains regions measured from magnetic resonance images in chimpanzees. See text for description of abbreviations. Positive AQ values reflect right hemisphere biases and negative values reflect left hemisphere biases.

the CS, hippocampus and inferior parietal lobe. The leftward asymmetries in the PT have similarly been reported in cadaver chimpanzee brains [Gannon et al., 1998].

In addition to the region of interest analyses, we have also examined petalia asymmetries in the shape of the brain [see Pilcher et al., 2001 for description of methods]. The human brain is not symmetrically shaped but rather is torqued such that the right frontal lobe protrudes rostrally and left occipital lobe protrudes caudally. The so-called right-frontal, left-occipital petalia asymmetry has been reported in measurements taken from tissue as well as from endocasts of skulls [reviewed by Falk et al., 1986; LeMay, 1982]. Previous studies of endocasts have reported a right-frontal and left-occipital asymmetries in apes [Holloway & de Lacoste-Lareymondie, 1982] and we have found a similar pattern of results in our chimpanzee samples when the volumes of the frontal and occipital poles were compared between hemispheres in a sample of 85 chimpanzees (see Figure 9). As with humans, the chimpanzees show a population-level right-frontal, left-occipital asymmetry, a pattern of results some have claimed is uniquely human [Crow, 2004].

Behavioral Correlates of Brain Asymmetries

The long-term behavioral data on handedness combined with the more recently obtained neuroanatomical asymmetry data have allowed for

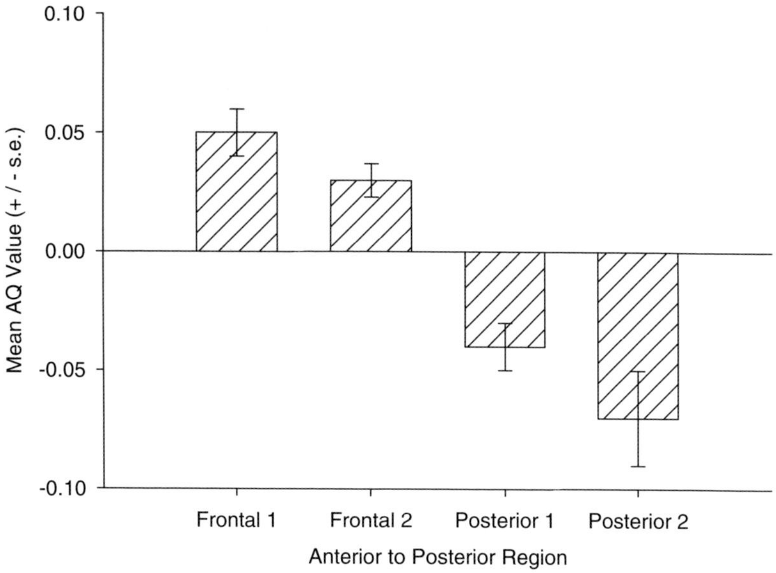

Figure 9. Mean asymmetry quotients (AQ) (±s.e.) of cerebral volumes from the frontal and occipital poles as measured from magnetic resonance images of chimpanzees. Frontal 1 = first 10% of the frontal pole for the length of the entire hemisphere. Frontal 2 = first 30% of the frontal pole for the length of the entire hemisphere. Posterior 1 = last 30% of the occipital pole for the length of the entire hemisphere. Posterior 2 = last 10% of the occipital pole for the length of the entire hemisphere. Positive AQ values reflect right hemisphere biases and negative values reflect left hemisphere biases.

examination of brain–behavior association in lateralized behavior. Of specific interest has been the association between hand preferences and asymmetries in the IFG, PT and KNOB. These regions were of primary interest because they have been implicated in human handedness [Foundas et al., 1995; Hammond, 2002]. For the TUBE task and bimanual feeding, significant associations have been found between preferences and the KNOB with right-handed subjects having a more leftward KNOB compared to left-handed subjects, a finding recently replicated in capuchin monkeys [Phillips & Sherwood, 2005]. We have also found some evidence that handedness for manual gestures correlates with asymmetries in the IFG [Hopkins & Cantalupo, 2003; Taglialatela et al., 2006]. Thus, hand preferences for different tasks correlate with asymmetries in different brain regions.

Lastly, a number of investigators working with human participants have examined the association between handedness and corpus callosum (CC) morphology. The CC is the tract of fibers that connects homotopic regions between the left and right cerebral cortex and it has been suggested that smaller

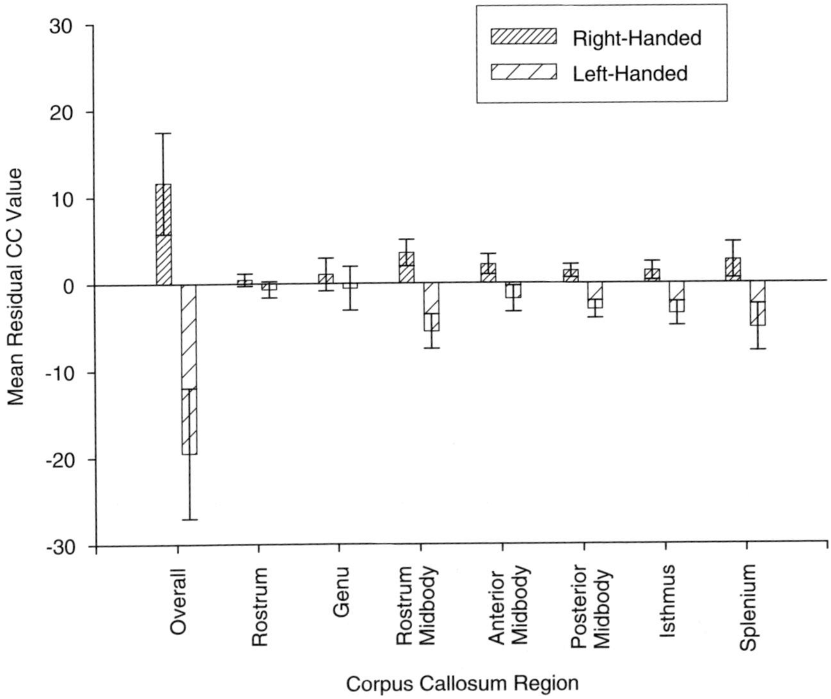

Figure 10. Mean residual corpus callosum values (±s.e.) for right- and left-handed chimpanzees [see Dunham & Hopkins, 2006 for description].

CC size is related to greater asymmetry in the brain. In humans, although not entirely consistent across studies, it has been generally claimed that left- or inconsistently-handed individuals have a larger CC compared to right- or consistently-handed individuals [Burke & Yeo, 1994]. Sex differences in the relative CC size have also often been reported in humans [Bishop & Wahlsten, 1997]. We have also examined the effect of handedness and sex on CC size in our chimpanzee sample [Dunham & Hopkins, 2006]. The CC was divided into seven sub-regions on the mid-saggital slice of the MRI scans according to the criteria employed by Witelson in human brains. Significant main effects for handedness were found for the overall CC size in the chimpanzees (see Figure 10), but particularly in the rostrum body, anterior-midbody, posterior-midbody, isthmus and splenium regions of the CC with right-handed subjects having a larger CC compared to left-handed subjects. These results are consistent with previous reports in rats and dogs but differ from reports in humans.

CONCLUSION

A summary of findings on handedness in chimpanzees over the past 15 years has revealed several significant and important findings. First, chimpanzees show task-specific population-level right-handedness and handedness appears to be a multidimensional trait in chimpanzees. Second, tasks that assess coordinated bimanual actions appear to be more sensitive in detecting population-level handedness in chimpanzees than unimanual tasks, especially simple reaching. Third, hand preferences are relatively consistent across rearing conditions and laboratories. Fourth, chimpanzees make fewer errors with the right compared to the left hand. Additionally, chimpanzees make more precision grips with the right compared to the left hand, suggesting that the left hemisphere is specialized for motor skill. Lastly, whether considering specific tasks or overall measures of handedness, chimpanzees are less lateralized than humans.

With respect to the evidence of population-level handedness, our results indicate that asymmetries can be found whether specific tasks are examined or whether handedness is derived from multiple measures of hand use. Thus, the claim that human handedness reflects true handedness while nonhuman primate handedness is task specific is not supported by our findings. Moreover, the claim that rearing history has a significant influence on handedness is not supported by our findings. No significant rearing effects were found and there is more than adequate statistical power to detect these effects. An examination of the mean *HI* scores for mother-reared, human-reared and wild-caught apes for each measure (see Table II) shows that the influence of rearing is randomly distributed across measures. Lastly, the claim that the findings of population-level handedness in chimpanzees are unique to the YERKES animals is not supported by the existing data [also see Hopkins, 2006]. In short, evaluating hand preferences in captive chimpanzees using multiple approaches reveals consistent evidence of population-level handedness.

Whether considering specific tasks or overall handedness, another consistent result that emerges is that chimpanzees are less lateralized than humans. Within specific tasks, such as gestures or throwing, there is a 2:1 or at best 3:1 ratio of right- to left-handed apes. This value is lower than the typical ratio of 8:1 or 9:1 reported for humans. The basis for the difference is not clear, but genetic factors might play a role and life history and socio-cultural explanations cannot be ruled out [Hopkins, 2006]. For example, it has been suggested that a mutation in the human genome caused the emergence of the pronounced manifestation of right-handedness [Corballis, 1997]. With detailed information about the human and chimpanzee genomes now available, the potential to identify candidate genes or potential mutations that might explain the discrepancy of human and nonhuman primate handedness is now at least feasible (but with much work to do). Alternatively, others have suggested that the biological and genetic basis for handedness in humans and apes are comparable but that the differences seen in the ratio of right- to left-handed individuals reflect the additive effect of socio-cultural adaptation unique to Hominid evolution [Hopkins, 2006]. Lastly, the differences may simply reflect differences in the measurement and assessment of handedness between human and nonhuman primates. Although we have attempted to compare the distribution of handedness between chimpanzees and humans in this chapter, these findings should be interpreted with caution. There are numerous problems in comparing human and nonhuman primate handedness, the least of which is the different scales of measurement used to evaluate handedness. Ultimately, it might be more fruitful to test human subjects on the types of measures employed with nonhuman primates rather than attempt to look for commonalities in motor and cognitive demands across different tasks. Moreover, observational data on hand use are also needed from human subjects rather than self-reported hand use. These data will provide a more level playing field in the comparison of direction and strength of handedness in human and nonhuman primates. Continued research should provide important and critical data on these three fundamental theoretical explanations.

The 2:1 or 3:1 ratio in right- to left-handed individuals also presents a pragmatic and statistical problem when evaluating handedness in nonhuman and human primates. The lower ratio of right- to left-handed subjects in chimpanzees indicates that handedness is not a robust behavioral trait and recently it has been shown that effect size for handedness in chimpanzees is small to moderate [Hopkins, 2006]. Pragmatically, this means that larger sample sizes are needed to detect population-level handedness and, without considering this possibility, this enhances the probability of Type II error (falsely accepting the null hypothesis). Compounding this potential problem is the almost uniform adoption of the use of z-scores to classify subjects as left-, right- or ambiguously handed and the subsequent determination of population-level handedness on the basis of χ^2 tests. Nonparametric tests are less powerful than inferential statistics. Therefore, investigators are adopting less powerful tests to detect an already small to moderate effect size. For these reasons, it is not surprising that many investigators have not reported evidence of population-level handedness, particularly in wild chimpanzees where sample sizes are often small [see Hopkins & Cantalupo, 2005, for further discussion].

There remains a paucity of data from wild chimpanzees in terms of sample sizes and range of behaviors measured for hand use. Arguably, in the studies by McGrew and Marchant [2001] and Marchant and McGrew [1996], there were many behaviors observed but one limitation of these studies is that none of the actions recorded revealed particularly strong hand preferences at the individual level, save the measures of tool use. In the absence of recording behaviors that elicit hand preference at the individual level, it is extremely difficult to compare the findings between captive and wild apes. For this reason, we believe the claims of discrepant findings between captive and wild apes are premature. Moreover, where comparisons can be made (see Figure 4), wild chimpanzees show good evidence of asymmetries. Lastly, reinforcing our argument, tool use does elicit strong asymmetries in wild chimpanzees relative to other manual actions and they show population-level handedness for activities involving tools [see Lonsdorf & Hopkins, 2005].

The findings presented here also provide key data relevant to various evolutionary models of handedness. Specifically, some have suggested that handedness may have evolved in the context of articulated speech [Annett, 2002], gestural language [Corballis, 2002], tool use including throwing [Calvin, 1982], coordinated bimanual actions [Wundrum, 1986] and bipedalism [see Bradshaw & Rogers, 1993]. Based on the data presented here and other data showing little evidence of speech and manual gesture capabilities in primates, it does not appear that speech is a necessary condition for the emergence of handedness. Similarly, tool use also does not appear to be a necessary condition for the emergence of population-level handedness because captive chimpanzees show population-level asymmetries for nontool use activities. Moreover, there is at least some evidence that nontool using species and species that do not manually gesture, such as baboons and capuchin monkeys, show population-level handedness for certain tasks, including manual threat gestures [Vauclair et al., 2005; Meguerditchian & Vauclair, 2006]. The data presented here most strongly support the view that either the need for coordinated bimanual actions or the need for increasing postural control was a strong selective component for handedness. This conclusion is primarily supported by the findings that coordinated actions most consistently reveal population-level right-handedness across different tasks (e.g., TUBE task, bimanual feeding, tool use). The exception is throwing, which is an unusual but extremely interesting lateralized behavior. Throwing is not a bimanual task but it does require coordination of the upper and lower limbs and there is a substantial amount of postural control that is necessary for this manual action, particularly for chimpanzees that opt to throw from an upright posture. Eighty percent of the chimpanzees that throw overhand, throw right-handed which is in contrast to only 50% of chimpanzees that throw from a tripedal posture (and throw underhand). Thus, in the case of throwing, posture has a dramatic effect on the expression of handedness. From the standpoint of evolution, right-handed throwing may have emerged from inherent postural systems already lateralized such as those used in leading limb in locomotion [see Heestand, 1986; Peters, 2005]. Thus, rather than throwing being an antecedent condition for the emergence of right-handedness, it might have been facilitated by the existence of population-level laterality in postural systems associated with locomotion or other functions.

In closing, we wish to make some recommendations for continued work on the topic of laterality in nonhuman primates. In our view, there is some continued need to document variation in handedness in different species. These data would be particularly useful for evaluating the role of biomechanical and ecological adaptations on the development of handedness. These would also be useful in evaluating the potential role that social systems might have on the development of behavioral asymmetries, as has been well-documented in fish [see Vallortigara & Rogers, 2005]. Second, the question of reduced fitness and handedness remains a central topic in the human literature on handedness and we believe nonhuman primate studies can contribute significantly to this issue. With respect to human handedness, many have argued that left-handedness reflects a maladaptive trait and is a consequence of birth trauma, birth stress and other perinatal factors [see Bailey & McKeever, 2004; Searleman et al., 1989, for reviews]. Others have argued that left-handedness has a potential genetic basis and has been maintained in the population for adaptive reasons [McManus, 2002; Raymond & Pontier, 2004]. Ultimately, the question is really whether variation in handedness has some fitness-related or clinical significance and very few studies have considered these types of associations in nonhuman primates. Recently, Westergaard et al. [2004] have reported associations between immunological and biochemical factors and handedness in macaques, and similar studies in other primates would be of interest. Lastly, this chapter has focused on handedness but there is clear evidence of other behavioral asymmetries in primates such as auditory processing, oro-facial expressions and other cognitive processes. To the extent that scientists can broaden the scope of behaviors as well as the levels of analysis, we will enhance our evolutionary understanding of the similarities and differences in laterality in human and nonhuman primates. It very well might be that there is greater continuity in certain behavioral asymmetries in primates, relative to handedness, and this will provide a better framework for considering the evolution and specific adaptations that have characterized the pronounced expression of handedness in humans.

ACKNOWLEDGMENT

This research was supported by NIH grants NS-36605, NS-42867, NS-29574, U42-RR-15090 and RR-00165 to the Yerkes National Primate Research Center or The University of Texas M. D. Anderson Cancer Center (UTMDACC). The Yerkes Center and the UTMDACC Department of Veterinary Sciences are fully accredited by the American Association for Accreditation of Laboratory Animal Care. American Psychological Association guidelines for the ethical treatment of animals were adhered to during all aspects of these studies. Correspondence concerning this article should be addressed to Dr. William D. Hopkins, E-mail: whopkin@emory.edu or whopkins@agnesscott.edu

REFERENCES

Annett, M. Handedness and Brain Asymmetry: The Right Shift Theory. Hove, Psychology Press, 2002.

Bailey, L. M.; McKeever, W. F. A large-scale study of handedness and pregnancy/birth risk events: Implications for genetic theories of handedness. Laterality 9: 175–188, 2004.

Biro, D.; Inoue-Nakamura, N.; Tonooka, R.; Yamakoshi, G.; Sousa, C.; Matsuzawa, T. Cultural innovation and transmission of tool use in wild chimpanzees: Evidence from field experiments. Animal Cognition 6: 213–223, 2003.

Bishop, K. M.; Wahlsten, D. Sex differences in the human corpus callosum: Myth or reality? Neuroscience and Biobehavioral Reviews 21: 581–601, 1997.

Boesch, C. Handedness in wild chimpanzees. International Journal of Primatology 6: 541–558, 1991.

Bradshaw, B.; Rogers, L. The Evolution of Lateral Asymmetries, Language, Tool-Use and Intellect. San Diego, Academic Press, 1993.

Burke, H. L.; Yeo, R. A. Systematic variations in callosal morphology: The effects of age, gender, hand preference, and anatomic asymmetry. Neuropsychology 8: 563–571, 1994.

Byrne, R. W.; Byrne, J. M. Hand preferences in the skilled gathering tasks of mountain gorillas (*Gorilla berengei*). Cortex 27: 521–536, 1991.

Calvin, W. H. Did throwing stones shape hominid brain evolution. Ethology and Sociobiology 3: 115–124, 1982.

Cantalupo, C.; Hopkins, W. D. Asymmetric Broca's area in great apes. Nature 414: 505, 2001.

Cantalupo, C.; Pilcher, D. L.; Hopkins, W. D. Are planum temporale and sylvian fissure asymmetries directly related? A MRI study in great apes. Neuropsychologia 41: 1975–1981, 2003.

Corballis, M. C. The Lopsided Brain: Evolution of the Generative Mind. New York, Oxford University Press, 1992.

Corballis, M. C. The genetics and evolution of handedness. Psychological Review 104: 714–727, 1997.

Corballis, M. C. From Hand to Mouth: The Origins of Language. Princeton, Princeton University Press, 2002.

Corey, D. M.; Hurley, M. M.; Foundas, A. L. Right and left handedness defined. A multivariate approach using hand preference and hand performance measures. Neuropsychiatry, Neuropsychology, and Behavioral Neurology 14: 144–152, 2001.

Corp, N.; Byrne, R. W. Sex difference in chimpanzee handedness. American Journal of Physical Anthropology 123: 62–68, 2004.

Crow, T. Directional asymmetry is the key to the origin of modern *Homo sapiens* (the Broca-Annett axiom): A reply to Rogers' review of The Speciation of Modern *Homo sapiens*. Laterality 9: 233–242, 2004.

Crow, T. J. Sexual selection, timing and the descent of man: A theory of the genetic origins of language. Current Psychology of Cognition 17: 1079–1114, 1998.

Dadda, M.; Cantalupo, C.; Hopkins, W. D. Further evidence of an association between handedness and neuroanatomical asymmetries in the primary motor cortex of chimpanzees (*Pan troglodytes*). Neuropsychologia 44: 2582–2586, 2006.

Dunham, L.; Hopkins, W. D. Sex and handedness effects on corpus callosum morphology in chimpanzees (*Pan troglodytes*). Behavioural Neuroscience 120: 1025–1032, 2006.

Fagot, J.; Vauclair, J. Manual and hemispheric specialization in the manipulation of a joystick by baboons (*Papio papio*). Behavioral Neuroscience 107: 210–214, 1993.

Falk, D.; Cheverud, J.; Vannier, M. W.; Conroy, G. C. Advanced computer-graphics technology reveals cortical asymmetry in endocasts of rhesus monkeys. Folia Primatologica 46(2): 98–103, 1986.

Finch, G. Chimpanzee handedness. Science 94: 117–118, 1941.

Foundas, A.; Leonard, C.; Heilman, K. Morphological cerebral asymmetries and handedness: The pars triangularis and planum temporale. Archives of Neurology 52: 501–508, 1995.

Freeman, H.; Cantalupo, C.; Hopkins, W. D. Asymmetries in the hippocampus and amygdala of chimpanzees (*Pan troglodytes*). Behavioral Neuroscience 118: 1460–1465, 2004.

Gannon, P. J.; Holloway, R. L.; Broadfield, D. C.; Braun, A. R. Asymmetry of chimpanzee planum temporale: Humanlike pattern of Wernicke's brain language area homologue. Science 279: 220–222, 1998.

Hammond, G. Correlates of human handedness in primary motor cortex: A review and hypothesis. Neuroscience and Biobehavioral Reviews 26: 285–292, 2002.

Healey, J. M.; Liederman, J.; Geschwind, N. Handedness is not a unidimensional trait. Cortex 22: 33–53, 1986.

Heestand, J. Behavioral Lateralization in Four Species of Ape. Unpublished Doctoral Dissertation, University of Washington, Seattle, 1986.

Hellige, J. B. Hemispheric Asymmetry: What's Right and What's Left. Cambridge, MA, Harvard University Press, 1993.

Holloway, R. L.; de Lacoste-Lareymondie, M. C. Brain endocast asymmetry in Pongids and Hominids: Some preliminary findings on the paleontology of cerebral-dominance. American Journal of Physical Anthropology 58(1): 101–110, 1982.

Hopkins, W. D. Hand preferences for bimanual feeding in 140 captive chimpanzees (*Pan troglodytes*): Rearing and ontogenetic factors. Developmental Psychobiology 27: 395–407, 1994.

Hopkins, W. D. Chimpanzee handedness revisited: 54 years since Finch (1941). Psychonomic Bulletin and Review 3: 449–457, 1996.

Hopkins, W. D. Heritability of hand preference in chimpanzees: Evidence from a partial interspecies cross-fostering study. Journal of Comparative Psychology 113: 307–313, 1999.

Hopkins, W. D. Comparative and familial analysis of handedness in great apes. Psychological Bulletin 132: 539–559, 2006.

Hopkins, W. D.; Bard, K. A.; Jones, A.; Bales, S. Chimpanzee hand preference for throwing and infant cradling: Implications for the origin of human handedness. Current Anthropology 34: 786–790, 1993.

Hopkins, W. D.; Cantalupo, C. Does variation in sample size explain individual differences in hand preferences of chimpanzees: A reply to Palmer (2002). American Journal of Physical Anthropology 121: 878–881, 2003.

Hopkins, W. D.; Cantalupo, C. Handedness in chimpanzees is associated with asymmetries of the primary motor cortex but not with homologous language areas. Behavioral Neuroscience 118: 1176–1183, 2004.

Hopkins, W. D.; Cantalupo, C. Individual and setting differences in the hand preferences of chimpanzees (*Pan troglodytes*): A critical analysis and some alternative explanations. Laterality 10: 65–80, 2005.

Hopkins, W. D.; Cantalupo, C.; Wesley, M. J.; Hostetter, A.; Pilcher, D. Grip morphology and hand use in chimpanzees (*Pan troglodytes*): Evidence of a left hemisphere specialization in motor skill. Journal of Experimental Psychology: General 131: 412–423, 2002.

Hopkins, W. D.; Dahl, J. F. Birth order and hand preference in chimpanzees (*Pan troglodytes*): Implications for pathological models of human handedness. Journal of Comparative Psychology 114: 302–306, 2000.

Hopkins, W. D.; Dahl, J. F.; Pilcher, D. Genetic influence on the expression of hand preferences in chimpanzees (*Pan troglodytes*): Evidence in support of the right shift theory and developmental instability. Psychological Science 12: 299–303, 2001.

Hopkins, W. D.; Pilcher, D. L. Neuroanatomical localization of the motor hand area using magnetic resonance imaging: The left hemisphere is larger in great apes. Behavioral Neuroscience 115: 1159–1164, 2001.

Hopkins, W. D.; Rilling, J. K. A comparative MRI study of the relationship between neuroanatomical asymmetry and interhemispheric connectivity in primates: Implication for the evolution of functional asymmetries. Behavioral Neuroscience 114: 739–748, 2000.

Hopkins, W. D.; Russell, J. L. Further evidence of a right hand advantage in motor skill by chimpanzees (*Pan troglodytes*). Neuropsychologia 42: 990–996, 2004.

Hopkins, W. D.; Russell, J.; Cantalupo, C.; Freeman, H.; Schapiro, S. Factors influencing the prevalence and handedness for throwing in captive chimpanzees (*Pan troglodytes*). Journal of Comparative Psychology 119: 363–370, 2005a.

Hopkins, W. D.; Russell, J.; Hook, M.; Braccini, S.; Schapiro, S. Simple reaching is not so simple: Association between hand use and grip preferences in captive chimpanzees. International Journal of Primatology 26: 259–277, 2005b.

Hopkins, W. D.; Russell, J.; Remkus, M.; Freeman, H.; Schapiro, S. J. Handedness and grooming in chimpanzees (*Pan troglodytes*): Comparative analysis between findings in captive and wild individuals. International Journal of Primatology, in press.

Hopkins, W. D.; Wesley, M.; Russell, J. L.; Schapiro, S. J. Parental and perinatal factors influencing handedness in captive chimpanzees. Developmental Psychobiology 48: 428–435, 2006.

Kimura, D. Sex And Cognition. Cambridge, MA, MIT Press, 1999.

Knecht, S.; Deppe, M.; Draeger, B.; Bobe, L.; Lohman, H.; Ringelstein, E. B.; Henningsen, H. Language lateralization in healthy right-handers. Brain 123: 74–81, 2000.

Lehman, R. A. W. Manual preference in prosimians, monkeys, and apes. Pp. 107–124 in J. P. Ward; W. D. Hopkins, eds. Primate Laterality: Current Behavioral Evidence of Primate Asymmetries. New York, Springer-Verlag, 1993.

LeMay, M. Morphological aspects of human brain asymmetry: An evolutionary perspective. Trends in Neurosciences 5: 273–275, 1982.

Lonsdorf, E. V.; Hopkins, W. D. Wild chimpanzees show population-level handedness for tool use. Proceedings of the National Academy of Sciences 102: 12634–12638, 2005.

Marchant, L. F.; McGrew, W. C. Laterality of function in apes: A meta-analysis of methods. Journal of Human Evolution 21: 425–438, 1991.

Marchant, L. F.; McGrew, W. C. Laterality of limb function in wild chimpanzees of Gombe National Park: Comprehensive study of spontaneous activities. Journal of Human Evolution 30: 427–443, 1996.

McGrew, W. C.; Marchant, L. F. On which side of the apes? Ethological study of laterality of hand use. Pp. 255–272 in W. C. McGrew; L. F. Marchant; T. Nishida, eds. Great ape societies. Cambridge, Cambridge University Press, Cambridge, 1996.

McGrew, W. C.; Marchant, L. F. On the other hand: Current issues in and meta-analysis of the behavioral laterality of hand function in non-human primates. Yearbook of Physical Anthropology 40: 201–232, 1997.

McGrew, W. C.; Marchant, L. F. Ethological study of manual laterality in the chimpanzees of the Mahale mountains, Tanzania. Behaviour 138: 329–358, 2001.

McGrew, W. C.; Marchant, L. F.; Wrangham, R. W.; Klein, H. Manual laterality in anvil use: Wild chimpanzees cracking *Strychnos* fruits. Laterality 4: 79–87, 1999.

McManus, I. C. Right hand, left hand: The origins of asymmetries in brains, bodies, atoms, and cultures. London, Weidenfeld & Nicolson, 2002.

Meguerditchian, A.; Vauclair, J. Baboons communicate with their right hand. Behavioural Brain Research 171: 170–174, 2006.

Palmer, A. R. Chimpanzee right-handedness reconsidered: Evaluating the evidence with funnel plots. American Journal of Physical Anthropology 118: 191–199, 2002.

Palmer, A. R. Reply to Hopkins and Cantalupo: Chimpanzee right-handedness reconsidered – sampling issues and data presentation. American Journal of Physical Anthropology 121: 382–384, 2003.

Papademetriou, E.; Sheu, C. F.; Michel, G. F. A meta-analysis of primate hand preferences for reaching and other hand-use measures. Journal of Comparative Psychology 119: 33–48, 2005.

Perelle, I. B.; Ehrman, L. The development of laterality. Behavioral Science 28: 284–297, 1983.

Perelle, I. B.; Ehrman, L. An international study of human handedness: The data. Behavior Genetics 24: 217–227, 1994.

Peters, H. Behavioral Lateralization in the Orang-utan (*Pongo pygmaeus pygmaeus*). Unpublished Doctoral Dissertation, University of New England, Australia, 2005.

Phillips, K.; Sherwood, C. C. Primary motor cortex asymmetry is correlated with handedness in capuchin monkeys (*Cebus apella*). Behavioral Neuroscience 119: 1701–1704, 2005.

Pilcher, D.; Hammock, L.; Hopkins, W. D. Cerebral volume asymmetries in non-human primates as revealed by magnetic resonance imaging. Laterality 6: 165–180, 2001.

Porac, C.; Coren, S. Lateral Preferences and Human Behavior. New York, Springer, 1981.

Rasmussen, T.; Milner, B. The role of early left-brain injury in determining lateralization of cerebral speech function. Annals of the New York Academy of Sciences 299: 355–369, 1977.

Raymond, M.; Pontier, D. Is there geographical variation in human handedness? Laterality 9: 35–51, 2004.

Rogers, L. J.; Andrew, J. R. Comparative Vertebrate Lateralization. Cambridge, Cambridge University Press, 2002.

Searleman, A.; Porac, C.; Coren, S. Relationship between birth order, birth stress, and lateral preferences: A critical review. Psychological Bulletin 105: 397–408, 1989.

Springer, S. P.; Deutsch, G. Left Brain, Right Brain. New York, Freeman, 1993.

Taglialatela, J.; Cantalupo, C.; Hopkins, W. D. Gesture handedness predicts asymmetries in the chimpanzee inferior frontal gyrus. Neuroreport 17: 923–927, 2006.

Tonooka, R.; Matsuzawa, T. Hand preferences in captive chimpanzees (*Pan troglodytes*) in simple reaching for food. International Journal of Primatology 16: 17–34, 1995.

Vallortigara, G.; Rogers, L. J. Survival with an asymmetrical brain: Advantages and disadvantages of cerebral lateralization. Behavioral and Brain Sciences 28: 574, 2005.

Vauclair, J.; Meguerditchian, A.; Hopkins, W. D. Hand preferences for unimanual and coordinated bimanual tasks in baboons (*Papio anubis*). Cognitive Brain Research 25: 210–216, 2005.

Ward, J. P.; Hopkins, W. D. Primate Laterality: Current Behavioral Evidence of Primate Asymmetries. New York, Springer-Verlag, 1993.

Warren, J. M. Handedness and laterality in humans and other animals. Physiological Psychology 8: 351–359, 1980.

Wesley, M. J.; Fernandez-Carriba, S.; Hostetter, A.; Pilcher, D.; Poss, S.; Hopkins, W. D. Factor analysis of multiple measures of hand use in captive chimpanzees: An alternative approach to the assessment of handedness in nonhuman primates. International Journal of Primatology 23: 1155–1168, 2002.

Westergaard, G. C.; Chavanne, T. J.; Houser, L.; Cleveland, A.; Snoy, P. J.; Suomi, S. J.; Higley, J. D. Biobehavioral correlates of hand preference in free-ranging female primates. Laterality 9: 267–285, 2004.

Wundrum, I. J. Cortical motor asymmetries in Hominid feeding strategies. Human Evolution 1: 183–188, 1986.

A family of Guinea baboons, *Papio papio*. (Photograph by Maxime Cauchoix.)

The Evolution of Hemispheric Specialization in Primates
W.D. Hopkins (Editor)

PERCEPTUAL AND MOTOR LATERALIZATION IN TWO SPECIES OF BABOONS

Jacques Vauclair and Adrien Meguerditchian
Department of Psychology, Research Center in Psychology of Cognition, Language & Emotion, University of Provence, Aix-en-Provence, France (JV, AM); Center of Primatology – CNRS, Rousset-sur-Arc, France (AM)

ABSTRACT

Hemispheric cerebral specialization is usually considered as a specific trait of the human species. However, the results of many studies on behavioral asymmetries in numerous animals seem to challenge this view and studying functional laterality in nonhuman primates remains an important way to investigate potential precursors of human hemispheric specialization. Here, we review available data on functional lateralization in baboons in the manual (handedness, bimanual collaboration and manual communicative gestures) and in the visual domains. These findings are discussed in a comparative framework and reveal some continuities between baboons, great apes and humans.

Keywords: baboons, perceptual asymmetries, mental rotation, handedness

INTRODUCTION

An increase in the study of handedness and more generally a renewed interest in lateralization processes in nonhuman primates was stimulated by the review paper and proposals presented by MacNeilage et al. [1987] in a Behavioral and Brain Sciences' target article. The novelty of this seminal paper was threefold. First, the paper challenged the historical view in the literature that there was only individual, and not population-level, hand preferences in nonhuman primates. Second, MacNeilage et al. [1987] proposed an evolutionary scenario according to which primate handedness patterns could have evolved with adaptations to posture and feeding. Third, MacNeilage et al. [1987] suggested that the handedness patterns of nonhuman primates may offer clues for understanding patterns of human hemispheric specialization. These ideas were further developed in the "postural origins" theory of primate motor asymmetries [MacNeilage, 1991; see below]. The senior author of this chapter has been involved (then with the collaboration of Joël Fagot and now with Adrien Meguerditchian, the junior author of the present chapter) in the study of hemispheric specialization in nonhuman primates (two species of baboons, the

ISSN: 1936-8526 DOI: 10.1016/S1936-8526(07)05007-5

Guinea baboon, *Papio papio* and the olive baboon, *Papio anubis*, and lowland gorillas) since the mid-1980. The initial investigations on handedness were carried out with baboons living in social groups in large compounds [Vauclair & Fagot, 1987a]. Subsequently, as our research interests moved from the study of motor asymmetries to that of perceptual lateralization, it became necessary to use an apparatus specifically designed to allow for presentation of visual stimuli in a divided field format. This latter system (a video formatted matching-to-sample task) could be used only in laboratory conditions. More recently, we have returned to the study of baboons in social groups raised in semi-natural conditions, in our most recent research on bimanual coordination and manual gestures.

The goal of the present chapter is to summarize available data on lateralization processes in baboons. This review will include additional datasets collected in our laboratory, when necessary, to provide a more comprehensive analysis and interpretation of existing data. The first part of the chapter is concerned with lateralization in the motor domain (handedness, bimanual collaboration and manual gestures). The second part will cover a range of studies concerned with functional asymmetries in the visual domain, including performance of the monkeys on mental rotation tasks. Finally, in the third part, we will attempt to synthesize the main findings obtained in the studies of the baboons and discuss them within a comparative framework. Our objective is to offer a perspective on the baboon functional hemispheric specialization patterns in relation to those described in other Cercopithecidae and in great apes.

HANDEDNESS STUDIES IN BABOONS LIVING IN SOCIAL GROUPS

The assessment of manual asymmetries represents an appropriate way to infer functional hemispheric specialization. In humans, the preferential use of one hand in most manual activities is linked to the dominance of the opposite cerebral hemisphere. Nearly 90% of humans are right-handed and this bias reflects a left-hemispheric lateralization for manual control [Annett, 1985]. From an evolutionary view point, there is a strong debate as to whether such a population-level right-handedness is unique and exclusively related to language or if it has precursors in nonhuman primates [Bradshaw & Rogers, 1993; Corballis, 2002]. Although there is substantial evidence of population-level right-handedness reported in apes and monkeys for manipulative activities [Hopkins, 1995; Hopkins & Pearson, 2000; Hopkins et al., 2001, 2003a, 2004; Spinozzi et al., 1998; Vauclair et al., 2005; Westergaard & Suomi, 1996], some authors have challenged the interpretation of the data and the debate remains intense [Hopkins, 1999; Hopkins & Cantalupo, 2003; Palmer, 2002; Marchant & McGrew, 1996; McGrew & Marchant, 1997]. From a comparative perspective, whether manual behaviors of nonhuman primates are consistently lateralized for a given task at the individual-level and at a group-level has been largely investigated with some contradictory results [see for reviews, Fagot & Vauclair,

1991; McGrew & Marchant, 1997; Papademetriou et al., 2005]. Here, we address this question by focusing on an Old World monkey species, the baboon. We review the data collected with various approaches and contexts including spontaneous simple manipulations, effect of different experimental tasks on handedness' distributions, assessment of group-level handedness in large samples for simple and complex tasks, and handedness in gestural communication (see Table 1 for a synopsis of our handedness data).

NATURAL SPONTANEOUS HAND USE

To our knowledge, the first investigation conducted on manual asymmetries in baboons has concerned a troop of 18 captive Guinea baboons, *P. papio* [Vauclair & Fagot, 1987a]. The authors assessed manual preferences of subjects by recording their natural unimanual and bimanual behaviors performed (a) with an object (stone, food, stick, etc.), (b) toward a social partner (touching) or (c) self-centered actions (like self-grooming, scratching). This study revealed that spontaneous hand use in baboons involved individual lateralization in 7 subjects but no significant group-level handedness: 5 subjects were right-handed, 2 left-handed and 11 ambiguously handed with little variation in laterality patterns when unimanual actions (4 right-, 2 left- and 12 ambiguously handed subjects) were distinguished from bimanual actions (3 right-, 3 left- and 12 ambiguously handed). Analysis of data showed no effect of sex on handedness but some age effects were found. Individual hand preferences and asymmetries of bimanual activities were more pronounced in adults than in juveniles, indicating that manual biases in baboons could develop during ontogeny via biological maturation and/or experience.

TASK EFFECTS ON HANDEDNESS

The second step in the investigations on baboons was to evaluate the effect of the complexity of a given task on hand preferences [Fagot & Vauclair, 1988a]. Three different tasks were used to assess manual preferences in six adult male Guinea baboons (*P. papio*) familiarized to the experimental conditions. The first task was a simple unimanual reaching of food placed on a board: a symmetrical distribution of individual handedness was described (3 left- and 3 right-handed subjects). The second task involved using a metal box with food inside. In an initial version of the task, the baboons could raise the whole box with one hand to take the food with the same or the other hand. In a second version of the box task, the use of an additional stop screw prevented the box from remaining open and required the subject to keep the box lifted with one hand in order to retrieve the food with the opposite hand. For both versions, the subjects presented the same individual hand preferences to open the metal box than for simple reaching, indicating no major difference between unimanual reaching and the box-opening tasks. The third experimental apparatus was a transparent Plexiglas panel with two small windows. The panel could slide horizontally in both

Table 1. Summary of handedness data reported in baboons (R: right-handed subjects; L: left-handed subjects; A: ambiguously handed subjects).

Type of manual task	Results	*N*	Population-level bias	Reference
Unimanual reaching food	7 R, 2 L, 2 A	11	No bias	Vauclair and Fagot [1987a]
Unimanual natural spontaneous actions	4 R, 2 L, 12 A	18	No bias	Vauclair and Fagot [1987b]
Bimanual natural spontaneous actions	3 R, 3 L, 12 A	18	No bias	Vauclair and Fagot [1987b]
Natural spontaneous actions (total)	5 R, 2 L, 11 A	18	No bias	Vauclair and Fagot [1987b]
Board task (unimanual reaching food)	3 R, 3 L	6	No bias	Fagot and Vauclair [1988a, b]
Unimanual box opening	3 R, 3 L	6	No bias	Fagot and Vauclair [1988a, b]
Horizontal sliding panel (adjusting apertures)	1 R, 5 L	6	Left	Fagot and Vauclair [1988a, b]
Vertical sliding panel (adjusting apertures)	6 L	6	Left	Fagot and Vauclair [1988a, b]
Delayed vertical sliding panel (adjusting apertures)	1 R, 5 L	6	Left	Fagot and Vauclair [1988a, b]
Hole task (unimanual reaching food)	3 R, 3 L	6	No bias	Fagot and Vauclair [1988a, b]
Tube (bimanual coordinated actions)	52 R, 33 L, 19 A	104	Right	Vauclair et al. [2005]
Unimanual reaching food	27 R, 25 L, 42 A	94	No bias	Vauclair et al. [2005]
Communicative Gesture	35 R, 10 L, 15 A	60	Right	Meguerditchian and Vauclair [2006]

directions in front of a metal board in which a median aperture was made to place a food reward inside. The task consisted of sliding the Plexiglas panel using a handle to align one of its windows with the aperture of the metal board in order to have access to the food. A vertical-sliding version of the panel with one window was also used to force the subjects to perform the task with both hands (one hand adjusted the aperture, while the other took the food). For both versions of the task, the distribution of individual hand preferences was asymmetrical and in favor of the left hand for alignment and adjustment of apertures (5 left-handed and 1 right-handed baboons for the horizontally sliding panel and 6 left-handed baboons for the vertically sliding panel task).

Two hypotheses were proposed from these results: (1) the left hand was preferred to move the Plexiglas panel, consequently inducing the use of the right hand to take the food, (2) the right hand was preferred to take the food and such a manual preference required using the left hand to slide the panel. Results of a novel experiment confirmed the first hypothesis. In effect, the authors modified the vertically sliding panel task in order to dissociate the adjusting action from the food reaching. In the first modified task, where the baboons had to align one of the apertures, the food fell down through a hole and was collected at the end of a tube. This procedure introduced a delay that was necessary to test independently the actions for adjusting the apertures. For such a task, 5 subjects were left-handed and 1 right-handed. In the second version of the task, to test only the food-reaching action, the authors used a vertical metal board with a hole in its center in which food was left. In this later case, there was no group-level handedness for taking the food in the aperture (3 right-handed and 3 left-handed).

This study revealed that individual handedness in baboons [and in other nonhuman primates species, see the review by Fagot & Vauclair, 1991] could depend on the complexity of motor and cognitive functions involved in performing a given manual task. Moreover, similar findings were reported with the same tasks (unimanual, box and sliding panel tasks) in captive lowland gorillas [Fagot & Vauclair, 1988b]. No group-level handedness for unimanual reaching (3 right-, 3 left- and 4 ambiguously-handed subjects) was observed, but a group-level left hand preference for the alignment of apertures in the horizontally sliding panel task was described (1 right- and 7 left-handed subjects). Thus, high-level manual activities are more effective behaviors than low-level manual actions for assessing handedness. High-level tasks induced an asymmetrical distribution of manual preferences, which can be related to the lateralization of the contralateral hemisphere, contrary to low-level tasks (e.g., simple reaching), which induced a symmetrical distribution of hand preferences. Such a dichotomy can explain why the previous handedness data in nonhuman primates, which were collected mainly with simple reaching tasks, were not consistent and often contradictory [see also Papademetriou et al., 2005, for a review]. From an evolutionary viewpoint, further investigations are necessary to predict the direction of group-level handedness of the species in large samples in order to better evaluate possible continuities with human patterns of handedness.

STUDY OF GROUP-LEVEL HANDEDNESS IN LARGE SAMPLES

One limitation of the early studies in baboons was sample size. Thus, in renewed research on the topic of handedness, we have sought to develop and use less complex tasks that still were considered an efficient high-level manual task. This was done to facilitate the generalization of handedness measures to large groups of baboons. For that purpose, we used the task designed by Hopkins to test handedness in captive chimpanzees, referred to as the "coordinated bimanual tube task" [Hopkins, 1995]. In this task, a PVC tube with peanut butter smeared on the inside was given to a focal subject and the hand used to remove the food was recorded while the opposite hand held the tube (see Figure 1a for an illustration). Since its first use by Hopkins [1995], the tube task has been employed in studies with several nonhuman primates such as capuchin monkeys [Spinozzi et al., 1998; Westergaard & Suomi, 1996], macaques [Westergaard & Suomi, 1996; Westergaard et al., 1997], gorillas and orangutans [Hopkins et al., 2003b]. Recently, the tube task was also used with olive baboons (*P. anubis*) in a large sample of 104 subjects [Vauclair et al., 2005]. A significant preference for the right hand was observed (52 right-, 33 left- and 19 ambiguously-handed subjects). Thus, among the 85 lateralized individuals, 61% were right-handed. In the same study, for comparison purposes, a unimanual reaching task (low-level) was also assessed. For this task, small grains of cereals were dispersed in the baboons' outside enclosures (see Figure 1b for an illustration of the reaching action). Ninety-four baboons were tested and, as in our earlier studies [Fagot & Vauclair, 1988a; Vauclair & Fagot, 1987a, 1987b], we obtained a symmetrical distribution of hand preferences (27 right-, 25 left- and 42 ambiguously-handed subjects), and no effect of sex on

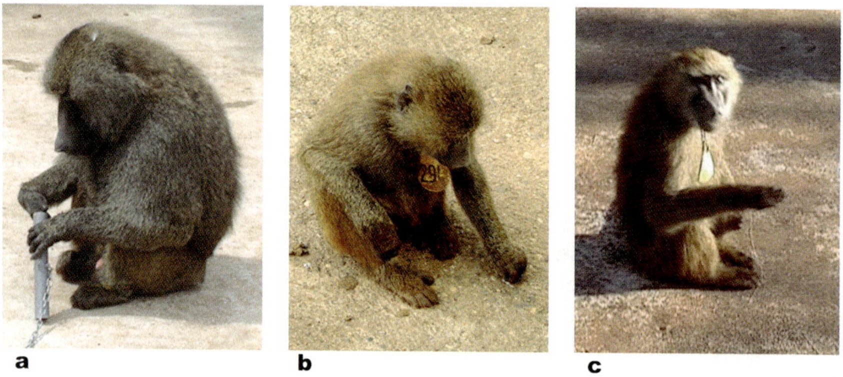

Figure 1. Photographs of different manual actions performed by baboons and used to assess their handedness. (a) Coordinated bimanual tube task in an adult male baboon; (b) unimanual simple reaching for grains by a young female baboon; (c) a young male baboon is starting a hand slapping ground gesture in order to intimidate the human observer. (Photos by Adrien Meguerditchian.)

hand biases was observed. However, contrary to the previous studies reviewed above, no effect of age on the direction and on the strength of manual asymmetries was found. Additionally, the results of hand preferences recorded for both tasks some months later, for a portion of the sample, showed a strong consistency of hand use across time for the tube task but not for the simple reaching task. These results support the hypothesis of "task effect" on hand use but also the fact that high-level tasks provide better measures than low-level tasks for measuring handedness.

Although the degree of group-level right-handedness in humans is higher than that obtained in other primates, the latest findings leave the door open for further investigations of continuity between human and nonhuman primates concerning hemispheric specialization and handedness. A prevalence for right-handedness has been shown several times for the tube task in large samples of captive chimpanzees [Hopkins, 1995; Hopkins & Pearson, 2000; Hopkins et al., 2001, 2003a, b, 2004] and also in smaller samples of gorillas [Hopkins et al., 2003b] and capuchin monkeys [Spinozzi et al., 1998]. Nevertheless, the origin and the phylogenetic evolution of handedness from nonhuman to human primates are still unclear, because other primate species tested with the tube task showed divergent laterality patterns. Thus, orangutans presented a group-level left-handedness [Hopkins et al., 2003b], while rhesus monkeys seem not to be lateralized at a group-level [Bennett et al., unpublished data].

HAND USE IN GESTURAL COMMUNICATION

On a theoretical level, there is a lively debate concerning the evolutionary relationship between the emergence of language and handedness [Corballis, 2003]. It is well-established that both production and perception of speech are mainly controlled by the left hemisphere and that this hemispheric specialization is related to right handedness [96% of right-handed people presented a left-hemispheric dominance for language functions, Knecht et al., 2000]. Moreover, it has been shown that (1) the activity of the right hand is predominant for manual movements when people are talking [Kimura, 1973] or signing [deaf people: Vaid et al., 1989], and (2) the right hand is preferred to the left hand in pointing gestures by infants during speech development [Blake et al., 1994]. Such reports and other considerations have led some authors [e.g., Corballis, 2002] to consider manual communicative behaviors as a phylogenetic mediator leading to right-handedness and speech, both characterized by left cerebral control. But, whether gestural communication in nonhuman primates involves similar left-hemispheric dominance is still unclear.

We have recently addressed this topic in captive olive baboons (*P. anubis*) living in social groups, by recording their manual preferences for a species-specific communicative gesture performed to threat or intimidate conspecifics or a human observer [Meguerditchian & Vauclair, 2006]. This intentional gesture consists of a repetitive slapping or rubbing of the hand on the ground toward a focused individual [Kummer, 1968; see Figure 1c for an illustration]. In a sample of 60 captive baboons, results revealed significant group-level right-handedness for both intra- and inter-specific threat gestures. In fact, among the 45 lateralized subjects, 78% were right handed, indicating a left-hemispheric specialization for the communicative functions of the hands. A similar preference of the right hand in gestural communication has been shown in chimpanzees for food-begging gestures produced to communicate with humans [Hopkins et al., 2005; see also Hopkins & Cantero, 2003; Hopkins & Leavens, 1998; Hopkins & Wesley, 2002]. The patterns of laterality revealed by the baboons' gestures differed from those reported in noncommunicative bimanual and unimanual motors' actions [Vauclair et al., 2005, see above] insofar as (1) hand preferences did not correlate within the same individuals who performed both types of manual actions and as (2) the degree of group-level right hand bias reported for communicative gestures was stronger than that obtained for noncommunicative actions, such as the tube task. Additionally, whereas no effect of age was shown on handedness data for the tube and the unimanual tasks (see above), the right hand biases for communicative gestures were significantly more pronounced in adults than in young baboons. Considering these findings, we suggest that the neural substrate of manual communication may be dissociated from the substrates involved in manipulative motor functions in nonhuman primates. The production of gestures, when considered as communicative signals, is strongly lateralized to the left hemisphere, whereas manual motor actions without communicative functions are less lateralized [Meguerditchian & Vauclair, 2006]. Moreover, a similar difference in the degree of right-handedness between communicative and manipulative tasks was observed in chimpanzees [Hopkins et al., 2005] and in children raised by deaf parents [Bonvillian et al., 1997]. The report of asymmetries in baboons' gestures suggests a continuity between this species, chimpanzees and humans. This continuity leads us to support the view that lateralization for language may have evolved from a gestural system of communication lateralized in the left hemisphere in their common ancestor at least 30 million years ago [Meguerditchian & Vauclair, 2006].

LABORATORY STUDIES ON THE LATERALIZATION OF VISUAL INFORMATION

In human studies, researchers interested in hemispheric lateralization have repeatedly used a tachistoscope [Hellige, 1993]. The use of the tachistoscope requires that subjects gaze on a central fixation point and view a short display (<150 ms) of a stimulus in either the left or the right peripheral visual half-field. Given the anatomy of the visual system, visual information presented with a sufficient degree of eccentricity (more than 2.5°) relative to the fixation point is directed to the hemisphere that is opposite to the stimulated half-field [Bryden, 1982]. It is thus possible with this method to selectively activate the left or the right hemisphere of the intact brain. Following Washburn et al. [1989], we developed for humans and baboons an automated test procedure that has several characteristics in common with the use of a tachistoscope. During this procedure, primates are initially trained to manipulate a joystick, so as to displace a cursor in any direction on a computer monitor (see Figure 2).

Subjects are then tested in a matching-to-sample task in which the sample stimulus was presented to either the left or the right of a fixation point for a duration shorter (e.g., 120 ms) than the latency to initiate a visual saccade [Wilde et al., 1994]. To ensure that the sample form was presented in one visual half-field only, fixation is imposed by having the subject place and maintain the cursor on the fixation point before the presentation of the sample stimulus. Therefore, this technique is analogous to the tachistoscopic mode of stimulus presentation [see Vauclair & Fagot, 1994, 1995 for additional information on the joystick system].

This system is advantageous because it can be adapted for testing both human and nonhuman subjects in strictly identical experimental conditions. Use of identical procedures does not equate all the factors that may affect the results (for instance, the attentional, memory and motivational factors). Nevertheless, testing humans and monkeys with similar procedures is a reasonable approach to assessing species differences and similarities in hemispheric processing. This

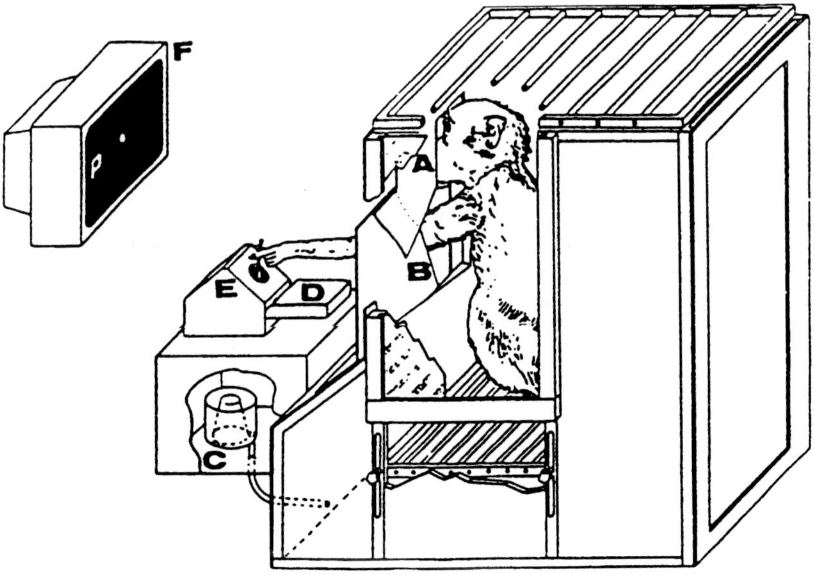

Figure 2. The testing environment used for the study of hemispheric specialization in baboons. Experimental cage fitted with a viewport (A), two hand ports (B), a food dispenser (C), a touch-sensitive pad (D), a joystick (E), and a 14-in color monitor (F).

joystick system was used for testing baboons' abilities to solve different pattern discrimination problems [e.g., categorization: Vauclair & Fagot, 1996].

Two sets of studies illustrate this approach and its potential heuristic value. In one study, the video task was used with the aim of assessing lateralization for the processing of spatial relations in humans and baboons. The human cerebral hemisphere appears to be specialized to process spatial relations [Kosslyn, 1987]. Studies involving normal and brain-damaged humans, or neural network simulations [see Jager & Postma, 2003, for a review] are convergent and show a left-hemisphere superiority to categorize topological relations of the above or below type, and a right-hemisphere advantage to evaluate distances. From an evolutionary standpoint, we questioned this left or right dissociation by using metric tasks. Indeed, nonhuman primates are capable of processing distances [e.g., Fujita, 1997]. However, hemispheric specialization in the processing of distances had not yet been investigated with nonhuman primates. In our study [Dépy et al., 1998], baboons and humans were tested with the divided-field procedure and were required to decide whether a dot was near or far from a horizontal line. Using an arbitrary matching-to-sample task, subjects had to select a red stimulus when the sample belonged to the "near" category and a green one when it belonged to the "far" category. Results of this experiment revealed a left-hemisphere advantage for the baboons for distance processing. Naive human participants exhibited a right-hemisphere advantage in the early trials of the task, but this effect disappeared with practice. It can be hypothesized from these results that in the early trials of the task, the metric aspects of the stimuli predominated and induced a right-hemisphere advantage. With practice, categorical aspects of the task increased and overrode the metric aspects, thus inducing a greater involvement of the left hemisphere. This study stresses the fact that hemispheric lateralization changes with practice. A similar shift has also been observed in a study with baboons in the processing of composite visual stimuli [Fagot & Vauclair, 1994]. In short, the dissociation proposed by Kosslyn [1987] of two kinds of spatial representations and of their underlying cerebral control was thus only partially verified [see Vauclair et al., 2006, for further discussions].

The second set of studies examined hemispheric specialization in mirror-image discrimination and mental rotation in humans and baboons. Mirror-image stimuli are reputed to be difficult to discriminate by nonhuman species [see Hopkins et al., 1993, for review]. The use of a matching-to-sample procedure within the framework of a divided field presentation of visual stimuli appeared to be an efficient tool for performing mirror-image discrimination. In one of the experiments [Experiment 3: Hopkins et al., 1993], we investigated the effect of bilateral versus unilateral visual input on accuracy and response times in the baboons. We found significantly lower accuracy and longer response times for bilateral compared with unilateral visual input. We were then in a position to study the phenomenon of mental rotation [Shepard & Metzler, 1971] in baboons [Vauclair et al., 1993]. The monkeys were taught a matching-to-sample task within the constraints of a divided visual field presentation. That is, once eye fixation on the center of the monitor was assured, the sample stimulus (the letter F or P) was presented for 150 ms to the right or to the left of the fixation point (in the left visual or right visual field). Then, two comparison stimuli were displayed on the monitor: one test stimulus matched the sample stimulus; the other was its mirror image (see Figure 3 for an example). In addition, both test shapes were oriented at 60°, 120°, 180°, 240°, 300° or 360°. Moving the cursor, by way of joystick manipulation, to the location of the matching comparison stimulus was reinforced with a food reward.

Baboons solved the mirror-image discrimination task with an accuracy averaging 77% correct for the left as well as right visual field presentations. Moreover, the analysis of response times for right visual field presentations revealed a mental rotation curve, suggesting that the left hemisphere of the baboon utilized a human-like mental rotation strategy. The right hemisphere of the baboon could also successfully discriminate mirror images, but apparently the monkeys used an as yet unknown strategy that differs from the typical mental rotation phenomenon. Overall reaction times for the baboons were roughly 2.5 times faster than that for human participants tested in the same conditions [Vauclair et al., 1993]. Explaining the differences in response times and mental rotation rates between humans and baboons requires follow-up investigations. At present, any explanation remains speculative. It is noteworthy, however, that the letter stimuli used in our study had a specific meaning or reinforcement history for human participants that did not exist for baboons. This fact might account for some of the differences in response times. In addition, it is also possible that the baboons performed a spatial transformation of one part of the object instead of rotating the whole object as human participants do. Alternatively, other factors cannot be ruled out such as the differences in overall size of the brain, overall differences in the sensorimotor systems or cognitive levels of processing between the two species.

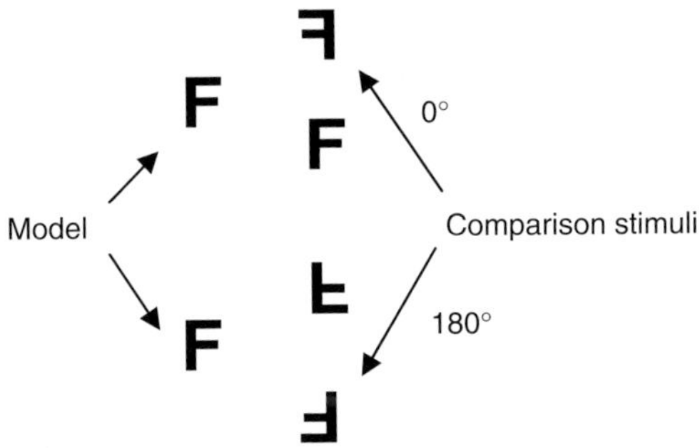

Figure 3. Two stimulus presentation conditions (0° and 180°) used in the mental rotation experiment with humans and baboons [Vauclair et al., 1993].

There was little evidence of mental rotation in animals before our study on baboons. Hollard and Delius [1982] found that pigeons, unlike humans, were able to discriminate between image and mirror image of rotated stimuli without delay, irrespective of their angular disparity in a time-independent rotational invariance process. One study performed with a single individual [a California sea lion: Mauck & Dehnhardt, 1997; see also Stich et al., 2003] has reported a linear increase of reaction times for discriminating (within a matching-to-sample paradigm) simple asymmetrical visual stimuli rotated between 0° and 120°, and a decrease for 150° and 180°. A striking aspect of the sea lion's performance is, however, the rather long reaction times measured (an average of 5 s for 0° rotation and 8 s for 120° rotation). More recently, a 5-year-old female lion-tailed macaque (*Macaca silenus*) was tested with a matching-to-sample task to discriminate visual stimuli (bilaterally presented) that compare in complexity with the forms we used with baboons [Burmann et al., 2005]. The monkey had great difficulties to discriminate mirror-image stimuli and the two stimuli rotated by 30°, 60°, 90° or 120° both clockwise and counterclockwise. Analysis of the macaque's reaction times revealed that they did not increase as a function of the angular disparity of the stimuli. Similar findings were obtained with three rhesus monkeys [Köhler et al., 2005].

The relative facility with which our monkeys solved the discrimination of mirror-image shapes and used a mental rotation strategy could be explained by the lateralization of the visual information inherent to the paradigm we used. As proposed by Noble [1966, 1968] [see also Rollenhagen & Olson, 2000], the difficulty with lateral mirror images reflects the reception in each hemisphere of antagonistic information regarding the horizontal extent of the visual field. The information received via the lateral geniculate nucleus conflicts with information received from the other hemisphere via homotopic interhemispheric connections. Lateralizing the visual information might have prevented the formation of the duplicate of the stimulus representation and thus might have facilitated mirror-image discrimination and use of a mental rotation strategy in baboons.

The same group of baboons has also been used to test the processing of global versus local information and its lateralization. The laterality for processing compound stimuli [Navon, 1977] has thus been studied by Deruelle and Fagot [1997] and Fagot and Deruelle [1997]. Baboons (and humans) showed a right-hemisphere advantage for global trials and a reversed but not significant left-hemisphere advantage for local trials. Hopkins [1997] found similar results in chimpanzees. In this latter study, compound stimuli made of letters at both the local and global levels were presented to chimpanzees in a divided field setup. There was no significant effect of lateralization for accuracy scores. For the speed of the response, however, there was a significant left-hemispheric advantage for local processing, but no significant advantage for global processing.

We can finally note that the video formatted task has also been used with success to investigate hemispheric lateralization for the control of spatial movements in the baboon. Thus, Vauclair and Fagot [1993] have tested the abilities of eight Guinea baboons to transfer abilities of each hand to reach (via joystick manipulation) a randomly moving target on a computer screen. Two interesting results were obtained: (1) a transfer effect from one hand to the other for mastering the task, and (2) a better performance of the left hand compared to the right hand. This latter performance was expressed by a shorter path made by the left hand to bring the cursor into collision with the target. This left hand advantage is compatible with the findings obtained in our research group according to which spatial tasks were better performed by the right hemisphere (see above). In addition, this kind of task allows researchers to shed light on the relationship between hand preference and hand performance.

CONCLUSION

As the study of perceptual asymmetries is no longer being pursued in our research group, the discussion below will mostly concern manual lateralization. In our work on perceptual asymmetries, the search for cognitive similarities between humans and nonhuman primates permits us to evaluate the dependence of cognitive functions vis-à-vis the functions that are considered as being typically human, such as language. Studies with nonhuman primates illustrate perfectly the importance of the comparative approach for understanding human lateralization. Following Broca's [1861] first observation of a case of lateralization, there was a pervasive view that lateralization was a typical characteristic of humans [Geschwind & Galaburda, 1985]. Our results clearly challenged this view by showing that lateralization is independent from language. Also interesting is that the study of animal lateralization showed important similarities between humans and other primates, for instance, in the processing of faces [e.g., Weiss et al., 2002] or in the treatment of the global or local characteristics of the visual forms [e.g., Deruelle & Fagot, 1997]. It is thus presumed that instead of being primarily linked to language, lateralization actually emerged as a solution to either perceptual or motor demands.

Concerning evidence of handedness in baboons, high-level tasks represent the best candidate for the evaluation of hand use in relation to hemispheric asymmetries compared to low-level tasks [Fagot & Vauclair, 1988a, 1991]. From an evolutionary viewpoint, the report of group-level right-handedness with the coordinated bimanual tube task (high-level task) on a sample of 104 baboons suggests the existence of a phylogenetic continuity with humans concerning hemispheric lateralization [Vauclair et al., 2005]. Notwithstanding, the phylogenetic emergence of right-handedness remains unclear. MacNeilage [1991] proposed a theory of a postural origin of handedness from arboreal primates (left hand preference for manual actions like grasping for food while the opposite hand is used to support the body in the tree), whereas more terrestrial primates like baboons would have developed a right-handedness for manual activities. Some of the results obtained with the tube task are consistent with this theory. Thus, a right hand preference was reported for terrestrial species such as chimpanzees [Hopkins, 1995], gorillas [Hopkins et al., 2003b] and baboons [Vauclair et al., 2005], while a bias in favor of the left hand was observed in an arboreal primate, the orangutan [Hopkins et al., 2003b]. However, the postural origin theory of handedness is not supported by other reports. For example, the capuchin monkey, a mostly arboreal species, displayed a preference for the right hand in the tube task [Spinozzi et al., 1998, but see Westergaard & Suomi, 1996] and no group-level handedness was seen in the rhesus monkey, a terrestrial primate [Bennett et al., unpublished data]. In fact, further comparative and extensive hand use data must be collected on large samples and on many more nonhuman primate species in order to clarify this question.

Concerning communicative gestures, as has been shown in chimpanzees [Hopkins et al., 2005], the laterality pattern in baboons for unimanual and bimanual manipulative motor actions differed from the one exhibited for communicative gestures [see above, and Meguerditchian & Vauclair, 2006]. This discrepancy can serve to support the view that the manipulative and communicative functions of the hands do not share the same cerebral system in nonhuman primates. This statement is reinforced by results of studies using MRI techniques in chimpanzees revealing that handedness for non-communicative motor actions correlated with asymmetries in the primary motor cortex but not with homologues of human language areas [Hopkins & Cantalupo, 2004]. This finding is opposite to the human pattern for which right-handedness is clearly linked to left-hemispheric dominance for language functions [e.g., Knecht et al., 2000]. However, recently a significant correlation was shown [Taglialatela et al., 2006] between right-handedness for manual communication in chimpanzees and the left-asymmetries reported in the inferior temporal gyrus [homologous of the language area involved in humans during pronunciation of deictic sentences, see Lvenbruck et al., 2005]. These findings strongly support the idea that lateralization for language may have evolved from a gestural system of communication lateralized in the left hemisphere in the common ancestor of chimpanzees and humans at least 5–7 million years ago. Further investigations, both behavioral and with MRI techniques, are necessary with baboons in order to test our hypothesis of a similar continuity with this Old World monkey species [Meguerditchian & Vauclair, 2006]. It would be, in particular, very important to investigate evidence for correlations between hand preferences for gestures and neurobiological asymmetries of homologous language regions in the baboon brain.

Within a phylogenetic framework, handedness expressed in non-communicative actions and handedness observed in manual communication could possibly be distinguished. It is conceivable that the cerebral circuit of human handedness results from an evolutionary convergence of these two different lateralized systems (communicatory and motor), which evolved independently in our distant ancestors. And, because of the mediation of lateralized manual communications involving likely precursors of language cerebral areas (see above), the lateralization of the motor manipulative systems could have been progressively associated with the communicatory manual system to become ultimately related to a left lateralization for language [Knecht et al., 2000].

From a different view, on the basis of the debatable and poor evidence that the left-hemisphere might control vocal functions in nonhuman primates [see Corballis, 2003, and commentaries], Corballis [2002, 2003] proposed the hypothesis of a progressive integration of this lateralized vocal system into a gestural communicatory system during the course of the evolution which might finally lead to right-handedness in humans. In chimpanzees, Hopkins and Cantero [2003] have demonstrated that right hand preferences for gesturing were the most pronounced when gestures are simultaneously associated with vocalizations. A similar relationship has been described for human infants. For example, the right hand is more strongly involved than the left hand when human infants produced repetitive vocal-motor activity during the babbling phase [Locke et al., 1995]. Moreover, the degree of right-handedness for pointing gestures has been shown to increase during speech development [Blake et al., 1994]. Returning to baboons, their hand slapping gestures which exhibited a strong group-level right-handedness [Meguerditchian & Vauclair, 2006] have never been observed with a simultaneous production of vocalizations. Thus, we believe that the emergence during primate phylogeny of a preference for the right hand cannot be explained only within the framework of the progressive integration of vocalizations into gestural communicative behaviors [e.g., Corballis, 2003].

To us, it is quite obvious that baboon species are interesting models for the study of perceptual and motor asymmetries as well as for brain lateralization. The results obtained on Guinea and olive baboons compared well with other monkey species and also with apes. We firmly believe that the behavioral data collected on baboons along with future brain imaging studies should significantly contribute to a better understanding of the roots of human handedness and communication.

ACKNOWLEDGMENTS

This work, as part of the European SCIENCE Foundation EUROCORES Programme OMLL, was supported by funds from the CNRS (OHLL Programme) and the EC Sixth Framework Programme under Contract no. ERAS-CT-2003-980409.

REFERENCES

Annett, M. Left, right, hand and brain: The right shift theory. Hillsdale, NJ, Erlbaum, 1985.

Blake, J.; O'Rourke, P.; Borzellino, G. Form and function in the development of pointing and reaching gestures. Infant Behavior and Development 17: 195–203, 1994.

Bonvillian, J. D.; Richards, H. C.; Dooley, T. T. Early sign language acquisition and the development of hand preferences in young children. Brain and Language 58: 1–22, 1997.

Bradshaw, J. L.; Rogers, L. The evolution of lateral asymmetries, language, tool-use and intellect. San Diego, CA, Academic Press, 1993.

Broca, P. Remarques sur le siège de la faculté du langage articulé suivies d'une observation d'aphémie. Bulletin de la Société d'Anatomie 6: 398–407, 1861.

Bryden, M. Laterality: Functional asymmetry in the intact brain. New York, Academic Press, 1982.

Burmann, G.; Dehnhardt, G.; Mauck, B. Visual information processing in the lion-tailed macaque (*Macaca silenus*): Mental rotation or rotational invariance? Brain, Behavior and Evolution 65: 168–176, 2005.

Corballis, M. C. From hand to mouth. The origins of language. Princeton, NJ, Princeton University Press, 2002.

Corballis, M. C. From mouth to hand: Gesture, speech, and the evolution of right-handedness. The Behavioral and Brain Sciences 26: 199–260, 2003.

Dépy, D.; Fagot, J.; Vauclair, J. Comparative assessment of distance processing and hemispheric specialization in humans (*Homo sapiens*) and baboons (*Papio papio*). Brain and Cognition 38: 165–182, 1998.

Deruelle, C.; Fagot, J. Hemispheric lateralization and global precedence effects in the processing of visual stimuli by humans and baboons (*Papio papio*). Laterality 2: 233–246, 1997.

Fagot, J.; Deruelle, C. Processing of global and local visual information and hemispheric specialization in humans (*Homo sapiens*) and baboons (*Papio papio*). Journal of Experimental Psychology: Human Perception and Performance 23: 429–442, 1997.

Fagot, J.; Vauclair, J. Handedness and manual specialization in the baboon. Neuropsychologia 26: 795–804, 1988a.

Fagot, J.; Vauclair, J. Handedness and bimanual coordination in the lowland gorilla. Brain, Behavior and Evolution 32: 89–95, 1988b.

Fagot, J.; Vauclair, J. Manual laterality in nonhuman primates: A distinction between handedness and manual specialization. Psychological Bulletin 109: 76–89, 1991.

Fagot, J.; Vauclair, J. Video-task assessment of stimulus novelty effects on hemispheric lateralization in baboons (*Papio papio*). Journal of Comparative Psychology 108: 156–163, 1994.

Fujita, K. Perception of the Ponzo illusion by rhesus monkeys, chimpanzees and humans: Similarity and difference in the three primate species. Perception and Psychophysics 59: 284–292, 1997.

Geschwind, N.; Galaburda, A. M. Cerebral lateralization: Biological mechanisms, associations and pathology. Cambridge, MA, MIT Press, 1985.

Hellige, J. B. Hemispheric asymmetry: What's right and what's left. Cambridge, MA, Harvard University Press, 1993.

Hollard, V.; Delius, J. D. Rotational invariance in visual pattern recognition by pigeons and humans. Science 218: 804–806, 1982.

Hopkins, W. D. Hand preferences for a coordinated bimanual task in 110 chimpanzees: Cross-sectional analysis. Journal of Comparative Psychology 109: 291–297, 1995.

Hopkins, W. D. Hemispheric specialization for local and global processing of hierarchical visual stimuli in chimpanzees (*Pan troglodytes*). Neuropsychologia 35: 343–348, 1997.

Hopkins, W. D. On the other hand: Statistical issues in the assessment and interpretation of hand preference data in nonhuman primates. International Journal of Primatology 20: 851–866, 1999.

Hopkins, W. D.; Cantalupo, C. Does variation in sample size explain individual differences in hand preference of chimpanzees (*Pan troglodytes*)? An empirical study and reply to Palmer (2002). American Journal of Physical Anthropology 121: 878–881, 2003.

Hopkins, W. D.; Cantalupo, C. Handedness in chimpanzees is associated with asymmetries in the primary motor cortex but not with homologous language areas. Behavioural Neuroscience 118: 1176–1183, 2004.

Hopkins, W. D.; Cantero, M. From hand to mouth in the evolution of language: The influence of vocal behavior on lateralized hand use in manual gestures by chimpanzees (*Pan troglodytes*). Developmental Science 6: 55–61, 2003.

Hopkins, W. D.; Fagot, J.; Vauclair, J. Mirror-image matching and mental rotation problem solving by baboons (*Papio papio*): Unilateral input enhances performance. Journal of Experimental Psychology: General 122: 1–12, 1993.

Hopkins, W. D.; Fernandez-Carriba, S.; Wesley, M. J.; Hostetter, A.; Pilcher, D.; Poss, S. The use of bouts and frequencies in the evaluation of hand preferences for a coordinated bimanual task in chimpanzees (*Pan troglodytes*): An empirical study comparing two different indices of laterality. Journal of Comparative Psychology 115: 294–299, 2001.

Hopkins, W. D.; Hook, M.; Braccini, S.; Schapiro, S. Population-level right handedness for a coordinated bimanual task in chimpanzees (*Pan troglodytes*): Replication and extension in a second colony of apes. International Journal of Primatology 24: 677–689, 2003a.

Hopkins, W. D.; Leavens, D. A. Hand use and gestural communication in chimpanzees (*Pan troglodytes*). Journal of Comparative Psychology 112: 95–99, 1998.

Hopkins, W. D.; Pearson, K. Chimpanzee (*Pan troglodytes*) handedness: Variability across multiple measures of hand use. Journal of Comparative Psychology 114: 126–135, 2000.

Hopkins, W. D.; Russell, J.; Freeman, H.; Buehler, N.; Reynolds, E.; Schapiro, S. J. The distribution and development of handedness for manual gestures in captive chimpanzees (*Pan troglodytes*). Psychological Science 6: 487–493, 2005.

Hopkins, W. D.; Stoinski, T. S.; Lukas, K. E.; Ross, S. R.; Wesley, M. J. Comparative assessment of handedness for a coordinated bimanual task in chimpanzees (*Pan troglodytes*), gorillas (*Gorilla gorilla*), and orangutans (*Pongo pygmaeus*). Journal of Comparative Psychology 117: 302–308, 2003b.

Hopkins, W. D.; Wesley, M. J. Gestural communication in chimpanzees (*Pan troglodytes*): The effect of situational factors on gesture type and hand use. Laterality 7: 19–30, 2002.

Hopkins, W. D.; Wesley, M. J.; Izard, M. K.; Hook, M.; Schapiro, S. J. Chimpanzees (*Pan troglodytes*) are predominantly right-handed: Replication in three populations of apes. Behavioral Neuroscience 118: 659–663, 2004.

Jager, G.; Postma, A. On the hemispheric specialization for categorical and coordinate spatial relations: A review of the current evidence. Neuropsychologia 41: 504–515, 2003.

Kimura, D. Manual activity during speaking. I. Right-handers. Neuropsychologia 11: 45–50, 1973.

Knecht, S.; Deppe, M.; Draeger, B.; Bobe, L.; Lohman, H.; Ringelstein, E. B.; Henningsen, H. Language lateralization in healthy right-handers. Brain 123: 74–81, 2000.

Köhler, C.; Hoffmann, K. P.; Dehnhardt, G.; Mauck, B. Mental rotation and rotational invariance in the rhesus monkey (*Macaca mulatta*). Brain, Behavior and Evolution 66: 158–166, 2005.

Kosslyn, S. M. Seeing and imagining in the cerebral hemispheres: A computational approach. Psychological Review 94: 148–175, 1987.

Kummer, H. Social organization of Hamadryas Baboons: A field study. Chicago, The University of Chicago Press, 1968.

Locke, J. L.; Bekken, K. E.; McMinn-Larson, L.; Wein, D. Emergent control of manual and vocal-motor activity in relation to the development of speech. Brain and Language 51: 498–508, 1995.

Lvenbruck, H.; Baciu, M.; Segebarth, C.; Abry, C. The left inferior frontal gyrus under focus: An fMRI study of the production of deixis via syntactic extraction and prosodic focus. Journal of Neurolinguistics 18: 237–258, 2005.

MacNeilage, P. F. The "postural origins" theory of primate neurobiological asymmetries. Pp. 165–187 in N. A. Krasnegor; D. M. Rumbaugh; R. L. Schiefelbusch; M. G. Studdert-Kennedy, eds. Biological and behavioral determinants of language development. Hillsdale, NJ, Erlbaum, 1991.

MacNeilage, P. F.; Studdert-Kennedy, M. G.; Lindblom, B. Primate handedness reconsidered. Behavioral and Brain Sciences 10: 247–303, 1987.

Marchant, L. F.; McGrew, W. C. Laterality of limb function in wild chimpanzees of Gombe National Park: Comprehensive study of spontaneous activities. Journal of Human Evolution 30: 427–443, 1996.

Mauck, B.; Dehnhardt, G. Mental rotation in a California sea lion. The Journal of Experimental Biology 200: 1309–1316, 1997.

McGrew, W. C.; Marchant, L. F. On the other hand: Current issues in and meta-analysis of the behavioural laterality of hand function in non human primates. Yearbook of Physical Anthropology 40: 201–232, 1997.

Meguerditchian, A.; Vauclair, J. Baboons communicate with their right hand. Behavioural Brain Research 171: 170–174, 2006.

Navon, D. Forest before trees: The precedence of global features in visual perception. Cognitive Psychology 9: 353–383, 1977.

Noble, J. Mirror-images and the forebrain commissures of the monkey. Nature 211: 1263–1265, 1966.

Noble, J. Paradoxical interocular transfer of mirror-image discriminations in the optic chiasm sectioned monkey. Brain Research 10: 127–151, 1968.

Palmer, A. R. Chimpanzee right-handedness reconsidered: Evaluating the evidence with funnel plots. American Journal of Physical Anthropology 118: 191–199, 2002.

Papademetriou, E.; Sheu, C. F.; Michel, G. F. A meta-analysis of primate hand preferences, particularly for reaching. Journal of Comparative Psychology 119: 33–48, 2005.

Rollenhagen, J. E.; Olson, C. R. Mirror-image confusion in single neurons of the macaque inferotemporal cortex. Science 287: 1506–1508, 2000.

Shepard, R. N.; Metzler, J. Mental rotation of three-dimensional objects. Science 171: 701–703, 1971.

Spinozzi, G.; Castorina, M. G.; Truppa, V. Hand preferences in unimanual and coordinated-bimanual tasks by tufted capuchin monkeys (*Cebus apella*). Journal of Comparative Psychology 112: 183–191, 1998.

Stich, K. P.; Dehnhardt, G.; Mauck, B. Mental rotation of perspective stimuli in a California sea lion (*Zalophus californianus*). Brain, Behavior and Evolution 61: 102–112, 2003.

Taglialatela, J. P.; Cantalupo, C.; Hopkins, W. D. Gesture handedness predicts asymmetry in the chimpanzee inferior frontal gyrus. Neuroreport 17: 923–927, 2006.

Vaid, J.; Bellugi, U.; Poizner, H. Hand dominance for signing: Clues to brain lateralization. Neuropsychologia 27: 949–960, 1989.

Vauclair, J.; Fagot, J. Spontaneous hand usage and handedness in a troop of baboons. Cortex 23: 265–274, 1987a.

Vauclair, J.; Fagot, J. Visually guided reaching in adult baboons. The Behavioral and Brain Sciences 10: 287, 1987b.

Vauclair, J.; Fagot, J. Manual and hemispheric specialization in the manipulation of a joystick by baboons. Behavioral Neuroscience 107: 210–214, 1993.

Vauclair, J.; Fagot, J. A joystick system for the study of hemispheric asymmetries in nonhuman primates. Pp. 69–75 in J. R. Anderson; J. J. Roeder;

B. Thierry; N. Herrenschmidt, eds. Current Primatology (Vol. III): Behavioral Neuroscience, Physiology and Reproduction. Strasbourg, Université Louis Pasteur, 1994.

Vauclair, J.; Fagot, J. Une méthode non verbale pour l'étude des asymétries hémisphériques visuo-spatiales chez le singe et l'homme. Revue de Neuropsychologie Française 5: 3–21, 1995.

Vauclair, J.; Fagot, J. Categorization of alphanumeric characters by baboons (*Papio papio*): Within and between class stimulus discrimination. Current Psychology of Cognition 15: 449–462, 1996.

Vauclair, J.; Fagot, J.; Hopkins, W. D. Rotation of mental images in baboons when the visual input is directed to the left cerebral hemisphere. Psychological Science 4: 99–103, 1993.

Vauclair, J.; Meguerditchian, A.; Hopkins, W. D. Hand preferences for unimanual and coordinated bimanual tasks in baboons (*Papio anubis*). Cognitive Brain Research 25: 210–216, 2005.

Vauclair, J.; Yamazaki, Y.; Güntürkün, 0. The study of hemispheric specialization for categorical and coordinate spatial relations in animals. Neuropsychologia 44: 1524–1534, 2006.

Washburn, D. A.; Hopkins, WD.; Rumbaugh, D. M. Automation of learning set testing: The video task paradigm. Behavior, Research Methods, Instruments & Computers 12: 38–45, 1989.

Weiss, D.; Ghazanfar, A.; Miller, C. T.; Hauser, M. D. Specialized processing of primate facial and vocal expressions: Evidence for cerebral asymmetries. Pp. 480–530 in L. Rogers; R. Andrews, eds. Cerebral Vertebrate Lateralization. New York, Cambridge University Press, 2002.

Westergaard, G. C.; Champoux, M.; Suomi, S. J. Hand preference in infant rhesus macaques (*Macaca mulatta*). Child Development 68: 387–393, 1997.

Westergaard, G. C.; Suomi, S. J. Hand preference for a bimanual task in tufted capuchins (*Cebus apella*) and rhesus macaques (*Macaca mulatta*). Journal of Comparative Psychology 110: 406–411, 1996.

Wilde, J.; Vauclair, J.; Fagot, J. Eye movements in baboons performing a matching-to-sample task presented in a divided-field format. Behavioural Brain Research 63: 61–70, 1994.

Hand in hand: *Cebus apella*, *Homo sapiens*. (Photograph by Valentia Truppa.)

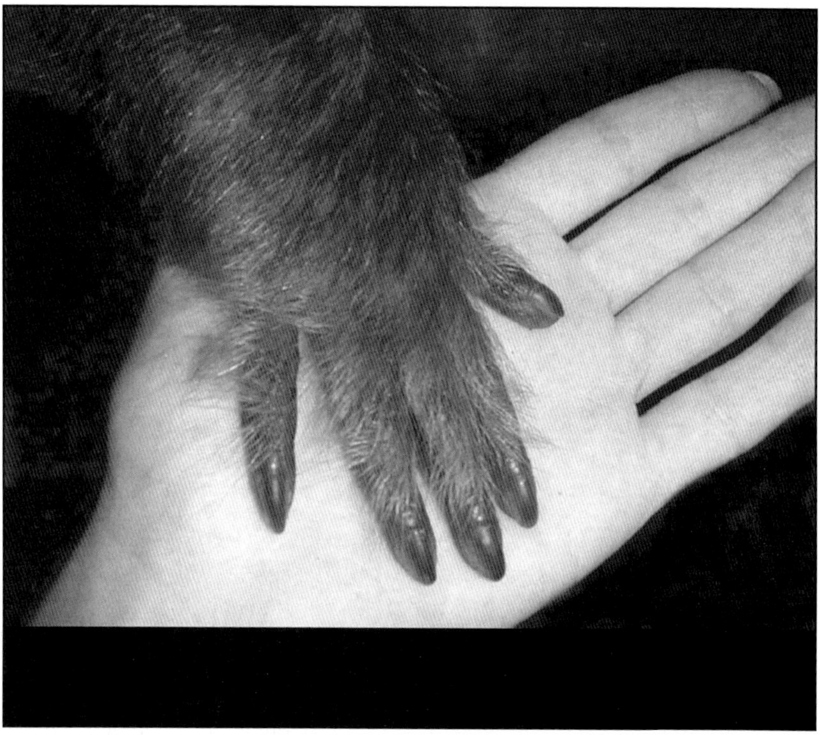

The Evolution of Hemispheric Specialization in Primates
W.D. Hopkins (Editor)

FACTORS AFFECTING MANUAL LATERALITY IN TUFTED CAPUCHINS (*CEBUS APELLA*)

Giovanna Spinozzi
Unit of Cognitive Primatology, Institute of Cognitive Sciences and Technologies, CNR, Rome, Italy

ABSTRACT

Capuchin monkeys are relatively large-brained New World monkeys and have been the subject of extensive studies on object manipulation, grasping morphology, and tool use. This chapter summarizes a series of studies in our laboratory that has examined the role of bimanual and postural demands on hand preference in capuchin monkeys. In addition, results on performance difference in grasp skill and grip morphology are presented and compared to existing results from other nonhuman primates. Parallels and differences in hand preferences and performance between capuchin monkeys and other nonhuman primate species are discussed.

Keywords: capuchin monkeys, grasping, motor skill, bimanual coordination, laterality, grip morphology

INTRODUCTION

Until quite recently it was generally argued that cerebral lateralization was a uniquely human characteristic [Warren, 1980]. Indeed, it has long been held that, in nonhuman species, the left and right sides of the brain are equivalent at both the structural and functional levels. This conviction derived from the emphasis on hemispheric asymmetry for language and the belief that language processes are uniquely human. However, the growing awareness that behavioral asymmetries in humans are not restricted to language led to a renewed interest in laterality in other species and to the use of animal models for the investigation of the mechanisms of hemispheric asymmetry. There is now a large body of evidence in favor of the presence of both functional and behavioral asymmetries in a wide range of animal species. For example, asymmetries for visual processing have been documented not only in mammals and birds, but also in lower vertebrates, such as fish, reptiles, and amphibians [Bisazza et al., 1998; Bradshaw & Rogers, 1993; Robins & Rogers, 2004; Vallortigara et al., 1999]. However, it is still unclear whether these asymmetries are homologous or

ISSN: 1936-8526 DOI: 10.1016/S1936-8526(07)05008-7

reflect independent evolutionary histories [Bisazza et al., 1998; see also Corballis, 1983].

Research addressing manual asymmetries and the way these asymmetries relate to brain organization has focused mainly on nonhuman primates because of their phylogenetic proximity to the human species, as well as their human-like prehensile hand and fine manipulative skills. Given that laterality in hand use is one of the most conspicuous manifestations of functional brain asymmetries in humans [Annett, 1985; Corballis, 1991; Kimura, 1979], comparative research on nonhuman primate laterality is thus critical to shed light on possible precursors of human hemispheric lateralization.

Early systematic research of manual laterality in nonhuman primates failed to yield population-level hand biases, although individual animals often showed consistent preferences for either hand across repetitions of the same task. This led several authors to conclude that nonhuman primate manual laterality was primarily the result of experience and was independent of any asymmetry of cerebral activity [Lehman, 1980; Warren, 1980]. However, as various researchers have recently argued [Fagot & Vauclair, 1991; Hopkins, 1996; Hopkins & Morris, 1993; MacNeilage et al., 1987], failure to detect lateral preferences arose from too simplistic an approach to animal handedness. For example, the majority of the early hand preference studies were carried out on samples composed of a very limited number of subjects, which did not allow the generalization of the results, and tended to rely on tasks that were too simple to elicit manual laterality. Furthermore, the possible effect of several factors on the expression of laterality, such as subjects' sex and age, posture adopted by the animals while performing specific activities, and intrinsic characteristics of the actions examined, was not taken into account.

In more recent years, the evaluation of these variables provided evidence for population-level lateral hand bias for different manual activities in a broad range of primate species, ranging from prosimians [Masataka, 1989; Ward et al., 1993] to New World and Old World monkeys [Diamond & McGrew, 1994; Fagot et al., 1991; Lacreuse & Fragaszy, 1996, 1999; Westergaard et al., 1997, 1998; Westergaard & Suomi, 1996], and great apes [Hopkins, 1995; Hopkins & Leavens, 1998; Hopkins & Pearson, 2000]. Notwithstanding this, the question of whether nonhuman primate asymmetries are analogous to or homologous with that of humans is still debatable [Corballis, 1997; Hopkins, 1999; McGrew & Marchant, 1996, 1997]. What remains unclear is whether human and nonhuman primates share similar mechanisms and factors affecting the expression of hand preference.

Among these factors, particular attention has been addressed to the level of skill required to execute a task [Fagot & Vauclair, 1991; Hopkins & Morris, 1993; Hopkins & Rabinowitz, 1997; MacNeilage et al., 1987]. Fagot and Vauclair [1991] argued that low-level tasks involving simple and familiar behaviors, such as single acts of reaching for and grasping food in an open space, would not necessarily be expected to induce consistent hand bias at group level.

In contrast, novel or cognitively complex tasks implying attentional processes and/or fine spatiotemporal discrimination or coordination would consistently elicit group-level handedness. Fagot and Vauclair [1991] claimed that in complex tasks, but not in simple or familiar tasks, preferences may demonstrate an underlying specialization of the hemisphere contralateral to the hand used, as reported in humans [Young et al., 1983]. There is evidence, in fact, that manual laterality in humans is a multidimensional phenomenon since it can vary as a function of the level of skill required to execute a movement. For example, activities involving fine sequential finger movements are more likely to produce lateral asymmetries than those implying simpler motor patterns [Healey et al., 1986; Steenhuis, 1996; Steenhuis & Bryden, 1989].

Several recent findings, however, have only partially confirmed the model proposed by Fagot and Vauclair [1991] since complex activities, like those defined by these authors, do not always induce consistent hand bias in nonhuman primate species. For example, Anderson et al. [1996] tested capuchins in a series of manipulative tasks of differing complexity and found no significant hand bias for any of the actions examined, although their data showed that the strength of hand preference was strictly related to the type of manipulation performed, with simple reaching tasks eliciting the weakest form of hand preference. Similar findings have been reported in chimpanzees by Colell et al. [1995], who used various complex tasks to determine the manual laterality of the subjects involved. They found that visuospatial adjustment tasks, requiring the use of one hand or either hands to perform sequential or both sequential and concurrent acts (such as aligning a sliding window with an aperture to obtain food) did not produce any significant group-level hand bias. In contrast, tasks involving single or multiple one-handed problem-solving acts for food retrieval (such as opening a box to grasp food) induced a significant right-hand preference for reaching or manipulating in a subgroup of adolescent and adult individuals. Overall, these findings stress the necessity to evaluate carefully the variables included in the hand preference tasks in order to understand more fully the relationship between task complexity and the expression of hand preference in nonhuman primates.

The present chapter illustrates the results of several experiments on hand preference in a sample of 26 captive tufted capuchins (*Cebus apella*) carried out in the past few years in our laboratory. The tasks my colleagues and I have employed to determine manual asymmetries differed in complexity with respect to both the type and number of actions necessary to execute them, and the availability of visual cues to obtain a goal. The purpose of the present paper is to examine the relationships between task complexity and hand use, and to highlight the factors that could possibly affect the expression of laterality in this nonhuman primate species.

Capuchins represent a very interesting species for the study of manual laterality. They have the highest degree of manual dexterity of all New World monkeys, and display a strong propensity, especially in captivity, to use tools in a variety of contexts and for many purposes [see Fragaszy et al., 2004, for a review]. Although their prehensile hand does not possess a fully opposable thumb typical of catarrhine primates, capuchins are able to perform relatively independent movements of the fingers in most of their spontaneous activities and use different forms of precision grip that mainly involve the lateral aspects of digits, for picking up small objects [Christel & Fragaszy, 2000; Costello & Fragaszy, 1988; Spinozzi et al., 2004].

MANUAL LATERALITY BY CAPUCHINS FOR DIFFERENT TASKS

Simple Tasks

Unimanual Reaching Tasks. Our first experiments with capuchins were designed to assess the effect of posture on hand preference for a simple, unimanual food reaching task [Spinozzi et al., 1998]. The selection of this measure was inspired in part by MacNeilage et al.'s [1987] postural origins theory of primate asymmetry, which postulates that handedness in primates evolved through selection pressures favoring lateral bias for postural support and feeding. Briefly, these authors argued that a left-hand specialization for visually guided reaching first evolved in prosimians, accompanied by right-hand specialization for postural support. This left-hand superiority remains in monkeys. However, with the change from vertical clinging to a quadrupedal locomotion in higher primates, this postural specialization evolved into a right-hand superiority for fine, sequential manipulation, particularly in the context of bimanual coordination. Other reports, however, failed to support the evolutionary model proposed by MacNeilage et al. [1987] since manual asymmetries for visually guided prehension have been shown to be weak and unstable and strongly influenced by situational factors. For example, a significant left-hand preference has been noted in squirrel monkeys when reaching for moving goldfish, but such a manual bias was lost when the animals reached for stationary food items both in a quadrupedal and a bipedal posture and changed to a significant right-hand food reaching preference when they hung in a vertical cling posture [King & Landau, 1993]. Moreover, the direction of hand preference does not appear to be uniform across species. For instance, a group-level left-handedness has been reported in several prosimian species when a bipedal, but not a quadrupedal posture was required for the subjects [Larson et al., 1989]. A bias toward use of the right hand has been noted in several great ape species when reaching from a species-atypical bipedal posture but not from a familiar quadrupedal stance [Hopkins, 1993; Hopkins et al., 1993; Olson et al., 1990].

Our first study was aimed at empirically investigating the direction and strength of hand preference in a captive group of 26 capuchins (13 females and 13 males, mean age 9.08 years, $SD = 5.06$) when performing a simple reaching task that involved the use of one hand for picking up a small food item from a large surface. Given the influence of posture, we incorporated this variable in our test, recording capuchins' responses both when standing quadrupedally and when standing bipedally. Of particular interest was whether the monkeys would exhibit a shift toward right or left preferential hand use for bipedal reaching when contrasted with quadrupedal reaching. Capuchins were tested individually in their indoor cage. To elicit quadrupedal reaching, a small piece of food (a slice of banana or apple) was placed on the cage floor. A trial was scored when the subject maintained both hindlimbs and one forelimb on the floor while reaching. To evoke bipedal reaching, the food was placed on a wooden shelf fixed to the cage wire mesh 25–30 cm above the floor. During reaching, the monkey adopted an upright stance, keeping both feet on the cage floor and using one forelimb to maintain postural support. Each monkey performed approximately 100 trials in each postural condition. In this task, as well as in the other tasks reported below, we used a directional handedness index (*HI*) to characterize the subjects' lateral bias. This measure is calculated using the formula $(R-L)/(R+L)$, in which R and L are the total number of right- and left-handed responses, respectively. The resulting values, ranging from $+1.0$ to -1.0, situate each monkey's hand preference on a continuum ranging from strongly right-handed ($+1.0$) to strongly left-handed (-1.0). The absolute value of the *HI*, *ABS-HI*, represents the strength of hand preference irrespective of its direction.

Table I presents the mean value of the *HI* observed in our capuchin sample as a function of posture (see *unimanual-quadrupedal* and *unimanual-bipedal*). It includes the *t*-values and the hand bias for the group. As can be seen, unimanual reaching from a bipedal posture elicited a significant right-hand bias at group level. By contrast, in quadrupedal reaching, no significant lateral hand bias emerged for the group. In this postural condition, the mean *HI* per subject did not deviate significantly from chance. Moreover, the change from quadrupedal to bipedal posture resulted in a significant shift toward a greater preferential use of the right hand: the mean *HI* scores were greater during upright reaching compared with quadrupedal reaching. However, the strength in hand preference did not differ between measures (mean *ABS-HI*: quadrupedal = .38, bipedal = .45). Overall, these results are consistent with those obtained for several species of apes, including gorillas, chimpanzees, and orangutans [Hopkins, 1993; Hopkins et al., 1993; Olson et al., 1990]. Moreover, they agree with prior work showing that, in capuchins, the adoption of bipedal posture for food reaching increased the use of the right hand [Westergaard et al., 1997, 1998]. It is noteworthy that none of the studies with capuchins showed a left-hand preference for single acts of visually guided prehension.

It is unclear what neural mechanisms actually may mediate the effect of posture on hand preference in nonhuman primates. Ward et al. [1993] posit that neuronal arousal activated by postural instability or readjustment may facilitate lateral biases. Others have suggested that nonhuman primates may have a mild or "soft" form of manual laterality that is intermediate between population-level handedness and absence of manual preference [Roney & King, 1993]. In this view, soft handedness may prevail when an increase of the postural

Table I. Mean *HI* scores and associated *t*-values for behavioral measures of handedness.

Task	Action	*N*	*HI*	*t*	*p*	Bias
1. Unimanual-quadrupedal	Reaching	26	.10	1.09	>.10	No bias
2. Unimanual-bipedal	Reaching	26	.27	2.91	<.01	Right
3. Tube-crouched	Reaching[a]	26	.44	3.01	<.01	Right
4. Tube-bipedal	Reaching[a]	26	.33	2.13	<.05	Right
5. Hole	Reaching	24	.33	2.31	<.05	Right
6. Horizontal panel	Moving	25	.13	0.82	>.10	No bias
	Reaching	25	.33	2.25	<.05	Right
7. Vertical panel	Lifting	23	.08	0.46	>.10	No bias
	Reaching[a]	23	.12	0.59	>.10	No bias
8. Box-2-act-solution	Lifting	18	.20	1.09	>.10	No bias
	Reaching	19	.29	2.27	<.05	Right
9. Box-3-act-solution	Lifting	23	.04	0.28	>.10	No bias
	Reaching[a]	23	.12	0.72	>.10	No bias
10. Visual reach	Reaching	26	.29	2.10	<.05	Right
11. Haptic discrimination	Reaching	26	−.30	2.17	<.05	Left
12. Visual-tactual reach	Reaching	26	−.36	2.52	<.05	Left

Note. N = number of subjects; HI = handedness index (HI scores refer to group-level hand bias); [a]Coordinated-bimanual act: “reaching” is concurrent with “holding.”

instability during manual reaching intensifies an already existing hand preference present during less posturally demanding reaching. On the other hand, Fagot and Vauclair [1991] claim that the adoption of an unstable posture causes strong spatiotemporal constraints since it requires a dynamic adjustment of the body in order to maintain balance, and movement regulation of the proximal–distal musculature to achieve the goal. In their view, bipedal reaching constitutes a high-level task for those animals for which bipedalism is not a species-typical trait.

Complex Tasks

Tube Tasks. It has been suggested that tasks requiring complex sensorimotor coordination, such as bimanual coordination, may be the most useful measures of lateral bias in primates. Performance in these tasks is less dependent on postural factors and may better reflect a specialization of the contralateral hemisphere to the preferred hand [Fagot & Vauclair, 1991; Hopkins, 1995]. To examine this issue, the same capuchins used in the previous experiments were tested in a task requiring the complementary use of both hands for food retrieval [Spinozzi et al., 1998]. To elicit bimanual coordination, a small piece of ripe banana was placed inside a vertical transparent tube through a hole made on its side. The tube was hung on a chain attached to the ceiling of the experimental cage. In each trial, we scored the hand used by capuchins to extract the reward from the tube. To assess whether the observed effects were caused by the motor requirements of the task or the monkey's posture, manual preferences were assessed in both a crouched and a bipedal posture. In the crouched condition (Figure 1a), the distance of the tube from the floor was about 5 cm, whereas in the bipedal condition the distance was about 25 cm (Figure 1b). As in the previous experiment, each monkey performed approximately 100 trials in each postural condition.

Table I (see *tube-crouched* and *tube-bipedal*) depicts the mean *HI* values for each postural condition of the task. A significant right-hand bias for the group was evident for coordinated-bimanual hand use in both crouched and upright posture. Moreover, in contrast to the previous findings on simple, unimanual reaching tasks, we noted that a change from a crouched to a bipedal stance did not affect the direction of hand preference: in this task, the mean *HI* scores did not differ significantly between the two postural conditions [see Spinozzi et al., 1998, for more details]. Likewise, the strength in hand use was not affected by the posture assumed by the subjects: the mean *ABS-HI* scores for crouched (.82) and bipedal (.81) postures were similar. However, comparison with the previous hand preference data on unimanual hand use revealed that capuchins were more lateralized for the bimanual tube tasks compared with the unimanual reaching tasks (mean *ABS-HI*: unimanual–quadrupedal = .38, unimanual–bipedal = .45). Therefore, coordinated bimanual actions elicited stronger hand preferences than simple acts of visually guided prehension.

Our finding of a significant right-hand bias for coordinated-bimanual hand use in capuchins confirmed that previously reported in chimpanzees [Hopkins,

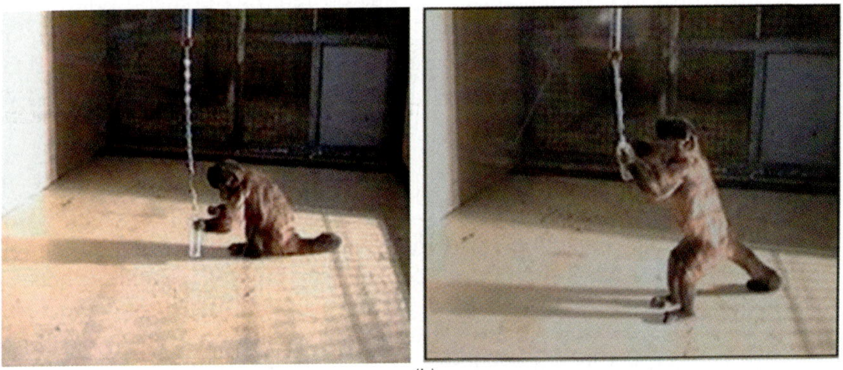

(a) (b)

Figure 1. An adult capuchin monkey performing the tube task from a *crouched* (a, left) and a *bipedal* (b, right) posture.

1995] and macaques [Westergaard & Suomi, 1996], but differed from those noted by Westergaard and Suomi [1996] in their capuchin sample. These authors found no group hand preference in their animals when performing a task designed to elicit the complementary use of hands to remove peanut butter smeared on the inside of a small tube. It is possible that the discrepancy between findings is because of differences in the manipulative demands involved in the tasks. For example, in Westergaard and Suomi's [1996] study, the baited tubes were placed on the monkeys' cage floor. By contrast, in our study the tube was hanging from a chain. This situation required the subjects to perform a continuous spatial adjustment of the hand in order to stabilize the tube and gain access to the reward. The greater demands of our task may have produced a group-level lateral bias in capuchins, eliciting a preferential use of the left hand to hold the tube and a consistent use of the right hand to extract the food.

Multicomponent Tasks. The tube-task experiments yielded evidence that reaching acts implying the complementary use of both hands elicited lateralized hand preferences in capuchins, independently of the posture adopted by the animals. A different picture, however, emerged from the data of two subsequent studies in which hand preference was tested when capuchins were performing several multicomponent tasks involving either sequential or both sequential and concurrent problem-solving acts.

In the first study [Spinozzi & Truppa, 1999], we examined and compared capuchin hand preferences in three reaching tasks – the hole, the horizontal panel, and the vertical panel task, which differed in both the number and the type of problem-solving acts necessary to obtain a reward. In the hole task, the monkeys were required to perform a single, one-handed reaching action to grasp a piece of peanut located inside a small aperture in a metal board fixed onto the animal's cage wall. In the horizontal panel task, capuchins were required to move a transparent sliding panel to the right or the left until one of the two apertures in it matched the central opening of the metal board, where the peanut was placed (Figure 2). This task involved two sequential, one-handed acts: moving the panel, and reaching for food. In the vertical panel task, the monkey had to lift the panel until the only aperture in it was aligned with the central opening of the metal board containing the peanut (Figure 2). This task compelled the capuchins to use both hands in a complementary role, with one hand holding the panel up and the other grasping the reward. The following three problem-solving acts were recorded: lifting the panel, holding the panel up, and reaching for food; performing the task thus implied both sequential (lifting, holding) and concurrent (holding, reaching) acts. The number of capuchins tested in the hole, the horizontal panel, and the vertical panel tasks was 24, 25, and 23, respectively. Further information concerning the procedure is provided in Spinozzi and Truppa [1999].

Table I summarizes the hand preference data obtained for each action performed in each task (see *hole*, *horizontal panel*, and *vertical panel*). As can be seen, the three apparatuses elicited different hand preference patterns.

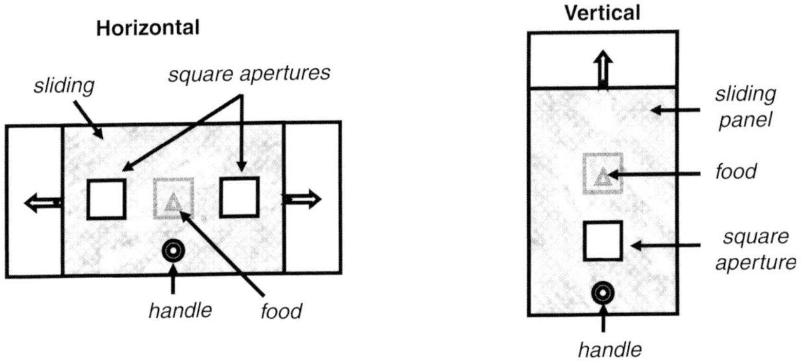

Figure 2. Illustration of the apparatuses used in the two sliding panel tasks. In the *Horizontal Panel* task (left), a transparent sliding panel with two symmetrical lateral apertures was mounted on a metal support with a square opening at its center. The monkey had to move the panel to the right or left (by means of a plastic handle) until one of the two apertures matched the square opening where a peanut was located. In the *Vertical Panel* task (right), a transparent panel with only one aperture was mounted on a metal support with a square opening at its center. The monkey had to lift the panel until the aperture in it was aligned with the square opening containing the peanut.

A significant group-level right-hand bias for reaching actions emerged in two one-handed tasks, that is the hole and the horizontal panel tasks, but not in the most complex task, the vertical panel, in which the food retrieval implied the complementary use of both hands. The latter result thus contrasts with the consistent right-hand preference previously observed in capuchins when performing the tube tasks, in which the food retrieval also entailed two different concurrent actions. Moreover, in the two sliding panel tasks, no significant hand bias emerged for the initial moving or lifting actions with high visuospatial components. Comparison of the strength of laterality across different reaching actions revealed that the mean *ABS-HI* scores for reaching in the vertical panel task (.93) were significantly higher than those for reaching in the horizontal panel (.70, $p < .05$) and the hole (.64, $p < .01$) tasks. Capuchins thus exhibited a stronger lateralization in hand use when reaching for food required bimanual cooperation than when it involved a unimanual act of visually guided prehension. Overall, these findings failed to confirm the claim that more complex motor tasks, such as those involving fine sensorimotor coordination, would consistently induce group-level hand asymmetries in nonhuman primates [Fagot & Vauclair, 1991; Hopkins & Rabinowitz, 1997]. Although the tasks used in our experiments differed with respect to the number and the complexity of the actions necessary to solve them, significant group-level hand biases emerged only for less demanding reaching actions.

The fact that capuchins exhibited a right-hand bias for bimanually coordinated reaching in the tube tasks, but not in the vertical panel task, might reflect differences in the tasks' manipulative demands. For example, the tube tasks may have required fractionated movements of the distal phalanges of the index and middle fingers, often used in combination with the fourth digit, to extract a sticky reward (a piece of banana) from the tube. These specific motor patterns, which are primarily under the control of the contralateral hemisphere [Brinkman & Kuypers, 1973], were not involved in the vertical panel task, in which a small piece of hard food (a piece of peanut) was used as a reward. Although we did not assess the gripping patterns used by our monkeys in this apparatus, there is evidence that this New World monkey species is capable of picking up very small objects using both precision and power grips [Spinozzi et al., 2004]. It should be noted, however, that capuchins exhibited reliable right-hand bias for reaching acts in both the hole and the horizontal panel tasks, which involved the same grip patterns as those used in the vertical panel apparatus. It seems, therefore, that differences in manipulative demands do not explain the discrepancy in hand preferences observed in the two sets of bimanual tasks.

Alternatively, it may be that the lack of a consistent hand bias for bimanual-coordinated action in the multicomponent task could be due to the presence of a problem-solving act, such as removing an obstacle (i.e., the sliding panel), preliminary to the attainment of the reward. This preliminary operation was not required in the tube tasks. One might suppose that in the vertical panel task the hand employed for the initial manipulatory activity might interfere with the division of roles between hands in performing two subsequent concurrent actions. This effect might not emerge in the horizontal panel task, which required the use of a single hand for each component act. In this task, which could be solved with either one hand or two hands acting independently, the inconsistency in directionality of the initial problem-solving act (moving the panel) could not affect the choice of the preferred hand for the final reaching action.

In order to evaluate the generality of this hypothesis, we carried out a second study [Spinozzi & Truppa, 2002] in which capuchins' lateral preferences were measured while performing two new multicomponent tasks. In both tasks, the monkeys had to open a small box by lifting its lid, before they could reach for a peanut placed inside the apparatus. The first task, Box 1, could entail either two sequential one-handed actions – lifting the lid, reaching for food – or three problem-solving actions – lifting the lid, holding the lid up, and reaching for food (Figure 3). This 3-act strategy thus involved both sequential (lifting, holding) and concurrent (holding, reaching) acts, and necessitated the use of both hands in a complementary role for the final food reaching action. In the second task, Box 2, only a 3-act solution was possible (Figure 3). Therefore, as in the two sliding panel tasks, also in the box-opening tasks, capuchins had to perform sequential multiple actions, which may imply bimanual coordination or not. A subgroup of 23 capuchins participated in the experiments. Each

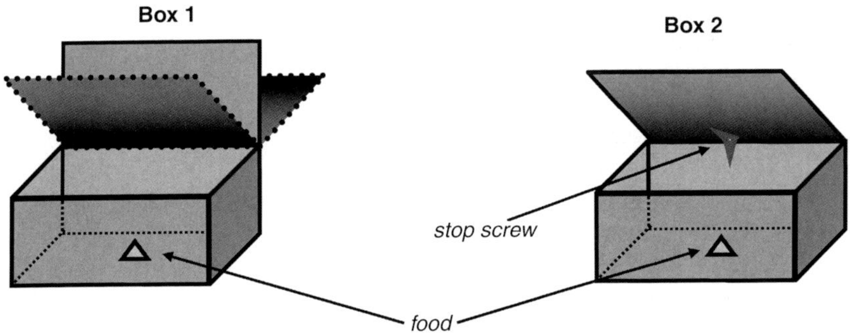

Figure 3. Illustration of the apparatus used in the two box-opening tasks. In both tasks, the apparatus consisted of a metal box, with one of the vertical walls in clear Plexiglas. The monkey could obtain the peanut placed inside the box by lifting a lid fixed to the back wall of the box. In the Box 1 (left), the lid remained raised once the capuchin lifted it by 90° or more. In this case, the task involved two sequential one-handed actions. However, if the monkey lifted the lid by less than 90°, it had to perform three problem-solving actions. In the Box 2 (right), the box included a stop screw on the back of the lid which caused the lid to fall shut if the subject did not hold it open. This task forced the monkeys to perform three problem-solving actions.

monkey performed 100 trials in each task [see, Spinozzi & Truppa, 2002, for more details].

Table I shows the hand preference data for each action performed in the tasks as a function of the type of solution (2-act and 3-act solution) adopted by the monkeys to achieve the goal. As can be seen, when the task solution involved two sequential, one-handed acts, capuchins exhibited a consistent right-hand bias in reaching for food, the final action of the sequence, but no significant preference for the initial lid-lifting action, a finding consistent with that previously noted in the horizontal panel task. In contrast, when the task solution required both sequential and concurrent acts (i.e., 3-act solution) implying the complementary use of the hands for the final food-reaching action, no asymmetrical hand usage emerged for any of them, a result that confirmed the one previously obtained with the vertical panel task. Comparison of the strength in hand preference between 2-act and 3-act solutions revealed that reaching involving bimanual coordination elicited stronger lateralization compared with reaching involving the use of only one hand (mean *ABS-HI*: 2-act-solution = .53, 3-act solution = .69, $p < .05$), a finding comparable to that noted in the sliding panel tasks.

Taken together, these findings seem to support the view that, in coordinated bimanual tasks, but not in tasks involving one-handed actions, the number of acts necessary to manage the task may be one of the variables affecting the expression of hand preferences in capuchins. In these monkeys, complex tasks involving the complementary use of both hands for food retrieval elicit significant hand asymmetries only when the reaching act is not preceded by any other manipulative operation required for solving the task.

The hand preference patterns shown by capuchins in the multicomponent tasks differ from those reported for other nonhuman primate species when solving similar apparatus problems. With respect to the two sliding panel tasks, Fagot and Vauclair reported that, when presented with the horizontal panel apparatus, their eight gorillas [1988a] and six baboons [1988b] showed a significant left-hand preference in the action of moving the panel, but a symmetrical distribution of the right and the left hand for one-handed reaching action. In the vertical panel task, baboons exhibited also a significant left-hand preference for lifting the panel but a consistent right-hand bias for the bimanually coordinated reaching act. On the basis of these results, these authors suggested that the left-hand preference for sliding panels, which involves a heavy spatial load, could be viewed as the expression of an underlying right-hemisphere specialization comparable to that of humans [Fagot & Vauclair, 1991]. Our findings for capuchins thus differ substantially from those reported for gorillas and baboons. Unlike the hypothesis that activities with higher visuo-spatial requirements would elicit a greater left-hand preference, there was no group-level hand bias in our monkeys when they acted on the panel. It is possible that these measures may not be good indicators of hemispheric specialization in this New World monkey species. However, as reported earlier, Colell et al. [1995] noted that, in the two sliding panel tasks, chimpanzees showed a symmetrical distribution of left and right hand use for each component act.

A similar discrepancy in the hand preference patterns across species comes from studies on the box-opening tasks. For example, Fagot and Vauclair [1988a] noted that their gorillas showed a left-hand bias for the initial lid-lifting action, but no significant preference for taking the food out of the box. The lack of hand asymmetry in the final action was evident whether the animals used one hand or both hands in complementary roles for food retrieval. On the other hand, their six baboons [Fagot & Vauclair, 1988b] showed a symmetrical distribution in the left or right hand use for each component act in both the two box-opening tasks. Hauser et al. [1991] found a different hand preference pattern in free-ranging rhesus macaques observed while lifting the lid of their chow dispenser and extracting food. The majority of individuals consistently lifted the lid with their left hand, a finding comparable with that reported for gorillas. However, the monkeys preferred to use their right hand to hold the lid open and the left hand to select chow. Our finding of a significant right-hand preference for the final one-handed reaching action in capuchins thus differs substantially from those reported for gorillas and baboons. Moreover, the absence of a significant hand bias for reaching involving bimanual coordination does not agree with the consistent left-hand preference observed in macaques. Overall, these findings show that, for similar multicomponent tasks, nonhuman primate species exhibit different hand preference patterns.

Haptic Tasks. The visually guided task experiments revealed that manual activities involving complex sensorimotor coordination did not always induce consistent hand biases in capuchins. The conflicting results emphasized the necessity to examine carefully all variables linked to the specific characteristics of the tasks, such as the number of acts necessary to obtain a goal, which might account for the lateral biases obtained. Further investigation of the relation between hand use and task complexity was carried out by means of a study aimed to evaluate the role of haptic perception in the expression of manual laterality in capuchins.

Haptic discrimination is a complex task requiring dynamic spatial and motor processing, and fine sensorimotor control of the digits and hand in searching for the perceptual attributes and the identity of the objects [Lederman & Klatzky, 1987]. Studies involving humans have usually shown a left-hand performance advantage, as measured by accuracy or response times, for haptic exploration of novel stimuli at least in right-handed people, suggesting a relative superiority of the right cerebral hemisphere in spatial-form perception [Fagot et al., 1997; Verjat, 1988].

In our study [Spinozzi & Cacchiarelli, 2000], we assessed and compared capuchin hand preference in three tasks – haptic discrimination, visual–tactual reaching, and the visually guided reaching tasks, which differed in sensorimotor demands and availability of visual cues. In all tasks, the apparatus consisted of a split-level box with the upper-level jutting out 5 cm at the back, at a height of 10 cm from the base. Subjects could access the contents of the box by inserting one hand only into a square aperture located at the center of the jutting part at the back of the box. In the haptic discrimination task, the monkeys were required to discriminate haptically between two pumpkin seeds and two tinfoil items stuck into a tray fixed to the lower-level floor of the box. Thus, continuous displacement and spatial adjustments of the digits were needed for locating stimuli in the tray and for recognizing food successfully in the absence of visual cues. In the visual–tactual reaching task, the tray was baited with two pumpkin seeds and fixed on the box floor. The monkey could see the food items through the transparent walls of the apparatus before reaching inside, but it could rely only on tactual cues during reaching since the location of the food inside the box prevented direct visual feedback while operating. Also, this task entailed fine digit movements precisely to locate the seeds in the tray in the absence of direct vision. In the visually guided reaching task, the baited tray was placed 5 cm behind the square opening of the transparent box. This ensured that the monkeys could rely on visual cues while reaching for and grasping food. This situation was thus less demanding in terms of spatio-temporal characteristics of the movements necessary to obtain the food. A total of 26 capuchins served as subjects. The sample included 21 individuals used in the experiments reported above, plus 5 new individuals. Each monkey performed approximately 50 trials in each task. In each trial, only the initial reaching action (i.e., grasping and taking the item out the box) was scored.

Table I summarizes the hand preference data for each task (see *visual reach*, *haptic discrimination*, and *visual–tactual reach*). As can be seen, capuchins exhibited reliable group-level hand biases for all reaching measures, but the direction of laterality was influenced by the sensorimotor and perceptual requirements of the task. A significant group-level left-hand bias was found in both the haptic discrimination and the visual–tactual tasks, which involved similar finely tuned motor patterns for locating stimuli (in both the tasks) and discriminating between edible and non-edible foods (in the haptic task) in the absence of visual feedback. These data are consistent with those previously reported for several New and Old World monkeys, such as capuchins [Lacreuse & Fragaszy, 1996, 1999; Parr et al., 1997], spider monkeys [Laska, 1996] and rhesus monkeys [Fagot et al., 1991], supporting the hypothesis of a greater involvement of the right hemisphere in processing spatial information during haptic exploration in these nonhuman primate species. Conversely, a consistent right-hand bias was found in the less demanding task, the visual reaching task, which did not require any exploratory food retrieval skills. The latter result is consistent with that previously noted in the hole task, in which the visually guided reaching act also entailed a precise movement of the forelimb toward a small opening where the reward was placed. There were no differences in the strength of hand preference across tasks; the mean *ABS-HI* scores for the visual reaching task (.66) were similar to those observed for haptic discrimination (.68) and visual–tactual (.72) tasks. Thus, it is the direction and not the degree of lateral bias that was largely influenced by the manipulo-spatial requirements of the tasks and the availability of visual cues. In the haptic discrimination task, besides the interest in the direction of laterality, we also considered the relationships between hand preference and manual performance through an analysis of the capuchins' accuracy in recognizing the seeds [see Spinozzi & Cacchiarelli, 2000, for more details]. Significant performance differences were found between hands in recognizing food successfully: capuchins' retrieval responses were more accurate with the left hand than with the right hand, a finding comparable to that observed in humans. Because the monkeys as a group preferred the left hand for performing this task, the observed left-hand superiority in food discrimination seems to reflect a common underlying neural mechanism involved in haptic processing.

Overall, these findings do not fully support the current thinking on the relationships between handedness and task difficulty [Hopkins, 1999; Fagot & Vauclair, 1991], which predicts that species-level handedness ought to occur under more difficult task conditions. In our study, significant group-level hand biases were evident in all reaching tasks, despite differences in the spatial demands of the movements for food retrieval. Furthermore, our data do not support MacNeilage et al.'s [1987] proposal that single acts of visually guided prehension should induce left-hand preference in nonhuman primates, whereas more complex, manipulative tasks should elicit greater use of the right hand.

Consistency Across Different Reaching Measures. It has been suggested that, although nonhuman primates can show population-level laterality for some selected activities, this asymmetry is not comparable to the human handedness [McGrew & Marchant, 1997]. In reviewing findings on laterality of hand use in nonhuman primates and confronting different tasks employed in handedness research, McGrew and Marchant [1997] argued that the population hand bias observed in some species reflects task specialization rather than humanlike true handedness. In their model, task specialization refers to situations in which the majority of individuals within a sample use the same hand for only one task, whereas true handedness occurs when the predominant use of one hand is consistent across most subjects and most tasks. To address this issue, the consistency of individual capuchin hand preferences across different reaching actions involved in the various tasks (see Table I) was measured by correlating the *HI* scores among all possible pairs of actions. As reported above, our capuchin sample showed consistent hand preference biases at group level for seven out of eight unimanual reaching measures and for two out of four reaching actions involving bimanual coordination (see Figure 4).

With regard to the unimanual measures, a within-subject consistency emerged for reaching in the quadrupedal (Q), the bipedal (B), the hole (Ho), and the visual reach (Vis.) tasks [Q/B, $r(26) = .59$, $p < .01$; Q/Ho, $r(23) = .49$, $p < .05$; Q/Vis., $r(21) = .47$, $p < .05$; B/Ho, $r(23) = .50$, $p < .05$; B/Vis., $r(21) = .50$, $p < .05$; Ho/Vis., $r(21) = .79$, $p < .001$]. All these measures involved one-handed acts of visually guided prehension and none of them was the final action of a sequence. Moreover, a significant positive correlation emerged

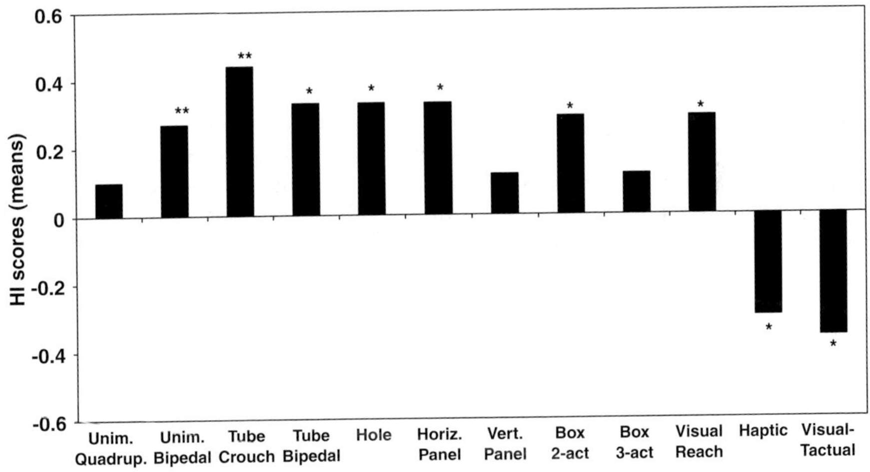

Figure 4. Mean *HI* scores for different reaching measures (see Table I and text for descriptions and tasks).

between the *HI* scores for reaching in the haptic and visual–tactual tasks [$r(26) = .61$, $p < .001$], which involved similar manipulative demands for locating food and the use of the sense of touch for guidance. Regarding the coordinated-bimanual reaching measures, a significant positive correlation was found between crouched and bipedal reaches in the tube tasks [$r(26) = .90$, $p < .001$]; both these actions, which entailed similar motor patterns for extracting food, were not preceded by any problem-solving acts. All the remaining correlations were not statistically significant.

Overall, these results seem to support the claim by McGrew and Marchant [1997] that evidence of handedness is virtually absent in nonhuman primates. Although our capuchin sample showed consistent hand biases for several tasks, only a few reaching measures were significantly and positively correlated with each other. The lack of a consistent individual hand use pattern across all measures thus should imply, on the basis of McGrew and Marchant's [1997] model, that capuchins' manual laterality is task-specific rather than being true handedness. However, the suggestion that only humans manifest "true handedness" – meaning the consistent use of the same hand across multiple measures – has no empirical support. There is in fact strong evidence that human hand use is not exclusive for any specific measure, but varies according to the nature of the task. For example, the ethological studies of Marchant et al. [1995] showed that, when a wide range of everyday behavioral patterns of hand use in traditional human societies was examined, the right-handedness appeared to be rather weak and did not show itself in all activities. The notable exception was the marked right hand preference shown by the individuals in tool-use activities, particularly those involving precision grip. Moreover, findings using principal components analysis have shown that human handedness is not a unidimensional trait but depends on the skill level of the task. For example, Healey et al. [1986] found that preferences for movements involving proximal (shoulder and arm) muscles differed from preferences for movements of the distal (hands and fingers) muscles. Likewise, Steenhuis and Bryden [1989] differentiated activities for which people report strong hand preference from those for which they report weak preference. From their analyses, these authors showed that certain questionnaire items load onto two hand preference factors that differed on the basis of the demands of motor skills involved in the tasks.

CONCLUSION

The results of our experiments show that group-level hand biases are evident in tufted capuchins in some reaching measures. The use of different tasks revealed the existence of certain factors that strongly affect either the direction or the degree of hand preference. Simple tasks involving single acts of picking up food in an open space, elicit weak forms of handedness: bias for the use of one hand seems to depend on the posture adopted by the animal. Conversely, tasks implying a precise movement of the forelimb toward a restricted space in which the reward is placed induce consistent group-level lateral biases, but the direction of hand preference is strongly affected by the spatio-temporal dimensions of the movement and the availability of the visual cues. Tasks involving the complementary use of both hands elicit a higher degree of laterality than those involving unimanual acts, but the emergence of a significant group-level hand bias for reaching is strongly related to the presence/absence of preliminary problem-solving acts. It seems clear from our findings that the level of task complexity is not sufficient to confer manual laterality in capuchins. The different hand preference patterns observed in our monkeys stress the necessity to consider carefully the requirements of the tasks in order to highlight the variables differently affecting manual asymmetries in nonhuman primates.

Although capuchins exhibit group-level manual asymmetries for different reaching actions, only a few measures are significantly and positively correlated with each other, with some degree of clustering depending on specific characteristics of the task. Current data from other nonhuman primate species have shown a similar dissociation among reaching measures requiring different hand use patterns [Colell et al., 1995; Hopkins & Pearson, 2000]. For example, Hopkins and Pearson [2000] examined the inter-task consistency in handedness across six reaching measures of hand use in a large sample of chimpanzees. They found that two measures involving bimanual coordination significantly correlated with each other, but neither of these measures correlated with the remaining unimanual measures. On the basis of these results, the authors hypothesized that, in nonhuman primates, handedness may not be a unidimensional phenomenon and that different measures of hand preference may involve different motor or neural systems, as has been proposed for humans [Healey et al., 1986; Steenhuis & Bryden, 1989]. A first indication supporting this hypothesis comes from a recent study by Hopkins and Cantalupo [2004] on chimpanzees. These authors correlated neuroanatomical asymmetries in three cortical regions – quantified from magnetic resonance imaging (MRI) – with three measures of handedness and found a significant relationship between hand preference for coordinated bimanual reaching and precentral gyrus morphology: specifically, right- and left-handed subjects differed with respect to asymmetry in this specific area of the brain. By contrast, handedness measures including simple reaching and feeding were not associated with neuranatomical asymmetries. A similar association between handedness for coordinated bimanual task and asymmetry of central sulcus depth has been recently reported by Phillips and Sherwood [2005] in a small group of male tufted capuchin monkeys.

Further research is needed in order to assess whether similar findings may be found in other nonhuman primate species. Early attempts to link hand preference to different brain regions in nonhuman primates have not produced reliable results [Ettlinger, 1988; Warren, 1980], leading some researchers to suggest that neurobiological substrates for hand preferences in humans are fundamentally different from those in other primates [Warren, 1980]. However, with the increasing use of non-invasive brain imaging techniques, further information on functional asymmetries in nonhuman primates can be acquired together with a better understanding of the neural mechanisms controlling the organization of motor patterns. This, in turn, should lead to new insight into the nature and evolutionary origins of handedness in primates, including humans.

ACKNOWLEDGMENT

The author thanks coauthors of the studies summarized in this chapter – Valentina Truppa, Tiziana Laganà, Maria Grazia Castorina, and Barbara Cacchiarelli – for their enthusiastic collaboration. Many of the studies conducted in the Unit of Cognitive Primatology, Institute of Cognitive Sciences and Technologies, CNR, were supported by the grant RBNE01SZB4 from the "Fondo per gli Investimenti della Ricerca di Base (F.I.R.B.), Ministero dell'Istruzione, dell'Università e dalla Ricerca (M.I.U.R.)".

REFERENCES

Anderson, J. R.; Degiorgio, C.; Lamarque, C.; Fagot, J. A multi-task assessment of hand lateralization in capuchin monkeys (*Cebus apella*). Primates 37: 97–103, 1996.

Annett, M. Left, Right, Hand and Brain: The Right Shift Theory. Hillsdale, NJ, Erlbaum, 1985.

Bisazza, A.; Rogers, L. J.; Vallortigara, G. The origins of cerebral asymmetry: A review of evidence of behavioral and brain lateralization in fishes, amphibians, and reptiles. Neuroscience and Behavioral Reviews 22: 411–426, 1998.

Bradshaw, J. L.; Rogers, L. The Evolution of Lateral Asymmetries, Language, Tool-Use and Intellect. San Diego, CA, Academic Press, 1993.

Brinkman, C.; Kuypers, H. G. J. M. Cerebral control of controlateral and ipsilateral arm, hand and finger movement in the split-brain rhesus monkey. Brain 96: 653–674, 1973.

Christel, M. I.; Fragaszy, D. M. Manual function in *Cebus apella*. Digital mobility, preshaping, and endurance in repetitive grasping. International Journal of Primatology 21: 697–719, 2000.

Colell, M.; Segarra, M. D.; Sabater-Pi, J. Manual laterality in chimpanzees (*Pan troglodytes*) in complex tasks. Journal of Comparative Psychology 109: 291–307, 1995.

Corballis, M. C. Human Laterality. New York, Academic Press, 1983.

Corballis, M. C. The Lopsided Ape. New York, Oxford University Press, 1991.

Corballis, M. C. The genetics and evolution of handedness. Psychological Review 104: 714–727, 1997.

Costello, M. B.; Fragaszy, D. M. Prehension in *Cebus* and *Saimiri*: I. Grip type and hand preference. American Journal of Primatology 15: 235–245, 1988.

Diamond, A. C.; McGrew, W. C. True handedness in the cotton-top tamarin (*Saguinus oedipus*)? Primates 35: 69–77, 1994.

Ettlinger, G. F. Hand preference, ability, and hemispheric specialization. How far are these factors related in the monkey? Cortex 24: 3389–3398, 1988.

Fagot, J.; Drea, C. M.; Wallen, K. Asymmetrical hand use in rhesus monkeys (*Macaca mulatta*) in tactually and visually regulated tasks. Journal of Comparative Psychology 105: 260–268, 1991.

Fagot, J.; Lacreuse, A.; Vauclair, J. Role of sensory and post-sensory factors on hemispheric asymmetries in tactual perception. Pp. 469–494 in S. Christman, ed. Cerebral Asymmetries in Sensory and Perceptual Processing. New York, Elsevier, 1997.

Fagot, J.; Vauclair, J. Handedness and manual specialization in the baboon. Neuropsychologia 26: 795–804, 1988a.

Fagot, J.; Vauclair, J. Handedness and bimanual coordination in the lowland gorilla. Brain, Behavior and Evolution 32: 89–95, 1988b.

Fagot, J.; Vauclair, J. Manual laterality in nonhuman primates: A distinction between handedness and manual specialization. Psychological Bulletin 109: 76–89, 1991.

Fragaszy, D. M.; Visalberghi, E.; Fedigan, L. The Complete Capuchin. The Biology of the Genus *Cebus*. Cambridge, University Press, 2004.

Hauser, M.; Perry, S.; Manson, J. H.; Ball, H.; Williams, M.; Pearson, E. It's all in the hands of the beholder: New data on free-ranging rhesus monkeys. Behavioral and Brain Sciences 42: 342–344, 1991.

Healey, J. M.; Liederman, J.; Geschwind, N. Handedness is not a unidimensional trait. Cortex 22: 33–53, 1986.

Hopkins, W. D. Posture and reaching in chimpanzees (*Pan troglodytes*) and orangutans (*Pongo pygmaeus*). Journal of Comparative Psychology 107: 162–168, 1993.

Hopkins, W. D. Hand preferences for a coordinated bimanual task in 110 chimpanzees (*Pan troglodytes*): Cross-sectional analysis. Journal of Comparative Psychology 109: 291–297, 1995.

Hopkins, W. D. Chimpanzee handedness revisited: 55 years since Finch (1941). Psychonomic Bullettin and Review 3: 449–457, 1996.

Hopkins, W. D. On the other hand: Statistical issues in the assessment and interpretation of hand preference data in nonhuman primates. International Journal of Primatology 20: 851–866, 1999.

Hopkins, W. D.; Bennett, A. J.; Bales, S. L.; Lee, J.; Ward, J. P. Behavioral laterality in captive bonobos (*Pan paniscus*). Journal of Comparative Psychology 107: 403–410, 1993.

Hopkins, W. D.; Cantalupo, C. Handedness in chimpanzees (*Pan troglodytes*) is associated with asymmetries of the primary motor cortex but not with homologous language areas. Behavioral Neuroscience 118: 1176–1183, 2004.

Hopkins, W. D.; Leavens, D. A. Hand use and gestural communication in chimpanzees (*Pan troglodytes*). Journal of Comparative Psychology 112: 95–97, 1998.

Hopkins, W. D.; Morris, R. D. Handedness in great apes. A review of findings. International Journal of Primatology 14: 1–25, 1993.

Hopkins, W. D.; Pearson, K. Chimpanzee (*Pan troglodytes*) handedness: Variability across multiple measures of hand use. Journal of Comparative Psychology 114: 126–135, 2000.

Hopkins, W. D.; Rabinowitz, D. M. Manual specialization and tool use in captive chimpanzees (*Pan troglodytes*): The effect of unimanual and bimanual strategies on hand preference. Laterality 2: 267–277, 1997.

Kimura, D. Neuromotor mechanisms in the evolution of human communication. Pp. 197–220 in H. D. Steklis; M. J. Raleigh, eds. Neurobiology of Social Communication in Primates: An Evolutionary Perspective. New York, Academic Press, 1979.

King, J. E.; Landau, V. I. Manual preference in varieties of reaching in squirrel monkeys. Pp. 107–124 in J. P. Ward; W. D. Hopkins, eds. Primate Laterality. Current Behavioral Evidence of Primate Asymmetries. New York, Springer-Verlag, 1993.

Lacreuse, A.; Fragaszy, D. N. Hand preferences for a haptic searching task by tufted capuchins (*Cebus apella*). International Journal of Primatology 17: 613–632, 1996.

Lacreuse, A.; Fragaszy, D. N. Left hand preferences in capuchins (*Cebus apella*). Role of spatial demands in manual activity. Laterality 4: 65–78, 1999.

Larson, C. F.; Dodson, D.; Ward, J. P. Hand preferences and whole-body turning biases of lesser bushbabies (*Galago senegalensis*). Brain, Behavior and Evolution 33: 261–267, 1989.

Laska, M. Manual laterality in spider monkeys (*Ateles geoffroyi*) solving visually and tactually guided reaching tasks. Cortex 32: 717–726, 1996.

Lederman, S.; Klatzky, R. Hand movements: A window into haptic object recognition. Cognitive Psychology 19: 342–368, 1987.

Lehman, R. A. W. Distribution and change in strength of hand preferences of cynomologus monkeys. Brain, Behavior and Evolution 17: 209–217, 1980.

MacNeilage, P. F.; Studdert-Kennedy, M. G.; Lindblom, B. Primate handedness reconsidered. Behavioral and Brain Science 10: 247–303, 1987.

Marchant, L. F.; McGrew, W. C.; Eibl-Eibesfeldt, I. Is human handedness universal? Ethological analysis from three traditional cultures. Ethology 101: 239–258, 1995.

Masataka, N. Population level asymmetry of hand preference in lemurs. Behavior 110: 244–247, 1989.

McGrew, W. C.; Marchant, L. F. On which side of the apes? Ethological study of laterality of hand use. Pp. 255–272 in W. C. McGrew; L. F. Marchant; T. Nishida, eds. Great Ape Societies. Cambridge, Cambridge University Press, 1996.

McGrew, W. C.; Marchant, L. F. On the other hand: Current issues in hand meta-analysis of the behavioral laterality of hand function in nonhuman primates. Yearbook of Physical Anthropology 40: 201–232, 1997.

Olson, D. A.; Ellis, J. E.; Nadler, R. D. Hand preferences in captive gorillas, orangutans and gibbons. American Journal of Primatology 20: 83–94, 1990.

Parr, L. A.; Hopkins, W. D.; deWaal, F. B. M. Haptic discrimination in capuchin monkeys (*Cebus apella*): Evidence of manual specialization. Neuropsychologia 35: 143–152, 1997.

Phillips, K. A.; Sherwood, C. C. Primary motor cortex asymmetry is correlated with handedness in capuchin monkeys (*Cebus apella*). Behavioral Neuroscience 119: 1701–1704, 2005.

Robins, A.; Rogers, L. J. Lateralized prey-catching responses in the cane toad, *Bufo marinus*: Analysis of complex visual stimuli. Animal Behaviour 68: 767–775, 2004.

Roney, L. S.; King, J. E. Postural effects on manual reaching laterality in squirrel monkeys (*Saimiri sciureus*) and cotton-top tamarins (*Sanguinus oedipus*). Journal of Comparative Psychology 107: 380–385, 1993.

Spinozzi, G.; Cacchiarelli, B. Manual laterality in haptic and visual reaching tasks by tufted capuchin monkeys (*Cebus apella*). An association between hand preference and hand accuracy for food discrimination. Neuropsychologia 38: 1685–1692, 2000.

Spinozzi, G.; Castorina, M. G.; Truppa, V. Hand preferences in unimanual and coordinated-bimanual tasks by tufted capuchin monkeys (*Cebus apella*). Journal of Comparative Psychology 112: 183–191, 1998.

Spinozzi, G.; Truppa, V. Hand preferences in different tasks by tufted capuchins (*Cebus apella*). International Journal of Primatology 20: 827–849, 1999.

Spinozzi, G.; Truppa, V. Problem-solving strategies and hand preferences for a multicomponent task by tufted capuchins (*Cebus apella*). International Journal of Primatology 23: 621–638, 2002.

Spinozzi, G.; Truppa, V.; Laganà, T. Grasping behavior in tufted capuchin monkeys (*Cebus apella*): Grip types and manual laterality for picking up a small food item. American Journal of Physical Anthropology 125: 30–41, 2004.

Steenhuis, R. E. Hand preference and performance in skilled and unskilled activities. Pp. 123–142 in D. Elliott; E. A. Roy, eds. Manual Asymmetries in Motor Performance. New York, CRC Press, 1996.

Steenhuis, R. E.; Bryden, M. P. Different dimensions of hand preference that relate to skilled and unskilled activities. Cortex 25: 289–304, 1989.

Vallortigara, G.; Rogers, L. J.; Bisazza, A. Possible evolutionary origins of cognitive brain lateralization. Brain Research Reviews 30: 164–175, 1999.

Verjat, I. La dissymmetrie fonctionnelle cerebrale dans la modalité tactilo-kinesthesique manuelle. L'Annéè Psychologique 88: 83–109, 1988.

Ward, J. P.; Milliken, G. W.; Stafford, D. K. Patterns of lateralized behavior in prosimians. Pp. 43–74 in J. P. Ward; W. D. Hopkins, eds. Primate Laterality: Current Behavioral Evidence of Primate Asymmetries. New York, Springer-Verlag, 1993.

Warren, J. M. Handedness and laterality in humans and other animals. Physiological Psychology 8: 351–359, 1980.

Westergaard, G. C.; Kuhn, H. E.; Lundquist, A. L.; Suomi, S. J. Posture and reaching in tufted capuchins (*Cebus apella*). Laterality 2: 65–74, 1997.

Westergaard, G. C.; Kuhn, H. E.; Suomi, S. J. Bipedal posture and hand preference in humans and other primates. Journal of Comparative Psychology 112: 55–64, 1998.

Westergaard, G. C.; Suomi, S. J. Hand preference for a bimanual task in tufted capuchins (*Cebus apella*) and rhesus macaques (*Macaca mulatta*). Journal of Comparative Psychology 110: 406–411, 1996.

Young, G.; Segalowitz, S. J.; Corter, C. M.; Trehub, S. E. Manual specialization and the developing brain: An overview. Pp. 3–12 in G. Young; S. Segalowitz; C. M. Corter; S. E. Trehub, eds. Manual Specialization and the Developing Brain. New York, Academic Press, 1983.

Moonstone, an adult male Garnett's bushbaby (*Otolemur garnettii*) surveys his home cage. (Photograph by David Hanbury.)

The Evolution of Hemispheric Specialization in Primates
W.D. Hopkins (Editor)

PROSIMIAN PRIMATES AS MODELS OF LATERALITY

Sheree L. Watson and David B. Hanbury
Department of Psychology, University of Southern Mississippi, Hattiesburg, Mississippi, USA

ABSTRACT

Prosimians are the most distantly related primates to humans and are unique subjects for studies of behavioral asymmetries. This chapter outlines the values of research from prosimians toward evolutionary models of hemispheric specialization in primates. We first review the evidence of functional asymmetries including handedness in prosimians. We next summarize the influence of posture and arousal on the strength and direction of hand preferences. In addition, the influence of age and sex on behavioral asymmetries and handedness are discussed in the context of lateralization being exhibited as a whole body phenomenon.

Keywords: individual differences, temperament, prosimians, posture and reaching

The decade from 1990 to 2000, by virtue of Presidential proclamation 1648, was designated the "Decade of the Brain". Almost as if by mandate, we witnessed an explosion in our understanding of neural structure, physiology, and function. One area, in particular, that enjoyed a surge of interest was the area of behavioral laterality. An unlikely catalyst in the surge of laterality research was the development of the prosimian model for the study of lateralized behavior. The investigations into prosimian laterality that began quietly in the late 1980s blossomed into a prolific undertaking that provided a foundation for examining the evolution of behavioral laterality from a novel and innovative vantage point.

Historically, behavioral laterality was considered a uniquely human phenomenon [MacNeilage et al., 1987]. For decades, researchers appeared not to notice the problem presented by denying the existence of evolutionary precursors of laterality during roughly 60 million years of prehominid and hominid development [MacNeilage et al., 1987]. Findings of reliable and consistent handedness in nonhuman primates, however, encouraged renewed inquiry into the evolutionary nature of hemispheric lateralization. Similar reliable findings of lateralized behavior in several prosimian species [e.g., Ward et al., 1993] secured acceptance of the idea that lateralized behavior is certainly a primate

ISSN: 1936-8526 DOI: 10.1016/S1936-8526(07)05009-9

character, if not a mammalian one. In this chapter, we will review the literature examining the existence of behavioral laterality on a nonvocal motor level in prosimian species, as well as the roles of several factors that may influence the expression of laterality. We will draw heavily on the work of Ward and colleagues conducted immediately prior to and during the "Decade of the Brain."

WHY STUDY PROSIMIANS?

Although the comparison of species offers a compelling approach to the understanding of adaptive development among related genera, most comparative studies either examine the adaptive strategies of a single species or seek to develop a nonhuman model of a human condition. Using the comparative approach to make direct comparisons of species with shared phylogenetic history, though, offers a unique opportunity to explore the similarities and differences that arise through adaptation of related species to diverse ecological demands [Ward, 1995]. Ward [1995] reminds us that the taxa best suited for comparative study are those that "contain a large number and variety of species and thereby offer many dimensions along which comparisons may be made" (p 293). The prosimian group offers just such diversity.

The prosimian primates are the most primitive group of species in the order Primates. The prosimians are comprised of three groups, lemurs, lorises, and tarsiers, which include 24 genera and 50 species [Martin, 1990]. There is great diversity within and among the prosimian groups in morphology, dietary habit, postural habit, ecological niche, and associated behavioral characters. They range in size from the tiny mouse lemur (*Microcebus murinus*) weighing about 109 g to the quite large indri (*Indri indri*) at 7,500 g [Rowe, 1996]. Circadian patterns may be nocturnal, diurnal, or cathermal [i.e., lacking clear circadian patterns; Tattersall, 1987] and social habits may range from solitary foragers to permanent group cohesiveness [Kappeler & van Schaik, 2002]. Prosimian ecological niches may range from mostly arboreal to mostly terrestrial and their diets may be frugivorous, folivorous, insectivorous, or omnivorous. The ecological feeding diversity is accompanied by diversity in locomotion modes, postural orientations, and manners of locomotion. This great diversity, especially in postural orientation and locomotion, allows for examination of many characters that may converge in the evolution of laterality. Also importantly, extant prosimian forms are believed to be similar in both appearance and habit to the ancestral prosimians and are, therefore, sometimes referred to as the "foundation stock" of the Primate order. It is this link to the evolutionary history of the modern primate that allows for inferences about the evolution of laterality in primates.

The prosimians occupy a unique taxonomic position in the primate phylogeny. If individual prosimians reliably show lateralized behavior, then we may infer that lateralization is a characteristic of the Primate order (and of the common primate ancestor). Finally, prosimians are neurologically and behaviorally less complex than anthropoids. This simplicity of organization in brain structure and function and use of hands allows for examination of the basic premises of hemispheric lateralization. As will be discussed below, the study of behavioral laterality in prosimian species has provided the basis for new, testable hypotheses in the evolution and adaptive significance of lateralized behavior.

WHY STUDY BEHAVIORAL LATERALITY?

Biological organisms are constructed such that congruence of structure and function is apparent. In other words, the anatomical structure of most organisms both supports and invites the associated function [Ward & Hopkins, 1993]. This design of nature gives rise to expectancies of both asymmetry and symmetry. The asymmetry of the dorsal–ventral plane and the anterior–posterior plane are reflected in structure and function. Following Ward and Hopkins [1993], we note that our fronts both look and perform differently from our backs, and our bottoms both look and perform differently from our tops. However, the distinction is less obvious in the lateral plane. Across the midline, there is a striking symmetry of structure and an apparent symmetry of function. On behavioral measures, though, the symmetry of function breaks down. The bilateral asymmetry of behavioral function has been the basis for an enduring and fascinating biological mystery. The essential role of behavioral adaptation in species survival adds to the allure of the laterality riddle. Indeed, even morphological modifications provide no adaptive advantage if the organism does not utilize them behaviorally to derive benefit from the environment. Thus, understanding the origins of laterality would seem inextricably linked to the study of behavior and its underlying motor systems [Ward & Cantalupo, 1997].

Accounts of human handedness, the most apparent manifestation of behavioral laterality, date back to Biblical times. Such references were often tinged with moral judgment and held the suggestion that lateral manifestation was a reflection of the individual's inherent goodness or evil. Broca's (circa 1861) discovery of specialization of cerebral hemispheres paved the way for more sophisticated theories of behavioral laterality, which peaked with the study of Sperry's "split brain" patients, to the fascination of the scientific and lay community alike [Sperry, 1968].

The study of behavioral laterality has been remiss, though, in its efforts to trace the evolutionary basis of the phenomenon. As mentioned above, behavioral laterality has generally been considered a uniquely human phenomenon, and has been considered, along with the capacity for speech, one of the characters that set humans apart from their nonhuman primate cousins. In the past two decades, this view has been called into question with the publication of reliable evidence of not only motoric laterality in prosimian and simian species [see Bradshaw & Rogers, 1993; Ward & Hopkins, 1993, for reviews], but also of the existence of cerebral functional specialization in several primate species [e.g., Hamilton & Vermeire, 1991; Morris & Hopkins, 1993; Ward & Hopkins, 1993).

THE ROLE OF POSTURE

The studies that initially demonstrated behavioral laterality in the prosimian consisted primarily of simple reaching tasks (see Table I). The first report of lateralized reaching behavior in a prosimian species is that of Subramoniam [1957], which reported a left hand bias for food reaching in the slender loris. Masataka [1989] found that 20 out of 22 lemurs (*Lemur catta*) exhibited a left hand preference for four simple behaviors (pick up food, push food into mouth, reach, and grasp), whereas the other 2 lemurs demonstrated a right hand preference. Similarly, of 33 black lemurs (*Lemur macaco*) 32 exhibited a statistically significant lateral bias for reaching for food; of those, 20 were left hand preferents [Forsythe & Ward, 1988].

Table I. Prosimian species have been widely demonstrated to exhibit lateral hand bias[a].

Study	Species	*N*	Bias
Sanford and Ward [1986]	*Galago senegalensis*	8	7L 1A
Forsythe and Ward [1988]	*Lemur macaco*	33	20L, 12R, 1A
Masataka [1989]	*Lemur catta*	22	20L, 2R
Ward et al. [1990]	*L. catta*	42	21L, 14R, 7A
	L. coronatus	11	6L, 4R, 1A
	L. macaco	29	19L, 7R, 3A
	L. mongoz	26	9L, 12R, 5A
	L. rubriventer	5	1L, 4R
	L. fulvus albifrons	15	6L, 5R, 4A
	L. f. collaris	24	15L, 5R, 4A
	L. f. fulvus	11	3L, 5R, 3A
	L. f. rufus	14	8L, 3R, 3A
	L. f. sanfordi	17	3L, 6R, 8A
Milliken et al. [1991]	*Otolemur garnettii*	23	7L, 7R, 9A[b]
Stafford et al. [1993]	*Hapalemur griseus*	13	8L, 5R
Ward and Cantalupo [1997]	*O. garnettii*	27	17L, 9R, 1A
Ward [1998]	*Galago moholi*	8	4L, 4R
Rigamonti et al. [2005]	*Indri indri*	16	5L, 2R, 9A[b]

Note: L = left hand preferent, R = right hand preferent, A = ambi-preferent.
[a]Based on measures of unilateral reaching.
[b]Preferences based on composite of multiple measures.

It became clear early on, however, that simple reaching interacts with posture to influence lateralized reaching. A sample of bushbabies (*Galago senegalensis*) exhibited a population-level left hand bias when reaching for mealworms if the animals reached from a bipedal posture but not from a quadrupedal posture [Sanford et al., 1984]. Forsythe et al. [1988] examined reaching behavior in five male black and white ruffed lemurs (*Varecia variegata variegata*) housed on a moated island at a zoological park. The study compared hand preference for reaching in three situations: free foraging, discrete food presentation on land, and discrete food presentation in the water. Although hand preference was not detected in the free foraging situation, all animals demonstrated a clear left hand preference in the two discrete food presentation conditions. The changes in degree of handedness paralleled changes in postural stability. In the free foraging condition, the food was scattered about such that the lemurs retrieved it at their convenience. Most of these food retrievals were made from a sitting or reclining position. In contrast, the discrete food presentations on land required the lemur to approach the food in a manner that involved whole body postural changes. That is, the lemur had to position its body such that it could reach the food. In this condition, lateral hand bias increased from an average of 56% for the overall sample in the free foraging condition to over 96% in the discrete food presentation (on land) condition. Finally, the discrete food presentation on the water required the greatest postural adjustment, in that the lemurs had to balance themselves with three limbs while reaching for the food with the fourth in a way that prevented falling into the moat. In this condition of extreme postural adjustment, use of preferred hand reached almost 100% (i.e., 99.8%).

More recently, Milliken et al. [2005] demonstrated the relationship between posture and laterality in the sifaka during arboreal feeding. In a study of postural stability and hand use, Milliken et al. [2005] videotaped three species of sifaka (*Propithecus verreauxi coquereli*, $n = 11$; *Propithecus verreauxi verreauxi* form: *majori*, $n = 2$; *Propithecus tattersalli*, $n = 3$) as they fed on leaves while perched on elevated branches. Sifakas characteristically assume a vertical posture while feeding in an arboreal environment. One hand is used to stabilize the animal on its perch and the other hand is used in concert with the mouth to feed on the leaves. Milliken et al. [2005] point out that the complexity of this context (i.e., a milieu of continuous access to food sources and perch configurations) would not be conducive to selection pressures for a lateral bias for postural stability and feeding. Accordingly, their results revealed modest evidence of lateralized hand preference. Nevertheless, the results indicated a tendency for dissociation between hand use for feeding and postural support. Nine of 16 sifakas exhibited significant consistent hand use for postural support and 7 of 16 showed a significant hand preference for both feeding and postural support. Moreover, five of those with hand preference for both feeding and postural support used the left hand for feeding and the right for maintaining stability [Milliken et al., 2005; MacNeilage, 1991]. Thus, Milliken et al. [2005] point out that the findings may reflect a "basic foundation from which stronger manual specializations could evolve" (p 436).

Rigamonti et al. [2005] examined arboreal foraging among indris (*Indri indri*) in their natural habitat. The feeding behavior was similar to that examined by Milliken et al. [2005] and the study yielded similar results. Specifically, Rigamonti et al. [2005] found only modest lateralization, in that 4 of 16 subjects showed a significant preference for feeding behavior (i.e., "branch-reach") and postural support (i.e., "higher support"). However, three of the four were lateralized in the same direction as the sifakas observed by Milliken et al. [2005] [i.e., left hand use for feeding and right hand use for postural support; MacNeilage, 1991].

Manipulation of prey visibility, angle of reaching, and postural elevation were further examined in an elegant study of food reaching in lesser bushbabies (*G. senegalensis*) [Larson et al., 1989]. Of the factors manipulated, only postural change (i.e., from a quadrupedal to bipedal stance) influenced laterality. The relevance of posture in lateralized behaviors of the bushbaby may not be surprising given the many circumstances in which this species adopts a bipedal posture. For example, bushbabies adopt an upright posture during the arboreal leaping and clinging among vertical supports [Napier & Napier, 1967], their terrestrial locomotion generally involves saltatory bipedalism [Doyle, 1974], and erect postures is assumed in "threat stance" [Andrew, 1964]. Perhaps most importantly for interpretation of Sanford et al. [1984] data, erect posture is adopted during the capture of flying insects, which comprise the primary diet of free ranging bushbabies [Doyle, 1974]. The role of posture thus became a central focus in the understanding of laterality in prosimians. Indeed, the influence of posture in the expression of laterality is so profound that many of Ward and colleagues' results have been consistent with and interpreted in light of MacNeilage's [1991] Postural Origins theory. The Postural Origins theory [MacNeilage, 1991] proposed that ancestral primates evolved a left hand (i.e., right hemisphere) specialization for visually guided movement to facilitate the ballistic grabbing characteristic of predation in arboreal insectivores. Such an adaptation would optimally be accompanied by a complementary right hand (i.e., left hemisphere) specialization for postural stabilization. Moreover, such an adaptation would have its origins in the early prosimians if the assumption is correct that it is in this group that the selective pressure for unimanual predation first arose [MacNeilage, 1991]. Relative to the importance of the prosimian model for understanding laterality in terms of the Postural Origins theory, MacNeilage [1991] quotes Eisenberg [1981, p 161], who stated that prosimians "... represent a conservative primate pattern carried forward into the present relatively unchanged." However, postural influences on handedness may be susceptible to acute environmental pressure. For example, although 16 out of 16 *Galago moholi* and 7 out of 8 *M. murinus* were lateralized in one study of food reaching, the degree or consistency of lateralized behavior was not affected by postural adjustment [Dodson et al., 1992]. In that study, however, the condition requiring a bipedal stance was altered somewhat by using a small platform to accommodate the small stature of the *G. moholi* and *M. murinus*, essentially overcompensating such that the postural adjustment did not have the degree of effect on laterality one might expect. This methodological accommodation accompanied by the already strong bias of the animal may have obscured the postural effect [Dodson et al., 1992]. However, the most consistently, though, the most consistently lateralized of the two species (*G. moholi*) is also the species that most consistently maintained an upright posture [Dodson et al., 1992].

THE ROLE OF CENTRAL NERVOUS SYSTEM AROUSAL

The effects of posture in laterality may relate to neural mechanisms that mediate arousal. By its very nature, bipedal posture involves a relative instability due to an elevated center of gravity. Sherrington [1906] suggested that the nervous system has an essentially integrative function. Many of its subsystems, in particular, those that support posture, locomotion, and movement patterns, are globally related. Bipedalism, as a species characteristic, results in a situation of chronic instability for the organism. This chronic instability may be a condition promoting increased strength of lateralization in action systems [Ward et al., 1993].

The amplification of lateral bias with postural instability may be accomplished through a general arousal system. Drawing on the work of Talbott [1979] and Ito and Sanada [1965], Ward and her colleagues make a compelling case for central nervous system (CNS) arousal as a mediating and a mitigating factor in the expression of lateralized behavior. Integration of overall postural mechanisms may be brought about by a system of visceral proprioception [Talbott, 1979]. Righting reflexes in monkeys respond to the gravitational pull of changes in body orientation through visceral afferents that project to several CNS locations, including primary and secondary somatosensory areas, as well as neocortex [Ito & Sanada, 1965]. Thus, Larson et al. [1989] applied Talbott's [1979] neuraxial model to postural effects on lateralization. Changing from a stable (i.e., quadrupedal) posture to an unstable (i.e., bipedal) one increases activity in the somatosensory system and "... may also act to increase the activation and integration of movement mechanisms in general since any movement of a body part would act to jeopardize the postural stability" [Larson et al., 1989, p 266].

The neuraxial system may relate other arousal effects to the strength of the lateral bias. For example, Ward noted that most of the prosimians were tested when moderately hungry and found that 80% had a lateral hand bias. Hunger may itself act as an arousing factor and may, as well, initiate a species-typical feeding system that reflects the ballistic "smash and grab" prey capture technique of the bushbaby [Doyle, 1974].

The black and white lemurs tested at the Memphis zoo [Forsythe et al., 1988] were tested according to the customary feeding schedule; thus, food presentation occurred at a time of day when the animals would have been relatively hungry. Although the presence of zoo patrons may have had an arousing effect, the lemurs were relatively relaxed during the free foraging phase and they demonstrated a correspondingly lower level of lateral bias. The condition of discrete food presentation on the water was likely the most arousing of the three conditions. In this condition, the consequences of failure to maintain stability would have been dire as the animals had to retrieve the food items without falling into the moat. This condition elicited almost 100% lateralized reaching.

ROLE OF AGE AND SEX

Handedness increases in strength from infancy to adulthood in humans and anthropoids [Stafford et al., 1990]. Similar ontogenetic increases have been documented in prosimians. Three studies of prosimian species, with large and varied subject populations, found either a putative shift in handedness or an increase in strength of handedness with increasing age. Among 33 black lemurs that ranged from age 1 to 18 years [Forsythe & Ward, 1988], left hand use in food retrieval was prevalent among the younger lemurs but the older lemurs tended to use the right hand more often. Forsythe and Ward [1988] speculated that maturational factors may be related to a shift in preferred hand for food retrieval. A second study of 194 individuals of five lemur species revealed a similar age and hand preference relation. Within-species comparisons of five lemur species and six subspecies revealed greater strength of lateral bias among the older animals [Ward et al., 1990]. Finally, similar age effects were detected in a study of *Otolemur garnettii* [Milliken et al., 1991a]. Not only were adult *O. garnettii* more likely than juveniles to use their hand rather than mouth when given the choice, they were also more consistent in use of preferred hand than juveniles.

The effects of age may also be related to a corresponding influence of sex. Male and female lemurs show different patterns of lateralized behavior and the influence of age appears to be more pronounced among females. Specifically, male lemurs tend to be left hand preferent across all ages. However, there is a small subset of males that are not left-preferent, and among these males there appears to be a role of age in stabilization of lateralized behavior. That is, with increasing age, the number of ambi-preferent males decreases and right-preferent males increases. Juvenile females, though, tend to be equally distributed across the three preference groups (i.e., left, right, ambi). As the females age the proportion of right-preferent females increases until, by adulthood, the majority of females are right-preferent. A study of *O. garnettii* showed a similar relationship between female sex and right hand bias, in that seven of nine females were strongly right hand preferent [Ward, 1991].

Milliken et al. [2005] found similar sex and age differences among sifaka. Males more consistently displayed left hand feeding biases and right hand biases for postural stability than did females. Moreover, two of the three juveniles were ambi-preferent [Milliken et al., 2005], mirroring Mason et al.'s [1995] finding that four of the four juvenile sifaka showed no hand preference for a food positioning task. Milliken et al. [2005] pointed out that sexual dimorphism in body size and, especially weight distribution, may contribute to differences in laterality between male and female sifakas. Excess weight in the thoracic region could render the male "top heavy" and thereby increase postural instability while feeding from an arboreal perch [Milliken et al., 2005].

What might account for these age and sex effects? Because the studies reviewed here are cross-sectional, the developmental sequences of laterality are unclear. Nevertheless, the findings are consistent with existing theories of hemispheric development, leading Ward et al. [1993] to offer two possible explanations. Geschwind and Galaburda [1985] proposed that during fetal development, the right hemisphere develops more rapidly than the left, resulting in a bias for using the left hand preferentially. Because the left hemisphere develops more slowly, it may be subject to an extended period of vulnerability to developmental insult. The increase with age in right hand preferent lemurs appears to be at the expense of their ambi-preferent counterparts. Possibly then, the right-preferent and ambi-preferent groups are essentially one group whose hand preference changes over time. The change may reflect the continuing development of the left hemisphere, indicating a progression from a period when both hemispheres are equally developed (ambi-preferent) to a later stage where the left hemisphere continues to develop until it attains functional dominance (i.e., lemur becomes right-preferent). Thus, there may actually be two lemur subpopulations: (1) an established left hand preferent group with stable right hemisphere dominance and (2) a group that is initially ambi-preferent but becomes right hand preferent with age and continued left hemisphere development [Ward et al., 1990].

A second possibility is that the development of the hemispheres reflects a physiological effect based in a functional specialization of the left hemisphere. Social competition, including competition in feeding, may activate such hemispheric development. Consistent with this latter possibility, Tucker and Williamson [1984] proposed a model of hemispheric specialization in which the left hemisphere is responsible for activation as it may be "integral to motor operations, supporting postural readiness, and motivationally directed action" (p 198). Many lemur species are matriarchal [Jolly, 1966]. In such a social system, adult females dominate feeding situations, while juveniles and some young adult males are tolerated but somewhat peripheralized. As fully adult males and dominant adult females become more assertive in feeding situations, there may be correspondingly greater activation of the left hemisphere and therefore the right hand. Geschwind and Galaburda [1985] proposed that testosterone is an endogenous factor that effects hemispheric development and results in comparably greater final level of development of the right hemisphere. The proposed role of arousal is consistent with this proposition, as the individuals engaged in feeding competition would likely be in a state of high arousal [Ward et al., 1990].

ROLE OF THE MOUTH

Lemur juveniles were significantly more likely than adults to pick up food items with the mouth instead of the hand [Ward et al., 1990]. Intrigued by relationships between mouth and hand, Ward and colleagues explored the role of the mouth in both the ontogenetic and phylogenetic development of lateral bias. Ontogenetically, there is a transition in food procurement from exclusively mouth-initiated to exclusively hand-initiated that is specific to primates. The transition may reflect developmental maturation of the neuromuscular control of the hand. Consistent with this possibility, Ward et al. [1993] also noted that infant lemurs displayed interest in their own hands in situations that were not related to feeding or food procurement. They noted that seven infant lemurs of five species spent long periods of time looking at and "mouthing" their own hands. This behavior is reminiscent of the behavior of human infants while "discovering their hands" at a few months of age. Exploration of hands among lemur infants is especially intriguing in view of the observation that the mother's infant-directed grooming tends to center on the infants' faces, heads, and bodies, though rarely, if ever, the hands.

Juvenile *O. garnettii* were also more likely to retrieve food with the mouth than were adult conspecifics [Milliken et al., 1991a]. However, the *O. garnettii*, as a species, were more mouth-oriented than lemur species. In fact, 17 of the 23 *O. garnettii* displayed such a marked mouth preference that it interfered with testing. The subjects had to be retested using a receptacle that was covered with the exception of a small opening ("port") that required the animal to place its hand into the port to obtain a food item. The relative tendency to use the mouth instead of the hand for food procurement may, then, reflect species differences. For example, *G. moholi* are strongly lateralized and show very little inclination to use the mouth for food retrieval. In contrast, *M. murinus* were so likely to prefer the mouth that a special apparatus had to be constructed to promote hand use in food retrieval tests. When the hands were utilized, however, *M. murinus* was much less strongly lateralized than *G. moholi* [Dodson et al., 1992].

Milliken et al.'s [2005] finding of only modest handedness in sifakas support the notion that the role of the mouth is a critical factor in the development of lateral hand bias. Sifakas are strongly mouth oriented and this characteristic seems to have attenuated strength of lateral hand bias, especially for feeding. So substantial is the role of the mouth by sifakas in food retrieval that Mason et al. [1995] examined hand preference by recording the hand with which the sifakas repositioned a food item that had initially been retrieved by mouth.

Mason et al. [1995] found that 10 of 11 sifakas displayed consistent hand bias on this task. Notably the food retrieval was from a terrestrial, rather than arboreal, substrate, a methodological difference between the studies of Mason et al. [1995] and Milliken et al. [2005] that may explain differences in number of animals that were found to have significant hand preference.

Laterality in the aye-aye (*Daubentonia madagascariensis*) was examined by Feistner et al. [1994]. They examined hand use in the performance of four behaviors: reach (for food item), hold (food item), tap (branch to determine whether grubs are present), and digit feed (eating food off the specialized digit). On three of these measures (hold, tap, and digit feed), modest laterality was observed. Of 11 aye-ayes, 6 (3 Right, 3 Left) were significantly lateralized on the "hold" measure, 5 (2 R, 3 L) on the "tap" measure, and 5 (2 R, 3 L) on the "digit feed" measure. However, on the measure "reach" too few data points were observed to perform meaningful analyses because the aye-aye almost always retrieves food items with the mouth rather than the hand. These results, too, are consistent with the notion of attenuated hand use bias in species that rely heavily on the mouth for food retrieval. However, it is noteworthy that Milliken [1995] reported a right hand preference among aye-ayes on tasks requiring high levels of skill and precision such as grub extraction [Milliken, 1995; Milliken et al., 1991b].

What species characteristic could explain this difference? The differences seem related to species-typical postural habit. *G. moholi* is more characteristically bipedal than *M. murinus*. Ward et al. [1993] hypothesized to the extent that a species typically consumes food with an upright (and, therefore, less stable) posture, use of the hand will be favored over use of the mouth for food procurement. If this is true, and posture is related to lateral bias, then there ought also to be an inverse relationship between mouth use and strength of lateral hand bias. Ward [1995] tested this hypothesis by comparing propensity for mouth use and strength of lateral bias for three species: Dwarf lemur (*Cheirogaleus medius*), Garnett's bushbaby (*O. garnettii*), and ringtail lemur (*L. catta*). Mouth use was inversely proportional to both strength of hand bias and characteristic species-typical postural habit (i.e., quadrupedal vs. bipedal). Thus, Ward's results imply that the relationship between hand, mouth, and lateral bias has its roots in phylogeny as well as ontogeny.

Not only does the role of mouth appear related to development of lateral bias, there may be lateralization of the mouth itself. The mouth is an interesting organ, in that it is a unitary organ, which lies in the lateral plane on the midline, yet is not bifurcated like the nose. Videotaped records of the gentle lemur (*Hapalemur griseus*) revealed a significant correlation between side of mouth and preferred hand for feeding the bamboo shoot into the mouth [Stafford et al., 1993]. *L. catta* displayed a similar "mouthedness" in favored side of mouth for biting off pieces of monkey chow [Ward et al., 1993]. In an another study of 21 *L. catta*, Bennett et al. [1995] found that all had a side of mouth preference when eating fruit and all except three had a side of mouth preference for both fruit and monkey chow. Although the lemurs in this study were preselected for strength of lateral hand bias, Bennett et al. [1995] did not find consistency between preferred hand and preferred side of mouth, suggesting that "mouthedness" may be a distinct phenomenon. The meaning and significance of this mouthedness in the larger scheme of behavioral laterality is unknown. Given that the mouth is used earlier and extensively in ontogeny, it may be that lateral bias of the mouth leads to lateral hand bias. It is equally plausible that side of mouth bias arises from lateral hand bias. Finally, the two may develop in unison. In any case, mouthedness may provide important clues in understanding the evolution of laterality. Mouthedness, such as observed above, would result in uneven tooth wear. Such asymmetry of tooth wear in the fossil record of ancestral primates would provide evidence that laterality is a conserved character from early primate forms [Covert & Kay, 1981; Ward et al., 1993].

OTHER EXPRESSIONS OF LATERALIZED BEHAVIOR

Most of the literature reviewed here has examined lateralized behavior related to food reaching. Indeed, the bulk of the research has examined this variable and it has been a useful vehicle for understanding many of the complexities of laterality. Other expressions of laterality, though, have demonstrated the ubiquity of the laterality phenomenon. In particular, studies of eye dominance and whole body turning have contributed to our understanding of laterality.

Eye Dominance

Brain lateralization has also been manifested in perceptual processing systems [e.g., Morris & Hopkins, 1993]. In macaque species, hemispheric differences exist with right hemisphere dominance in face discrimination [Hamilton & Vermeire, 1988] and production of facial expressions [Hauser, 1993], whereas there is left hemisphere dominance for distinguishing line orientation [Hamilton & Vermeire, 1988] and processing of species-specific vocalizations [Petersen et al., 1978]. Because the contralateral optic fibers have larger diameter and faster conduction than the ipsilateral optic fibers, eye dominance may indicate lateralization in the processing of visual information. *O. garnettii* appear also to exhibit stimulus-specific lateralization in processing of visual information [Rogers et al., 1994]. Four adult females and two infants (one male, one female) were allowed to process visual information by looking monocularly through a grid with five vertical slits in it. A minimum of 100 bouts of peering through the grid were recorded for the adults and a minimum of 50 were recorded for the infants. The visual stimuli were a food item (mealworm) and the experimenter. Three of the females had given birth within the past month and for these their own infant also served as a visual stimulus. One female was also presented with a toy monkey and a rubber snake to assess the response to novel stimuli.

The results indicated that the bushbabies did, indeed, exhibit lateralized eye preference and, moreover, the preferred eye may be stimulus-specific. Although the grid was designed such that the bushbabies' head could have been tilted to allow binocular viewing through the grid openings, only one bushbaby (an infant) did so, and then on only 3% of the trials. Thus, monocular viewing appears to provide sufficient visual input for processing of relevant information.

All bushbabies exhibited significant left eye preference when viewing the experimenter or the food item. The bias was present in both adults and infants, and was present in the bushbaby that did not have an infant as well as those that did have infants. Eye preference was not correlated with handedness either for food reaching or food holding. However, there was a clear relationship between type of stimulus and eye preference. Characteristic viewing of the experimenter or food item could be distinguished from viewing the toy monkey, the rubber snake, and the infant bushbabies. Although only one subject viewed the snake, she did so for longer durations (compared to viewing the food item or experimenter) and preferentially used the right eye. The same bushbaby shifted from left to right eye preference when viewing the toy monkey. Similarly, when viewing their infants, each of the mothers' eye preference shifted toward greater use of the right eye. Rogers et al. [1994] proposed that the differences resulted from the extent to which the stimuli were differentially arousing to the animals. That the various stimuli were differentially arousing to the subjects is supported by the observation that viewing the presumed arousing (and right eye preferent) stimuli was accompanied by other indicators of arousal, like growling and posturing, which did not accompany viewing the presumed nonarousing (and left eye preferent) stimuli. These results may serve to extend the hypothesis that behavioral laterality derives from the arousal of postural instability and suggest that CNS arousal may support hemispheric specialization in a comprehensive manner.

Whole Body Turning

Whole body turning may be the most ubiquitous and basic of measures of behavioral laterality, having been documented in a variety of species, including fish [Bisazza et al., 1997], toads [Bisazza et al., 1996], lizards [Deckel, 1995], chicks [Rogers, 1995], rats [Denenberg, 1981], and parrots [Harris, 1989]. Ward and her colleagues examined the direction and strength of whole body turning in several prosimian species.

In a study examining the effects of various conditions (posture, visibility, angle of reaching) on hand preference, Larson et al. [1989] also had the opportunity to assess the presence of turning bias in bushbabies. Specifically, each of the 480 reaching trials was a discrete trial, which concluded with the bushbaby returning the start box. This required that the bushbaby make a 180° turn. Nine out of 10 bushbabies (*G. senegalensis*) had a significant turning bias. Of those, eight displayed a left turning bias.

With turning again defined as execution of a turn of 180° about the long axis of the body, Dodson et al. [1992] found that 100% (16 of 16) of bushbabies (*G. moholi*) tested displayed a significant turning bias as did 75% (6 of 8) of mouse lemurs (*M. murinus*). Of the bushbabies, 15 had a left turning bias, whereas of the mouse lemurs, all 6 of the animals exhibiting a turning bias exhibited a left turn bias. Notably, the bushbabies, which are the more consistently bipedal of the two species, were more strongly lateralized on both turning measures and measures of lateralized food reaching.

The left rotational bias in whole body turning was further supported by additional data from Ward's laboratory demonstrating that 14 out of 17 bushbabies made whole body turns to the left on an average of 88.4% of the trials (range 62–99%) with a sample of 240 trials per subject [Ward, 1991]. Moreover, a comparable rotational measure was used to further examine turning in a sample of eight mouse lemurs and the average left turn bias was 58.7% (range 48.4–73.5%) on an average of 466 trials (range 225–1,271). Again, the more bipedal of the two species (the bushbaby) was more strongly lateralized than the less bipedal (mouse lemur).

Interaction of Systems

Because multiple motor systems must coordinate effectively to facilitate even basic behaviors [Ward & Cantalupo, 1997], it is implausible that the lateralized systems described here (i.e., hand preference, eye preference, and rotational bias) operate independently. Nevertheless, early studies that have employed multiple measures of laterality have not found consistent patterns of interaction or association among systems [Dodson et al., 1992; Larson et al., 1989; Stafford et al., 1993]. In two studies of multiple lateralized systems, Ward and Cantalupo uncovered information that may illuminate a portion of the laterality mystery. In two studies employing a total of 47 *O. garnettii*, Ward and Cantalupo tested the animals for reaching preference, eye dominance, and whole body turning [Ward & Cantalupo, 1997; Cantalupo & Ward, 2000]. The most salient emergent finding, evident in both studies, was that the bushbabies could be separated into two distinct groups that exhibited clearly identifiable, but different, patterns of lateralized behavior. One group could be labeled STABLE ([Cantalupo & Ward, 2000]; in Ward and Cantalupo [1997] this group was labeled SHIFTER). The other group was labeled UNSTABLE (or NONSHIFTER). As implied by the labels, animals in the STABLE group maintained use of the preferred hand (whether they be right- or left-preferent) in both bipedal and quadrupedal postures. In the first of these studies [$N = 27$; Ward & Cantalupo, 1997], it was apparent that the weakly lateralized SHIFTERS were more vulnerable than their NONSHIFTER counterparts to the demands of the task (i.e., bipedal vs. quadrupedal reaching). In quadrupedal stance, SHIFTERS demonstrated positive correlation between hand and eye use but in bipedal stance the correlation was between hand and rotational bias.

In addition, SHIFTERS engaged in more bimanual reaching. Task demands did not influence strength of lateralized reaching among NONSHIFTERS.

The second study [$N = 20$; Cantalupo & Ward, 2000] further examined the interaction of hand, eye, and rotational bias. Across all animals, there was increased use of the right eye when reaching from a bipedal stance. This result supports the proposal of Rogers et al. [1994] that eye dominance shifts toward use of the right eye as the situation becomes increasingly arousing. Also consistent with prior studies [see Ward et al., 1993], subjects were more strongly lateralized (regardless of direction) in the bipedal stance and females were more strongly lateralized than males. And again, the STABLE and UNSTABLE groups emerged. In this study though, the pattern of interactions was especially interesting. Clear and significant correlations emerged indicating that the STABLE group had a significant positive correlation between preferred hand and preferred eye. In this group of 11 bushbabies, 8 were female. In contrast, the UNSTABLE animals ($n = 9$) had a significant negative correlation between preferred hand and turning bias. In this group, males outnumbered females (7 males, 2 females). The results of these two studies taken together affirm that lateralized systems interact at multiple levels and in a variety of poorly understood ways. Nevertheless, it is clear that, for some individuals, there is close coordination between the motor system involved in reaching and the visual system that guides reaching. In other individuals, there appears to be greater interdependence between the motor systems involved in reaching and postural control [Cantalupo & Ward, 2000]. Whether the individual's sex influences these interrelationships is a corresponding question of great interest, given the prior research, indicating a role for sex in both direction and strength of lateralized behavior [see Ward et al., 1993; Ward, 1995].

CONCLUSION

Throughout the "Decade of the Brain", Ward and her colleagues brought to the study of laterality, science in its finest form. With far reaching questions and elegant methodologies, Ward and others established the importance of the prosimian model in investigating the breadth and complexity of hemispheric lateralization. This work both supported and complemented the simultaneous study of laterality in anthropoid primates being conducted by such notable scholars as Hopkins [see Ward & Hopkins, 1993], King [1995], MacNeilage [1987], and others [for a more comprehensive review, see Ward & Hopkins, 1993]. The potential for future studies examining such factors as the roles of ecological niche, maturational factors, and interactions between lateralized systems, including lateralization of vocal behaviors [e.g., Hook-Costigan & Rogers, 1998] will undoubtedly inspire behavioral scientists into the new millennium [Drea, 1996]. As shown in Table II, lateralized hand preference may be influenced by such factors as type of task relative to species-typical foraging behaviors and nature of the test environment. Systematic study of these

Table II. Hand preference by species and task for selected studies. The type of task may influence strength and expression of lateral hand bias.

Authors	Species	Task and outcome	Test environment
Milliken et al. [2005]	*Propithecus* species	Postural support 2L, 7R, 7A Leaf eating 5L, 2R, 9A	Arboreal feeding in naturalistic environment[a]
Mason et al. [1995]	*P. verreauxi coquereli*	Reposition food item following mouth retrieval 6L, 4R, 5A	Captive; food retrieval from terrestrial substrate
Feistner et al. [1994]	*Daubentonia madagascariensis*	Holding food item 3L, 3R, 5A Tap 3L, 2R, 6A Digit-feed 3L, 2R, 6A	Naturalistic environment[a]
Rigamonti et al. [2005]	*Indri indri*	Arboreal leaf feeding 5L, 2R, 9A Postural support 0L, 4R, 12A	Natural habitat
Forsythe and Ward [1988]	*Lemur macaco*	Reaching for food item on cage floor or in metal pan 20L, 12R, 1A	Captive

Forsythe et al. [1988]	*L. macaco*	Free foraging for food item on ground 1L, 0R, 4A Discrete food presentation on ground 5L, 0R, 0A Discrete food presentation on water (moat) surface 5L, 0R, 0A	Naturalistic zoo environment
Sanford et al. [1984]	*Galago senegalensis*	Reaching from bipedal stance 14L, 5R, 6A	Captive
Larson et al. [1989]	*G. senegalensis*	Whole body turning 8L, 1R, 1A	Captive
Masataka [1989]	*Eulemur catta*	Reach, grasp, pick up food item 20L, 2R, 0A	Captive

[a]Naturalistic environment refers to captive environments that have been environmentally enriched to stimulate the animal's species-typical behavioral and foraging activities.

variables may unravel much of the remaining mystery of lateralized behavior. Moreover, vocal laterality has not yet been investigated in prosimian species. Study of vocal lateralization may shed light on the interactions of motoric behaviors and communication strategies.

Many of the early studies conducted consisted of simple food reaching tasks in captive settings [e.g., Sanford et al., 1984; Larson et al., 1989]. More recently, prosiman species have been tested either in their natural environment [e.g., Mason et al., 1995] or in a naturalistic setting [e.g., Milliken et al., 2005]. Moreover, studies have begun to incorporate multiple tasks that allow examination of the integrated functioning of multiple lateralized systems [Chapman, 2005]. Given the likelihood of selective pressures acting on these systems holistically rather than independently, studies using multiple measures may further our understanding of the ways that lateralized motor systems promote adaptation in the organism's ecological niche. The groundbreaking work of Cantalupo and Ward [Cantalupo & Ward, 2000; Cantalupo et al., 2002] exploring the interactions of lateralized systems speaks to complexity of the laterality puzzle and the need for further investigation. Given the findings of ambi-preference among juveniles of several species on a variety of measures [e.g., Milliken et al., 2005; Feistner et al., 1994; Milliken et al., 1991a, b], there is a dire need for longitudinal studies of behavioral laterality. Longitudinal studies, especially those employing multiple measures, may be instructive in illuminating the relations between neural maturation, motor maturation, and integration of systems.

Ward and colleagues changed our understanding of laterality in nonhuman primates. Once considered a uniquely human character, Ward's work has solidified the notion that laterality is certainly a primate character, if not a more evolutionarily ancient one. We are reminded that we, as human primates, have a long evolutionary history that has molded and shaped our behavioral characters. Rather than conceptualizing laterality exclusively as a population-level characteristic, we now recognize that lateralized behavior occurs at the level of the individual. The roles of sex, posture, arousal, and age add to the intricacy and remind us of the many subtle threads of selective pressure that are woven into the tapestry of behavioral laterality. Ward's work has brought attention to the great complexity in relations among multiple lateralized systems. Indeed, the work reviewed here has addressed the centuries old enigma of human handedness and extended the allure of the mystery to a new generation of research on its nonhuman correlates, its associated variables, and its evolutionary history.

REFERENCES

Andrew, R. J. The displays of the primates. Pp. 227–309 in P. Buettner-Janush, ed. Evolutionary and Genetic Biology of Primates. New York, Academic Press, Vol. 2, 1964.

Bennett, A. J.; Ward, J. P.; Milliken, G. W.; Stafford, D. K. Analysis of lateralized components of feeding behavior in the ring-tailed lemur (*Lemur catta*). Journal of Comparative Psychology 109: 27–33, 1995.

Bisazza, A.; Cantalupo, C.; Robins, A.; Rogers, L. J.; Vallortigara, G. Right-pawedness in toads. Nature 379: 408, 1996.

Bisazza, A.; Cantalupo, C.; Vallortigara, G. Lateral asymmetries during escape behavior in a species of teleost fish (*Jenynsia lineata*). Physiology and Behavior 61: 31–35, 1997.

Bradshaw, J. L.; Rogers, L. J. The Evolution of Lateral Asymmetries, Language, Tool Use, and Intellect. San Diego, CA, Academic Press, 1993.

Broca, P. Remarks on the seat of the faculty of articulate language, followed by an observation of aphemia. Pp. 49–72 in G. von Bonin, ed. The Cerebral Cortex. Springfield, IL, C.C. Thomas, 1861.

Cantalupo, C.; Ward, J. P. Interaction between lateralized systems: Exploring the complexity of lateralized behavior. Brain and Cognition 43: 73–78, 2000.

Cantalupo, C.; Ward, J. P.; Franceschetti, D. R. Lateralized reaching as a dynamic symmetry breaking in the bushbaby (*Otolemur garnettii*): Preliminary evidence. Brain and Cognition 48: 297–304, 2002.

Chapman, K. M. Multiple measures of handedness and laterality in three species of *Lemur*: *Lemur catta, Eulemur mongoz and Eulemur fulvus rufus.* A Poster Session Presented at The 2005 Conference on Comparative Cognition, Melbourne Beach, Florida, 2005.

Covert, H. H.; Kay, R. F. Dental microwear and diet: Implications for determining the feeding behaviors of extinct primates, with a comment on the dietary pattern of *Sivapithecus*. American Journal of Physical Anthropology 7: 1–38, 1981.

Deckel, A. W. Laterality of aggressive responses in *Anolis*. Journal of Experimental Zoology 272: 194–200, 1995.

Denenberg, V. H. Neonatal postural asymmetry and sex differences in the rat. Brain Research 254: 417–419, 1981.

Dodson, D. L.; Stafford, D. K.; Forsythe, C.; Seltzer, C. P.; Ward, J. P. Laterality in quadrupedal and bipedal prosimians: Reach and whole-body turn in the mouse lemur (*Microcebus murinus*) and the galago (*Galago moholi*). American Journal of Primatology 26: 191–202, 1992.

Doyle, G. A. Behavior of prosimians. Pp. 155–353 in A. Schrier; F. Stollnitz, eds. Behavior of Nonhuman Primates. New York, Academic Press, Vol. 5, 1974.

Drea, C. M. Primate handedness: An emerging discipline? Contemporary Psychology 41: 681–683, 1996.

Eisenberg, J. G. The Mammalian Radiations: An Analysis of Trends in Evolution, Adaptation, and Behavior. Chicago, University of Chicago Press, 1981.

Feistner, A. T. C.; Price, E. C.; Milliken, G. W. Hand preference patterns for tapping, digit feeding, and food holding in captive aye-ayes. Folia Primatologica 62: 136–141, 1994.

Forsythe, C.; Milliken, G. W.; Stafford, D. K.; Ward, J. P. Posturally related variations in the hand preferences of the ruffed lemur (*Varecia variegata variegata*). Journal of Comparative Psychology 102: 248–250, 1988.

Forsythe, C.; Ward, J. P. Black lemur (*Lemur macaco*) hand preference in food reaching. Primates 29: 369–374, 1988.

Geschwind, N.; Galaburda, A. M. Cerebral lateralization: Biological mechanisms, associations, and pathology: I–III. A hypothesis and a program for research. Archives of Neurology 42: 428–459, 1985.

Hamilton, C. R.; Vermeire, B. A. Complimentary hemispheric specialization in monkeys. Science 242: 1691–1694, 1988.

Hamilton, C. R.; Vermeire, B. A. Functional lateralization in monkeys. Pp. 19–34 in F. L. Kitterle, ed. Cerebral Laterality: Theory and Research. Hillsdale, NJ, Lawrence Erlbaum, 1991.

Harris, L. J. Footedness in parrots: Three centuries of research, theory, and mere surmise. Canadian Journal of Psychology 43: 369–396, 1989.

Hauser, M. Right hemisphere dominance for the production of facial expressions in monkeys. Science 261: 475–477, 1993.

Hook-Costigan, M. A.; Rogers, L. J. Lateralised use of the mouth in production of vocalization by marmosets. Neuropsychologia 36: 1265–1273, 1998.

Ito, T.; Sanada, Y. Location of receptors for righting reflexes acting upon the body in primates. Japanese Journal of Physiology 15: 235–242, 1965.

Jolly, A. Lemur Behavior. Chicago, Chicago University Press, 1966.

Kappeler, P. M.; van Schaik, C. P. Evolution of primate social systems. International Journal of Primatology 23: 707–740, 2002.

King, J. E. Laterality in hand preferences and reaching accuracy of cotton-top tamarins (*Saguinus oedipus*). Journal of Comparative Psychology 109: 34–41, 1995.

Larson, C. F.; Dodson, D.; Ward, J. P. Hand preferences and whole-body turning biases of lesser bushbabies (*Galago senegalensis*). Brain, Behavior and Evolution 33: 261–267, 1989.

MacNeilage, P. The evolution of hemispheric specialization for manual function and language. Pp. 285–309 in S. P. Wise, ed. Higher Brain Functions: Recent Explorations of the Brain's Emergent Properties. Oxford, England, Wiley, 1987.

MacNeilage, P. The "Postural Origins" theory of primate neurobiological asymmetries. Pp. 165–188 in N. Krasnegor; D. Rumbaugh; R. Schiefelbusch; M. Studdert-Kennedy, eds. Biological and Behavioral Determinants of Language Development. Hillsdale, NJ, Lawrence Erlbaum, 1991.

MacNeilage, P.; Studdert-Kennedy, M.; Lindblom, B. Primate handedness reconsidered. Behavioral and Brain Sciences 10: 247–303, 1987.

Martin, R. D. Primate Origins and Evolution: A Phylogenetic Reconstruction. Princeton, NJ, Princeton University Press, 1990.

Masataka, N. Population level asymmetry of hand preference in lemurs. Behaviour 110: 244–247, 1989.

Mason, A. M.; Wolfe, L. D.; Johnson, J. C. Hand preferences in the sifaka (*Propithecus verreauxi coquereli*) during feeding in captivity. Primates 36: 275–280, 1995.

Milliken, G. W. Right hand preference and performance biases in the foraging behavior of the aye-aye. Pp. 261–292 in L. Alterman; G. A. Doyle; M. K. Izard, eds. Creatures of the Dark: The Nocturnal Prosimians. New York, Plenum Press, 1995.

Milliken, G. W.; Kraiter, K. S.; Ferra, G.; Ross, C. L. Reach and posture hand preferences during arboreal feeding in sifakas: A test of the postural origins theory of behavioral lateralization. Journal of Comparative Psychology 119: 430–439, 2005.

Milliken, G. W.; Stafford, D. K.; Dodson, D. L.; Pinger, C. D.; Ward, J. P. Analyses of feeding lateralization in the small-eared bushbaby (*Otolemur garnettii*): A comparison with the ring-tailed lemur (*Lemur catta*). Journal of Comparative Psychology 105: 274–285, 1991a.

Milliken, GW.; Ward, J. P.; Erickson, C. J. Independent digit control in foraging by the Aye-Aye (*Daubentonia madagascariensis*). Folia Primatologica 56: 219–224, 1991b.

Morris, R. D.; Hopkins, W. D. Perception of human chimeric faces by chimpanzees (*Pan troglodytes*): Evidence for a right hemisphere asymmetry. Brain and Cognition 21: 111–122, 1993.

Napier, J. R.; Napier, P. H. A Handbook of Living Primates. New York, Academic Press, 1967.

Petersen, M.; Beecher, B.; Zoloth, S.; Moody, D.; Stebbins, W. Neural lateralization of species-specific vocalizations by Japanese macaques (*Macaca fuscata*). Science 202: 324–327, 1978.

Rigamonti, M. M.; Spiezio, C.; Poli, M. D.; Fazio, F. Laterality of manual function in foraging and positional behavior in wild indri (*Indri indri*). American Journal of Primatology 65: 27–38, 2005.

Rogers, L. J. The evolution and development of brain asymmetry, and its relevance to language, tool use and consciousness. International Journal of Comparative Psychology 8: 1–15, 1995.

Rogers, L. J.; Ward, J. P.; Stafford, D. K. Eye dominance in the small-eared bushbaby, *Otolemur garnettii*. Neurosychologia 32: 257–264, 1994.

Rowe, N. The Pictorial Guide to the Living Primates. New York, Pogonias Press, 1996.

Sanford, C.; Guin, K.; Ward, J. P. Posture and laterality in the bushbaby (*Galago senegalensis*). Brain, Behavior and Evolution 25: 217–224, 1984.

Sanford, C. G.; Ward, J. P. Mirror image discrimination and hand preference in the bushbaby (*Galago senegalensis*). Psychological Record 36: 439–449, 1986.

Sherrington, C. S. The Integrative Action of the Nervous System. New Haven, Yale University Press, 1906.

Sperry, R. W. Hemispheric disconnection and unity in conscious awareness. American Psychologist 23: 723–733, 1968.

Stafford, D. K.; Milliken, G. W.; Ward, J. P. Lateral bias in feeding and brachiation in *Hylobates*. Primates 31: 407–414, 1990.

Stafford, D.; Milliken, G.; Ward, J. Patterns of hand and mouth lateral biases in bamboo leaf shoot feeding and simple food reaching in the Gentle Lemur (*Hapalemur griseus*). American Journal of Primatology 29: 195–207, 1993.

Subramoniam, S. Some observations on the habits of the slender loris, *Loris tardigradus L.* Journal of the Bombay Natural History Society 54: 386–398, 1957.

Talbott, R. E. Ferrier, the synergy concept, and the study of posture and movement. Pp. 1–12 in R. E. Talbott; D. R. Humphrey, eds. Posture and Movement. New York, Raven Press, 1979.

Tattersall, I. Cathemeral activity in primates: A definition. Folia Primatologica 49: 200–202, 1987.

Tucker, D. M.; Williamson, P. A. Asymmetric neural control systems in human self-regulation. Psychological Review 91: 185–215, 1984.

Ward, J. P. Prosimians as animal models in the study of neural lateralization. Pp. 1–17 in F. Kitterle, ed. Cerebral Laterality: Theory and Research. Hilldale, NJ, Lawrence Erlbaum, 1991.

Ward, J. P. Laterality in African and Malagasy prosimians. Pp. 293–309 in L. Alterman; G. Doyle; M. K. Izard, eds. Creatures of the Dark: The Nocturnal Prosimians. New York, Plenum Press, 1995.

Ward, J. P. Left hand advantage for prey capture in the galago (*Galago moholi*). International Journal of Primatology 11: 173–184, 1998.

Ward, J. P.; Cantalupo, C. Origins and functions of laterality: Interactions of motoric systems. Laterality 2: 279–303, 1997.

Ward, J. P.; Hopkins, W. D. (eds). Primate Laterality: Current Behavioral Evidence of Primate Asymmetries. New York, Springer-Verlag, 1993.

Ward, J. P.; Milliken, G.; Dodson, D.; Stafford, D.; Wallace, M. Handedness as a function of sex and age in a large population of *Lemur*. Journal of Comparative Psychology 104: 167–173, 1990.

Ward, J. P.; Milliken, G.; Stafford, D. K. Patterns of lateralized behavior in prosimians. Pp. 43–74 in J. P. Ward; W. D. Hopkins, eds. Primate Laterality: Current Evidence of Primate Asymmetries. New York, Springer-Verlag, 1993.

Group of rhesus monkeys on Morgan Island, South Carolina.

The Evolution of Hemispheric Specialization in Primates
W.D. Hopkins (Editor)

THE BIOLOGICAL CORRELATES OF HAND PREFERENCE IN RHESUS MACAQUES

Sue Howell, J. Dee Higley and Gregory C. Westergaard
The Mannheimer Foundation, Inc., Clewiston, FL, USA, (SH); Laboratory of Clinical Studies, National Institute on Alcohol Abuse and Alcoholism, USA (JDH); Alpha Genesis, Inc., Yemassee, SC, USA (GCW)

ABSTRACT

The rhesus monkey is one of the most widely studied primates in behavioral and biomedical research. Our laboratory has been interested in comparative differences in hand preference among related macaque species as well as biological correlates of individual differences in handedness in rhesus monkeys. This chapter summarizes work in our laboratory that has correlated the association between specific neurotransmitters, stress, and stress hormones with handedness. In general, significant associations are found among these variables but the causal link between these factors remains to be tested.

Keywords: handedness, rhesus monkeys, immunology, stress, simple reaching

INTRODUCTION

This chapter details the association between lateralized behavior patterns and biological characteristics for nonhuman primates and presents current evidence for the physiological correlates of lateralized behavior. We focus on rhesus macaques (*Macaca mulatta*) as they are frequently used to model human biobehavioral processes [Coe, 1993; Kalin et al., 1998; Lubach et al., 1996]. Over the past 10 years, we examined relationships among handedness, plasma cortisol levels, cerebrospinal fluid (CSF) monoamine metabolite concentrations, and testosterone, and considered the relationships between handedness, physiological variables, and multiple indices of sociality. Early research was conducted at the neonatal nursery at the National Institutes of Health Animal Center and later research was conducted among captive rhesus macaques at Yemassee Primate Center and free-ranging rhesus macaques at Morgan Island, South Carolina. Our goal has been to test the hypothesis that individual differences in handedness are associated with biological functions and particular behavioral profiles.

ISSN: 1936-8526 DOI: 10.1016/S1936-8526(07)05010-5

Lateralized behavior and its corresponding morphological asymmetries have been noted in numerous phylogenetically diverse species [Bradshaw & Rogers, 1993; MacNeilage et al., 1987; also see Warren, 1980]. Approximately 90% of the Western human population is right-handed and the corresponding left-hemisphere specialization for manual control is thought to have contributed to the evolution of several species-typical characteristics including language and other higher-order cognitive processes. Because of the presumed importance of hemispheric specialization to human evolution and development, a great deal of research has been directed toward understanding factors associated with lateral asymmetries and handedness. Human research has included genetic models [Corballis, 2000; Corballis et al., 1996; Harada et al., 2004], environment models [Tang & Verstynen, 2000], and combination models that suggest a gene culture/environment interaction for handedness determination [Laland et al., 1995]. That is, variation in handedness may be influenced by genetics but also by developmental factors that may include birth order [Annett & Ockwell, 1980; Tan & Nettleton, 1980]; prenatal stress or pregnancy complications [Bailey & McKeever, 2004; Schwartz, 1988]; and birth stress [Bakan et al., 1973; Coren, 1995; Laland et al., 1995; Levander et al., 1989; McManus, 1981; Tan & Nettleton, 1980; van Strien et al., 1987]. In particular, theorists including Bakan et al. [1973] and Coren [1995] view right-handedness as a universal human characteristic, due to a polygenic influence, and that left-handedness is influenced by pregnancy and birth-stress factors.

The predominance of right-handedness among humans is shared with nonhuman primates including chimpanzees [Hopkins et al., 2004, 2005a, b; Hopkins & Russell, 2004] and adult rhesus monkeys [Westergaard & Lussier, 1999]. As a result, nonhuman primates are often used as a model for improved understanding of the evolution of handedness and morphological asymmetries [Hopkins et al., 2000; Hopkins & Rilling, 2000]. A great deal of research has been conducted to consider the evolutionary importance of handedness to bimanual tasks, gestures, throwing, and grip strength [Hopkins et al., 2002, 2003, 2005a, b, c; Hopkins & Wesley, 2002; Lonsdorf & Hopkins, 2005; Vauclair et al., 2005; Westergaard et al., 1998a, b; Westergaard & Suomi, 1993, 1994, 1996a, b; Westergaard, 1991, 1993]. Recent magnetic resonance imaging (MRI) studies have also improved our understanding of brain lateralization [human: Foundas et al., 1998; Pujol et al., 1999; nonhuman primates: Cantalupo et al., 2003; Halpern et al., 2005; Hopkins & Pilcher, 2001; Freeman et al., 2004; Hopkins & Cantalupo, 2004; Pilcher et al., 2001]. A third productive area of research on handedness focuses on the proximate factors that may influence handedness including maternal age, parity, reproductive outcome, infant nipple preference, and maternal cradling [Jaffe et al., 2006]. Of particular relevance to our research has been the association between stress and lateral asymmetries.

A review of the literature revealed that pregnant female rats placed in two hours of daily restraint one week prior to parturition produce offspring that are biased toward right-sided turning in a T-maze compared to offspring of nonhandled female controls [Alonso et al., 1991]. Merely handling pregnant female rats produces sexually dimorphic brain asymmetries among offspring; the right cortex is significantly thicker than the left among males of restrained females with an opposite pattern for nonhandled rats. This study is supported by nonhuman primate research that suggests high cortisol levels are associated with high right-frontal electroencephalographic activity [Kalin et al., 1998] and the strength of lateral bias [Westergaard et al., 2001] and are consistent with human research that demonstrates a preferential role for the right hemisphere in emotion-induced cortisol release [Henry, 1993; Wittling & Pfluger, 1990].

Cortisol is released in the body during agitated states and helps regulate blood pressure and cardiovascular function. When cortisol is released, muscle proteins break down and lead to amino acid release into the bloodstream. Amino acids are then used by the liver to synthesize glucose (i.e., gluconeogenesis). This results in increased blood sugar levels for increased energy to combat the potential stressor. Cortisol release is controlled by a complex feedback system. The pituitary gland, a small gland at the base of the brain, makes and secretes adrenocorticotrophic hormone (ACTH). ACTH signals the adrenals to increase cortisol production and secretion. In turn, the pituitary receives signals from the hypothalamus via corticotrophin releasing hormone (CRH) that signals the pituitary to release ACTH. A stressful event results in the release of ACTH and CRH and right-frontal electroencephalographic activity [Kalin et al., 1998]. We hypothesize that increased stress and the concomitant increase in plasma cortisol may be associated with hand preference development.

Over the last 10 years, our research team has used the rhesus macaque as a nonhuman primate model to investigate the relationship between CSF 5-hydroxyindole acetic acid (5-HIAA) and impulsivity in order to better understand the developmental psychobiology of these pathological conditions [Higley et al., 1992, 1996a, b, c; Mehlman et al., 1994, 1995, 1997; Westergaard et al., 1999]. Prospective research shows that free-ranging rhesus males with low CSF 5-HIAA concentrations exhibit several deleterious behavior patterns analogous to those seen in their human counterparts.

Considerable evidence has also emerged that indicates a relationship between lateral asymmetries and psychopathology in the human population as numerous empirical and theoretical reports have suggested an increased incidence of left- and mixed-handedness among alcoholics, schizophrenics, and their first-degree relatives [Lesch et al., 1990; Orr et al., 1999; Sperling et al., 2000]. Although anomalies in the laterality of neurological and cognitive dimensions associated with psychopathology have been documented, it is unclear what role these anomalies play in the genesis and early development of major psychiatric disorders. Animal models have revealed a great deal concerning processes that underlie normative functional and structural asymmetry in humans and, in recent years, animal models have been used to investigate factors associated with developmental learning disabilities including dyslexia and other language disorders [Cowell et al., 1999; Galaburda, 1994]. The animal literature suggests that, as in humans, both normal and disrupted laterality patterns in nonhuman species have their basis in early development [Cowell et al., 1999]. Data from nonhuman primates have provided evidence consistent with the view that negative effect is associated with reactivity and hemispheric specialization in developing animals, as research with young rhesus monkeys indicates that fearful temperament is correlated with circulating cortisol levels and with high right-frontal hemispheric activity [Kalin et al., 1998]. Whether the relationship with hemispheric specialization, hormonal stress functioning, and disturbed effect is developmentally continuous through adolescence and into adulthood remains to be determined.

In our recent research, we also considered the association between physiology, hand preference, and psychopathology. In keeping with human studies [e.g., Cowen, 2002], we hypothesized that elevated cortisol levels may lead to lower brain serotonin (5-HIAA) function and this in turn may lead to abnormal behavioral states including increased aggression and behavioral reactivity [Heinz et al., 2002]. This is consistent with the finding that human patients with major psychopathologies exhibited decreased brain serotonin (5-hydroxytryptamine, 5-HT) function and elevated cortisol secretion [van Praag, 1996], and with rodent studies that suggest chronic activation of the hypothalamic–pituitary–adrenal (HPA) axis results in attenuated serotonin responsiveness [van Riel, 2003].

The HPA axis includes parts of the hypothalamus, the anterior lobe of the pituitary gland, the adrenal cortices, hormones, systems that transport hormones and feedback mechanisms that transport cortisol from adrenal glands back to the hypothalamus and to other parts of the brain. When we experience a stressful event, the level of glucocorticoids rises via receptors in the hippocampus, activating the hippocampus to secrete CRH. CRH, in turn, causes the pituitary gland to release ACTH into the bloodstream from which it enters the adrenal glands and causes them to secrete cortisol. This process creates a negative feedback loop in which excess cortisol activates the brain's glucocorticoid receptors and suppresses the production of CRH. In psychopathology (e.g., depression), this loop no longer works properly and results in excess production of cortisol. Many seriously depressed patients have high plasma cortisol levels brought on by chronic stress. Monoamine neurotransmitters, especially dopamine, serotonin, and norepinephrine (noradrenaline) [van Praag, 1996] may play a prominent role in regulation of the HPA axis. As such, we sought to consider their role in handedness for rhesus macaques.

In the studies that follow, we examine the relationship between hand preference, cortisol, and CSF monoamine metabolite levels, and multiple indices of sociality in rhesus macaques. We hypothesize that increased stress and the concomitant increase in plasma cortisol may be associated with hand preference development. Moreover, serotonergic functioning may be associated with individual differences in hand preference and manifested by a positive correlation between CSF 5-HIAA (the monoamine metabolite of serotonin) and the frequency of right- or left-hand preference. We also sought to determine whether hand preference is associated with a behavioral profile that places individuals at increased risk for violent trauma and premature death as has been shown for male rhesus macaques with low CSF 5-HIAA concentrations maintained in the present free-ranging population [Higley et al., 1996b]. We reasoned that if left-handed macaques were more socially isolated and subject to more frequent and violent attacks from conspecifics relative to their right-hand counterparts, then we might be able to identify a biobehavioral mechanism to decrease prevalence of left-hand preference that has been noted in the aged rhesus population [Westergaard & Lussier, 1999].

PLASMA CORTISOL AND HANDEDNESS IN INFANT MONKEYS

Our first study considered the association between plasma cortisol levels and handedness among infant rhesus macaques [Westergaard et al., 2001]. We tested the hypothesis that differences in hormone functioning were correlated with hemispheric specialization and manifested by a positive correlation between cortisol and the frequency of right- versus left-hand use. We used a within-subjects design and sampled cortisol from subjects at 1, 3, and 5 months of age. We recorded hand preference when these same animals were aged 4–11 months, a period theoretically equivalent in maturation to 1- to 4-year old human children. In humans, this period of development represents a transition in manual action, reflected in increased fine motor manipulation in skilled tasks and emergence of consistent adult-like handedness [Harris & Carlson, 1993]. In this initial study, we evaluated the relationship between plasma cortisol levels and handedness in infant rhesus macaques in order to extend our knowledge of cerebral development and manual laterality.

Subjects included 18 healthy infant rhesus macaques (9 males; 9 females) housed in the neonatal nursery at the National Institutes of Health Animal Center. Infants were cared for by human caregivers, received regular behavioral assessments, and were socialized with peers. Blood samples were collected at 1, 3, and 5 months of age and were intended to measure response to a moderately challenging condition. At month 1 of age, blood was collected following a 30-min assessment of orientation, temperament, and motor maturity that involved handling and restraint (holding) by a human caregiver [Schneider & Suomi, 1992]. At 3 and 5 months of age, samples were collected following a 30-min period of separation from social partners and housing in an isolated single cage. Hazelton-Washington Laboratories, Vienna, VA, performed radioimmunoassay for plasma cortisol.

Hand preference was scored during a bimanual task. We placed a plastic tube that contained raisin paste on the cage floor and recorded the hand (right or left) that was used to extract the raisin paste from the tube. Subjects held the tube in one hand and extracted the food with the preferred hand. A trial occurred when an animal inserted its finger into an opening of the tube and removed the food. We recorded a minimum of 10 trials per session, and a maximum of 25 trials per subject. Each subject was tested twice over a 2-week period.

We used two measures to characterize hand preference. First, we calculated a handedness index (*HI*) score for each subject [(R−L)/(R+L)] where "R" equals the frequency of right-hand responses and "L" equals the frequency of left-hand responses to the retrieval task. This measure quantifies lateral bias along a continuum from strongly left-handed to strongly right-handed. Positive values indicate a right-hand bias while negative values indicate a left-hand bias. Absolute values of each subject's *HI* score (*ABS-HI*) were used to characterize hand preference strength independent of direction. That is, a subject with a *HI* score of +0.90 would have the same *ABS-HI* score as subject with a *HI* score of −0.90. A Pearson correlation coefficient (r) was used to determine the direction and degree of association between lateral bias and cortisol level. Alpha level was set to 0.05, two-tailed, for all analyses. We also used a between-groups analysis to examine differences in hand preference between subjects assigned to high, middle, or low cortisol groups based on mean values for each of the three sampling periods (1, 3, and 5 months of age). Subjects in the upper third of the sample were categorized as "high", subjects in the middle third of the sample were categorized as "medium", and subjects in the lower third of the sample were categorized as "low". We conducted analysis of variance (ANOVA) and post hoc tests that were appropriate to determine the extent to which the two extreme-groups differed from each other and from the middle responding group. We conducted the analysis with both *HI* and *ABS-HI* data, where we set the alpha level at 0.05, two-tailed, for all analyses.

Results indicate significant positive correlations between *HI* scores and levels of plasma cortisol sampled at 1 and 3 months of age (Pearson r $(16) = 0.66$ and 0.48, respectively, $p<0.05$; Figure 1), but not at 5 months of age (r $(16) = 0.21$). Regarding strength of hand preference (independent of direction), we noted a significant negative correlation between *ABS-HI* scores and cortisol levels sampled at 5 months (r $(16) = -0.47$, $p<0.05$), but not at 1 and 3 months of age (r $(16) = 0.04$ and −0.18, respectively). A between-groups ANOVA indicated a significant main effect of cortisol group on *HI* scores (F $(2, 15) = 5.02$, $p<0.03$). Post hoc comparisons (Fisher's protected least squares significant difference test) revealed that the high cortisol group had a significantly higher mean *HI* score than did the low and middle cortisol groups ($p<0.05$). Further analysis indicated no significant main effect of cortisol group on *ABS-HI* scores.

Results of this study suggest a relationship between cortisol and hand preference in infant rhesus macaques and are consistent with research that indicates the early developmental influence of stress reactivity on the emergence of hemispheric specialization [Bradshaw & Rogers, 1993]. Results reveal significant positive correlations between *HI* scores and levels of plasma cortisol

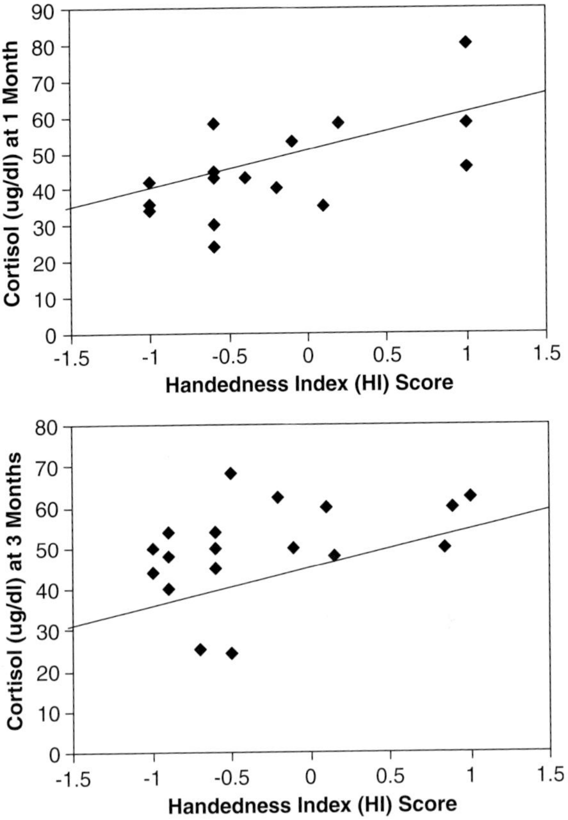

Figure 1. Scatterplots showing the relationship between handedness index (*HI*) scores and plasma cortisol at one (A) and three (B) months of age. A positive *HI* score indicates a right-hand bias and a negative *HI* score indicates a left-hand bias.

among very young infants although we found no significant correlation between cortisol level and the strength of hand preference for these young animals. We found these results keenly interesting to understand the biology of lateralized behavior. We continued our research to focus on later-age periods to pursue greater understanding of the correlation relationship between cortisol and hand preference in rhesus macaques.

PLASMA CORTISOL IN ADULT MALE RHESUS MONKEYS

In this study, we sampled cortisol from 34 adolescent males between 1 and 5 years of age and recorded hand preferences from these animals as adults aged 8–12 years [Westergaard et al., 2000]. We obtained physiological data in a fashion consistent across all studies to be described hereafter in this chapter.

Unlike Study 1, we conducted this study on subjects in a free-ranging locale at Morgan Island, South Carolina. Morgan Island is a 475 acre barrier sea island that is home to more than 4,000 rhesus monkeys that reside in 35 social groups with an approximate ratio of 2.5 adult females to each adult male. Four times per year (2 Fall, 2 Spring), monkeys were trapped for biannual exam processing. During this exam, we collected plasma samples for cortisol assay.

Following capture, subjects were anesthetized with ketamine hydrochloride (10 mg/kg). A blood sample was immediately gathered from each subject via venipuncture and placed on wet-packed ice. Samples were assayed for cortisol by a commercial laboratory via radioimmunoassay. We recorded the time of day and the number of minutes that elapsed from the time we entered the corral to the time when the blood was drawn to control for possible confounds on hormone values.

We obtained hand-preference data in a fashion consistent across all studies described in this chapter. We evoked quadrupedal reaching by throwing dried corn kernels in the vicinity of a focal subject's social group. The subject moved to the location to pick up the food. An observer blind to the physiological data and the specific study hypotheses, recorded the hand (left or right) that the subject used to retrieve the corn kernels. We counted a trial only when a subject maintained both hind limbs and one forelimb on the ground while reaching for the food. To insure independence of data points, we required subjects to move to the food site before each new reach to ensure postural repositioning. If locomotion did not occur before an animal picked up a food item, we did not consider the reach in the tally. We conducted 50 trials per subject across a 6-month period. To establish inter-observer reliability, two observers conducted simultaneous observations at the start, midpoint, and end of the study. Cohen's Kappa values were greater than 0.90 and indicated high inter-observer reliability.

Similar to the previous study, we used two measures to characterize hand preference. First, we calculated an *HI* score for each subject [(R−L)/(R+L)] where "R" equals the frequency of right-hand responses and "L" equals the frequency of left-hand responses to the corn kernel retrieval task. Absolute values of each subject's *HI* score (*ABS-HI*) were used to characterize hand preference strength independent of direction. We used multiple regression analyses (with age at physiological sampling and age at hand preference sampling held statistically constant) to examine the direction and degree of association between *HI* and *ABS-HI* scores and levels of plasma cortisol and ACTH. We also conducted linear analyses (Pearsons *r*) to evaluate relationships between *HI* and *ABS-HI* scores and absolute hormone levels. We set the alpha level at 0.05, two-tailed, for all analyses.

Eighteen of 34 subjects (53% of the subject sample) exhibited a positive *HI* score (indicative of right-hand bias) and 16 animals (47% of the subject sample) exhibited a negative *HI* score (indicative of a left-hand bias). A single-sample *t*-test indicated that the mean *HI* score per subject (−0.10) did not differ significantly from a null hypothesis distribution with a mean of 0, indicating a lack of population-level bias in the direction of hand preference in adult male rhesus macaques (t (33) = 1.40, $p>0.17$). Regarding the relationship between hand preference direction and strength, we noted a significant negative correlation between *HI* and *ABS-HI* scores indicating that left-hand biased animals exhibited greater strength of preference than did right-hand biased animals (Pearson r (32) = −0.70, $p<0.0001$).

Multiple regression analysis indicated a significant positive relationship between plasma cortisol levels and *HI* scores (t (3/31) = 2.62 $p< = 0.02$; Figure 2) and a significant negative correlation between plasma cortisol levels and *ABS-HI* scores (t (3/33) = −2.31, $p<0.03$). Linear analysis indicated a significant positive correlation between *HI* scores and plasma cortisol levels (r (32) = 0.42, $p<0.02$) and a near significant positive correlation between *ABS-HI* scores and plasma cortisol levels (r (32) = −0.32, $p<0.07$). Post hoc comparisons (Fisher's Protected Least Significant Difference test) revealed a significant lower mean *HI* score for the low cortisol group (−0.52) than for the middle and high cortisol groups (0.02 and 0.03, respectively), which did not differ significantly from each level ($p = 0.05$). Further analysis indicated a group-level left-hand bias among animals in the low cortisol group (single-sample t-test, t (7) = 3.89, $p<0.006$). Animals in the middle and high extreme group did not show a significant group-level bias toward either the left or right hand (for the middle cortisol group, t (7) = 0.29). Extreme-groups analysis of data reflecting strength of hand preference indicated no significant main effect of cortisol on *ABS-HI* scores (F (2/31) = 2.50).

Our results reveal a lack of population-level bias in the direction of hand preference for adult male rhesus. However, our results demonstrate a significant relationship between hand preference and levels of plasma cortisol in adult male rhesus macaques. Results suggest that left-hand biased animals exhibited

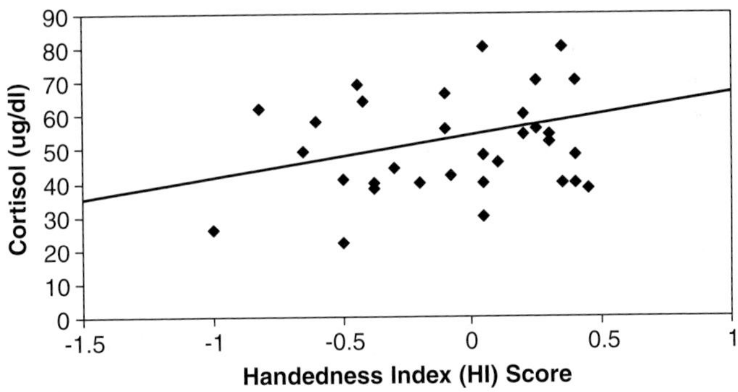

Figure 2. Scatterplot showing the relationship between plasma cortisol and handedness index (*HI*) scores for male rhesus. A positive *HI* score is indicative of right-hand bias and a negative *HI* score is indicative of left-hand bias.

greater strength of preference than did right-hand biased animals. The extreme-groups analysis indicated that hand preference direction was due primarily to the behavior of subjects in the low cortisol group as these animals were more biased toward use of their left hand than were animals in the middle or high cortisol groups. This result led us to conduct a similar study for female rhesus macaques.

PLASMA CORTISOL IN ADULT FEMALE RHESUS MONKEYS

In this study, we used three measures to characterize hand preference. We focused on the association between cortisol levels and hand preference for 41 young rhesus females aged 1.8–3.7 years (mean age = 2.9 years) at Morgan Island, South Carolina [Westergaard et al., 2004]. Plasma sampling and hand preference data collection methods were comparable to the male study described above. To evaluate hand preference for individual subjects we calculated a binomial *z*-score for each subject based on the number of right- versus left-hand responses. We next calculated an *HI* score for each subject using the formula [(R−L)/(R+L)] and used the absolute value of each subject's hand preference index score (*ABS-HI*) to characterize hand preference strength independent of the direction of lateral bias in a similar manner to males, described above. We used a Pearson product-moment correlation coefficient (r) to examine *HI* and *ABS-HI* data as a function of biological and behavioral measures. We set the alpha level at 0.05 for the analysis.

Examination of binomial *z*-score data indicated that 15 of the 41 animals in the overall subject sample (37%) exhibited significant preference toward use of either the right or left hand. Further examination of these data indicated that two animals exhibited a significant right-hand preference and that 13 animals exhibited a significant left-hand preference. Twenty-six animals did not exhibit a significant preference toward use of either hand. A binomial test indicated that the number of right- versus left-handed animals varied significantly from a random null hypothesis sample distribution with a bias toward left-hand use in the overall population ($t\ (40) = -3.63$, $p<0.001$).

Results demonstrate a significant positive correlation between *HI* scores and plasma cortisol concentrations ($r\ (40) = 0.45$, $p<0.003$). Regarding strength of hand preference, independent of direction, we noted a significant negative correlation between *ABS-HI* scores and plasma cortisol concentrations ($r\ (40) = -0.33$, $p<0.04$).

Our cortisol studies described herein were consistent, in that right-hand bias (>*HI* scores) was associated with relatively higher plasma cortisol levels and left-hand bias (<*HI* scores) was associated with relatively lower plasma cortisol levels. Although these studies do not confirm an etiological relationship between stress functioning and hemispheric specialization, they do provide evidence that is consistent with the notion that stress functioning is associated with lateral bias.

HAND PREFERENCE, MONOAMINE METABOLITE CONCENTRATIONS, AND BEHAVIORAL PATHOLOGIES

In this study, we examined the relationship between hand preference, cortisol and CSF monoamine metabolite levels, and multiple indices of sociality in free-ranging male rhesus macaques. We hypothesized that individual differences in serotonergic functioning are associated with individual differences in hand preference and are manifested by a positive correlation between CSF 5-HIAA concentrations and the frequency of right- or left-hand use. We also sought to determine whether hand preference is associated with a behavioral profile that places individuals at increased risk for violent trauma and premature death as has been shown for male rhesus macaques with low CSF 5-HIAA concentrations maintained in the present free-ranging population [Higley et al., 1996b]. We hypothesized that if left-handed macaques were more socially isolated and subject to more frequent and violent attacks from conspecifics relative to their right-hand counterparts, then we may be able to identify a biobehavioral mechanism for decreased prevalence of left-hand preference that has been noted in the aged rhesus population [Westergaard & Lussier, 1999].

Subjects included adult male rhesus macaques aged 6–10 years (mean age = 8 years) at study onset. Subjects were maintained at the free-ranging Morgan Island site described earlier in the chapter. Hand preference and physiological sampling methods were consistent to those described earlier in the chapter. We collected behavioral samples from 24 subjects, including 15 subjects for which we also collected physiological samples. Two observers collected behavioral data over a 6-month period during four 30-min sessions per subject. Data collection sessions were counterbalanced by the time of day. The observers used a focal animal sampling method and randomized data collection sessions for subject order and time of day. Social behavior categories followed those used for rhesus monkeys and other macaques in previous investigations [Altmann, 1962; Mehlman et al., 1994]. The observers recorded the following behaviors: (1) proximity; (2) groom; (3) submission, and (4) aggression. We initially defined and categorized aggressive behaviors by intensity into displacements, stationary threats, chases, and physical assaults, and then further categorized aggressive episodes as either 'low-intensity' (episodes involving only displacements or threat) or 'high-intensity' (episodes involving chases or physical assaults). Each occurrence of each behavior was treated as an event with a corresponding duration.

To establish inter-observer reliability, two observers conducted simultaneous observations at the beginning of the study, the midpoint, and the end of the study. The reliability of all recorded behavioral acts was shown by Cohen's Kappa values greater than 0.70. The reliability of all recorded hand preference acts was shown by Cohen's Kappa values greater than 0.90. Both observers were blind to all physiological data and to the specific hypotheses under investigation.

We attempted to control for potentially confounding variables inherent to the field conditions of this research by first transforming the absolute concentration of each metabolite and hormone into a standard (z) score. For values that exceeded ± 2 standard deviations from the mean, the subjects' records were examined to determine whether there were any technical or methodological errors that could have produced these extreme results. None were found. Second, we correlated each metabolite and hormone measure with: (1) the time of the day that the sample was taken and (2) the elapsed time between the subject's entrance into the corral and the time of sampling. We found no significant correlations among these variables. Further tests revealed no significant correlations between age and weight and any of the biological or behavioral measures.

We used three measures to characterize hand preference. To evaluate hand preference for individual subjects, we calculated a binomial z-score for each subject based on the number of right- versus left-hand responses. Subjects with a binomial z-score ≥ 1.96 were considered to be significantly right-handed, subjects with a binomial z-score ≤ -1.96 were considered to be significantly left-handed, and subjects with a binomial z-score < 1.96 and > -1.96 were considered to lack a significant preference toward use of either hands. We next calculated an *HI* score for each subject using the formula $[(R-L)/(R+L)]$, a standard method for calculating lateral bias in human and nonhuman primate research. Finally, we used the absolute value of each subject's *HI* score (*ABS-HI*) to characterize hand-preference strength independent of the direction of lateral bias. We used the Pearson product-moment correlation coefficient (r) to examine *HI* and *ABS-HI* data as a function of biological and behavioral measures. We set alpha at 0.05 for all analyses.

Examination of binomial z-score data indicated that 18 of the 33 animals in the overall subject sample (55%) exhibited significant preference toward use of either the right or left hand. Further examination of these data indicated that 8 animals exhibited a significant right-hand preference and that 10 animals exhibited a significant left-hand preference. Fifteen animals did not exhibit a significant preference toward use of either hands. A binomial test indicated that the number of right- versus left-handed animals did not vary significantly from a random null hypothesis sample distribution. This finding was further supported through examination of *HI* scores, as a single sample t-test indicated that mean *HI* score per subject (−0.11) did not vary significantly from a null hypothesis population with a mean of 0 (t (23) $= -1.19$, $p = 0.24$).

Examination of hand preference as a function of monoamine metabolite measures among the 15 subjects for which both hand preference and physiological data were available, indicated significant positive correlations between *HI* scores and CSF 5-HIAA concentrations (r (15) $= 0.46$, $p < 0.03$). Correlations between *ABS-HI* scores and monoamine metabolite measures were not statistically significant. Examination of hand preference as a function of hormonal measures indicated a significant positive correlation between *HI* scores and plasma cortisol concentrations (*r* (15) = 0.41, $p<0.05$). The correlation between *ABS-HI* scores and plasma cortisol concentrations was not statistically significant.

Examination of hand preference as a function of behavioral measures among the 15 subjects for which both hand preference and behavioral data were available, indicated significant positive correlations between *HI* scores and the frequency of social proximity bouts (*r* (15) = 0.45, $p<0.03$) and between *HI* scores and the frequency and duration of bouts in which animals received grooming (for the frequency of bouts in which animals received grooming, *r* (15) = 0.41, $p<0.05$; for the duration of bouts in which animals received grooming, *r* (15) = 0.41, $p<0.05$). We noted significant negative correlations between *HI* scores and the frequency of social submission bouts (*r* (15) = −0.59, $p<0.003$), and between *HI* scores and the frequency of bouts in which animals received aggression (*r* (15) = −0.58, $p<0.003$). Further examination indicated that the correlation between *HI* scores and the frequency with which animals received low-intensity aggression was not statistically significant (*r* (15) = −0.19, $p<0.38$), but that the negative correlation between *HI* scores and the frequency of bouts in which animals received high-intensity aggression was statistically significant (*r* (15) = −0.70, $p<0.0001$). Correlations between *HI* scores and other behavior measures were not statistically significant.

Regarding strength of hand preference, independent of direction, we noted significant positive correlations between *ABS-HI* scores and the frequency of submission bouts (*r* (22) = 0.70, $p<0.0001$), and between *ABS-HI* scores and the frequency of bouts in which animals received aggression (*r* (22) = 0.67, $p<0.002$). Further examination indicated significant correlations between *ABS-HI* scores and the frequency of bouts in which animals received both low-intensity and high-intensity aggression (for low-intensity aggression, *r* (22) = 0.46, $p<0.03$; for high-intensity aggression, *r* (22) = 0.66, $p<0.0003$). Correlations between *ABS-HI* scores and other behavioral measures were not statistically significant.

Results suggest left-hand preference is associated with a low incidence of positive social interaction and an increased risk of aggressive attacks from conspecifics. This is similar to results noted for adult male rhesus with low CSF concentrations of the serotonin metabolite 5-HIAA. Further research is needed in this area and we intend to continue to follow subjects over time to evaluate future rates of traumatic injury and premature deaths and the association between cortisol, 5-HIAA concentration and behavioral profiles.

CONCLUSION

At least two important issues, one procedural and the second methodological, remain to be resolved before we can definitively conclude whether stress functioning is causally related to hand preference or whether the relationship is produced by some third, as yet undefined, variable. One issue concerns whether the correlation between hand preference and plasma cortisol concentrations noted in our research indicates a true reduction in basal cortisol functioning or whether our results indicate a reduced sensitivity to the stress of the physical capture and blood sampling procedures. A similar procedural confound has been noted by Tiefenbacher et al. [2000] with regard to the correlation between low plasma cortisol concentrations and self-injurious behavior in adult male rhesus macaques. These authors caution that blood sampling under brief chemical restraint, such as ketamine, can result in mild activation of the pituitary–adrenocortical system [Elvidge et al., 1976]. We are currently attempting to address this issue in our field setting by assessing salivary cortisol under nonstressed conditions to determine whether there is a true difference in basal HPA activity as a function of hand preference. This procedure will also help us to evaluate the potentially confounding effects of ketamine as salivary cortisol can be obtained from unanesthetized animals.

A second issue to resolve concerns whether we can accurately infer that high or low cortisol concentrations in adult animals result from relatively high or low cortisol concentrations during earlier stages of development, when hand preference is more labile than in adulthood, such as during the pre-natal period or early infancy. The relevant studies to demonstrate life-span developmental continuity in stress hormone functioning have yet to be undertaken, although it is known that cortisol levels remain stable over shorter periods of time and that there is a significant genetic contribution to cortisol production in adulthood. It therefore seems reasonable to hypothesize that individual differences in cortisol concentrations remain stable across broadly separate lifespan periods [Moffat & Hampson, 1996].

We also believe it would be premature at this time to infer a developmental relationship among hand preference, stress reactivity, and the physiological and behavioral measures noted in the present research. We will however, continue to monitor these relationships in our ongoing and future studies. However, the results we have conducted across the past 10 years are intriguing, as hand preference is associated with an array of biological and behavioral processes among rhesus macaques.

The finding that left-hand use is correlated with low CSF 5-HIAA in our free-ranging male primate subject population is intriguing, given the relationship that has been noted in the human population between left-hand preference and various forms of psychopathology, including alcoholism and schizophrenia [Lesch et al., 1990; Orr et al., 1999; Sperling et al., 2000; also see Klar, 1999; Satz & Green, 1999]. Although it is difficult to note with precision the specific characteristics that are associated with the development of the major psychiatric disorders, it is generally agreed that the most powerful models combine both genetic predispositions and evidence of early neurological damage. One potential predisposing factor to pathological behavior that combines both genetics and evidence of early neurological damage is anomalous hemispheric

specialization expressed phenotypically as left- or mixed-hand preference. Sperling et al. [2000] examined the relationship between alcoholic subtypes and hand preference and found that left- or mixed-hand preference and numerous developmental risk factors are more prevalent in male alcoholics than in alcoholic women and nonalcoholic controls. A detailed comparison of alcoholic subtypes revealed a significant increased in vulnerability only among Type II alcoholics classified according to the Cloninger et al. [1988] model. Moreover, other researchers have reported a relationship between left- or mixed-hand preference and developmental risk factors among schizophrenics [Geschwind & Behan, 1984; Lewis, 1987; London, 1987; Mednick & Machon, 1988; Lesch & Walter, 1996]. It must be noted that among both alcoholics and schizophrenics the most significant increase in prenatal developmental risk factors occurs during the most vulnerable second trimester of pregnancy which in schizophrenics has been linked to disorders of neuronal migration [Bogerts et al., 1990; Beckmann & Jacob, 1994]. Of related interest in the context of the present research is the apparent influence of serotonin on motor output. In a functional magnetic resonance imaging (fMRI) study on the role of serotonin in regulating cerebral motor activity, Loubinoux et al. [1999] examined the effects of administering the monoamine neurotransmission enhancers fluoxetine and fenozolone to healthy right-handed male and female volunteers who performed a motor task with their preferred (right) hand. In response to both the drugs the authors found that subjects showed an increase in evoked signal intensity in the ipsilateral cerebellum, and a parallel reduction in the primary and secondary motor cortices. These results suggest a more focused activation of the contralateral sensorimotor area, greater involvement of posterior supplementary motor area, and a widespread decrease of bilateral cerebellar activation, and indicate involvement of monoamines and serotonin in the facilitation of cerebral motor activity. The prospects for conducting parallel fMRI studies in nonhuman primates are now actually quite promising and such studies should provide additional valuable avenues for exploring relationships among serotonin, psychopathology, and lateralized manual behavior.

The finding that left-hand use is associated with low plasma cortisol concentrations is consistent with the hypothesis that hemispheric specialization is linked to stress-related products of the HPA system [Geschwind & Galaburda, 1985a, b]. This is of particular interest in view of evidence indicating that prior stress or traumatic experience can lead to hypoactivity of the HPA axis [Heim et al., 2000]. Low basal cortisol levels can be induced in monkeys through manipulation of rearing condition, as has been shown in the infants of mothers subjected to an unpredictable variable foraging demand [Smith et al., 1997]. It has been hypothesized that the suppressive effects of cortisol early in development act predominantly on the right hemisphere, thereby inhibiting left-hand use [Moffat & Hampson, 1996]. In our own research with infant rhesus we found positive correlations between right-hand use and plasma cortisol concentrations providing indirect evidence in support of this view. In previous

studies, we have compared cortisol levels sampled during the juvenile period of development to hand preference collected during adulthood. Unfortunately, this research design has limited our ability to determine whether the relationship between cortisol and hand preference noted among infants is developmentally continuous later in life or whether the two processes develop independently subsequent to the first postnatal year. The results of the present study, in which we collected hand preference and cortisol data over the same period of time, suggest parallel development of hand preference and plasma cortisol levels in males through middle age. Further, it must be noted that Kalin et al. [1998] found higher cortisol levels in monkeys that exhibited more right-frontal hemispheric activity. However, this measure may not be indicative of overall hemispheric dominance and its relationship with hand preference has yet to be determined.

Behavioral correlates of hand preference noted in our macaque subject sample suggest that left-hand preference is associated with a low incidence of positive social interaction and an increased risk of violent attack from conspecifics. We have noted a similar behavioral profile in male rhesus macaques with low CSF 5-HIAA concentrations (irrespective of hand preference) and in addition, among these animals we have noted a disproportionately high rate of premature mortality as a result of violent trauma [Higley et al., 1996b]. A number of studies have indicated that the prevalence of left-hand preference declines with increased age in the human population [Coren, 1994; Porac, 1993; Porac & Coren, 1981], and at least one study has shown a similar age-based decline in the prevalence of left-hand preference among rhesus macaques [Westergaard & Lussier, 1999]. Specifically, the prevalence of left-hand preference in humans has been estimated to decline from approximately 15% in the teen-age population to about 1% in the 80-year-old population [Porac & Coren, 1981]. In rhesus macaques, the incidence of left-hand preference has been reported to decline from 62% in the young-adult population to 11% in the aged population [Westergaard & Lussier, 1999]. At present, it is unclear whether this decreased prevalence of left-hand preference occurs as a result of changes in hand use among individuals or whether the decreases are attributable to a decline in the number of left-handed individuals through selective mortality. Although the present study does not directly evaluate hand preference as a function of mortality, our data do suggest that left-hand preference is associated with social isolation and an increased incidence and severity of attacks from other animals. Right-hand preference, on the other hand, is associated with heightened social standing as reflected in proximity to other group members and the frequency with which animals receive grooming, reliable measures of social status, and integration within rhesus macaque society. We hypothesize that the biobehavioral profile of low serotonergic functioning and frequent left- versus right-hand use places male rhesus macaques at increased risk for traumatic injury, and perhaps as a consequence, premature death. We intend over the next several years to continue following the hand preference status of our

aging adult male population, and to evaluate evidence in support of or contrary to both the right-shift and selective mortality hypotheses.

Future research will focus on left-hand preference as it has potential to identify individuals at increased risk for dysregulation of the serotonergic and HPA systems, and for impaired social functioning, including a disproportionately low incidence of positive social interaction and an increased risk for violent attack. In future, we intend to examine further hypotheses linking left-hand preference to pathological behavior as well as stressful prenatal and early rearing experience. It is apparent that several issues remain to be resolved in order to understand more fully the psychobiology of hand preference, including issues pertaining to lifespan continuity, cause and effect relationships between physiology and behavior, and potential sex differences. We are currently seeking to address these issues in our continuing longitudinal research.

ACKNOWLEDGMENTS

This research was conducted with support from NIAAA Contracts N01AA42002 and N01AA02018. The subjects for this research are part of a breeding colony supported by Alpha Genesis, Inc. under Food and Drug Administration Contract 223-92-1101. Alpha Genesis, Inc. is an AAALAC accredited institution. The Alpha Genesis, Inc. Animal Care and Use Committee approved a research protocol for this study in accordance with and as required by the Animal Welfare Act. The authors also wish to thank Dr. Phillip Snoy (FDA) for authorization to conduct this research.

REFERENCES

Alonso, J.; Castellano, A.; Rodriguez, M. Behavioral lateralization in rats: Prenatal stress effects of sex differences. Brain Research 539: 45–50, 1991.

Altmann, S. A. A field study of the sociobiology of rhesus monkeys, *Macaca mulatta*. Annals of the New York Academy of Sciences 102: 338–345, 1962.

Annett, M.; Ockwell, A. Birth order, birth stress and handedness. Cortex 16(1): 181–187, 1980.

Bailey, L. M.; McKeever, W. F. A large-scale study of handedness and pregnancy/birth risk events: Implications for genetic theories of handedness. Laterality 9(2): 175–188, 2004.

Bakan, P.; Dibb, G.; Reed, P. Handedness and birth stress. Neuropsychologia 11(3): 363–366, 1973.

Beckmann, H.; Jacob, H. Prenatal development disorders of brain structures in schizophrenic patients. Nervenartz 65: 1435–1440, 1994.

Bogerts, B.; Falkai, P.; Haupts, M. Post-mortem volume measurements of limbic system and basal ganglia structures in chronic schizophrenia. Schizophrenia Research 3: 295–301, 1990.

Bradshaw, J.; Rogers, L. The Evolution of Lateral Asymmetries, Language, Tool-Use, and Intellect. New York, Academic Press, 1993.

Cantalupo, C.; Pilcher, D. L.; Hopkins, W. D. Are planum temporale and sylvian fissure asymmetries directly related? A MRI study in great apes. Neuropsychologia 31(14): 1975–1981, 2003.

Cloninger, C. R.; Sigvardsson, S.; Gilligan, S. B.; Von Knorring, A. L.; Bohman, M. Genetic heterogeneity and the classification of alcoholism. Advances in Alcohol and Substance Abuse 7: 3–16, 1988.

Coe, C. L. Psychosocial factors and immunity in nonhuman primates: A review. Psychosomatic Medicine 55: 298–308, 1993.

Corballis, M. C. Is the handedness gene on the X chromosome? Comment on Jones and Martin. Psychological Review 108(4): 805–810, 2000.

Corballis, M. C.; Lee, K.; McManus, I. C.; Crow, T. J. Location of the handedness gene on the X and Y chromosomes. American Journal of Medical Genetics 67(1): 50–52, 1996.

Coren, S. The diminished number of older left-handers: Differential mortality or social-historical trend? International Journal of Neuroscience 75: 1–8, 1994.

Coren, S. Family patterns in handedness: Evidence for indirect inheritance mediated by birth stress. Behavioral Genetics 25(6): 517–524, 1995.

Cowell, P. E.; Fitch, R. H.; Denenberg, V. H. Laterality in animals: Relevance to schizophrenia. Schizophrenia Bulletin 25: 41–62, 1999.

Cowen, P. J. Cortisol, serotonin and depression: All stressed out? The British Journal of Psychiatry 180: 99–100, 2002.

Elvidge, H.; Challis, J. R.; Robinson, J. S.; Roper, C.; Thorborn, G. D. Influence of handling and sedation on plasma cortisol in rhesus monkeys (*Macaca mulatta*). Journal of Endocrinology 70: 325–326, 1976.

Foundas, A. L.; Eure, K. F.; Luevano, L. F.; Weinberger, D. R. MRI asymmetries of Broca's area: The pars triangularis and pars opercularis. Brain and Language 64: 282–296, 1998.

Freeman, H. D.; Cantalupo, C.; Hopkins, W. D. Asymmetries in the hippocampus and amygdala of chimpanzees (*Pan troglodytes*). Behavioral Neuroscience 118(6): 1460–1465, 2004.

Galaburda, A. M. Developmental dyslexia and animal studies: At the interface between cognition and neurology. Cognition 50: 133–149, 1994.

Geschwind, N.; Behan, P. Left-handedness: Association with immune disease, migraine and developmental learning disorders. Proceedings of the National Academy of Science, USA 79: 5097–5100, 1984.

Geschwind, N.; Galaburda, A. M. Cerebral lateralization: Biological mechanisms, associations, and pathology: II. A hypothesis and a program for research. Archives of Neurology 42: 521–552, 1985a.

Geschwind, N.; Galaburda, A. M. Cerebral lateralization: Biological mechanisms, associations, and pathology: III. A hypothesis and a program for research. Archives of Neurology 42: 634–654, 1985b.

Halpern, M. E.; Gunturkun, O.; Hopkins, W. D.; Rogers, L. J. Lateralization of the vertebrate brain: Taking the side of model systems. Journal of Neuroscience 25(45): 10351–10357, 2005.

Harada, Y.; Hosoiri, Y.; Kuroda, R. Isolation and evaluation of dextral-specific and dextral-enriched cDNA clones as candidates for the handedness-determination gene in freshwater gastropod, *Lymnaea stagnalis*. Developmental Genetics and Evolution 214(4): 159–169, 2004.

Harris, L. J.; Carlson, D. F. Hand preferences for visually-guided reaching in human infants and adults. Pp. 285–306 in J. P. Ward; W. D. Hopkins, eds. Primate Laterality: Current Behavioral Evidence of Primate Asymmetries. New York, Springer-Verlag, 1993.

Heim, C.; Ehlert, U.; Hellhammer, D. H. The potential role of hypocortisolism in the pathophysiology of stress-related bodily disorders. Psychoneuroendocrinology 25: 1–35, 2000.

Heinz, A.; Jones, D. W.; Bissette, G.; Hommer, D.; Ragan, P.; Knable, M.; Wellek, S.; Linnoila, M.; Weinberger, D. R. Relationship between cortisol and serotonin metabolites and transporters in alcoholism. Pharmacopsychiatry 35: 127–134, 2002.

Henry, J. P. Psychological and physiological responses to stress: The right hemisphere and the hypothalamo-pituitary-adrenal axis, an inquiry into problems of human bonding. Integrative Physiological and Behavioral Science 28: 369–387, 1993.

Higley, J. D.; King, S. T.; Hasert, M. F.; Champoux, M.; Suomi, S. J.; Linnoila, M. Stability of interindividual differences in serotonin function and its relationship to aggressive wounding and competent social behavior in rhesus macaque females. Neuropsychopharmacology 14: 67–76, 1996a.

Higley, J. D.; Mehlman, P. T.; Higley, S. B.; Fernald, B.; Vickers, J.; Lindell, S. G.; Taub, D. M.; Suomi, S. J.; Linnoila, M. Excessive mortality in young free-ranging male nonhuman primates with low cerebrospinal fluid 5-hydroxyindoleacetic acid. Archives of General Psychiatry 53: 537–543, 1996b.

Higley, J. D.; Mehlman, P. T.; Poland, R. E.; Taub, D. T.; Vickers, J.; Suomi, S. J.; Linnoila, M. CSF testosterone and 5-HIAA correlate with different types of aggressive behaviors. Biological Psychiatry 40: 1067–1082, 1996c.

Higley, J. D.; Mehlman, P.; Taub, D.; Higley, S. B.; Vickers, J. H.; Suomi, S. J.; Linnoila, M. Cerebrospinal fluid monoamine and adrenal correlates of aggression in free-ranging rhesus monkeys. Archives of General Psychiatry 49: 436–441, 1992.

Hopkins, W. D.; Cantalupo, C. Handedness in chimpanzees (*Pan troglodytes*) is associated with asymmetries of the primary motor cortex but not with homologus language areas. Behavioral Neuroscience 118(6): 1176–1183, 2004.

Hopkins, W. D.; Cantalupo, C.; Wesley, M. J.; Hostetter, A. B.; Pilcher, D. L. Grip morphology and hand use in chimpanzees (*Pan troglodytes*): Evidence

of a left hemisphere specialization in motor skill. Journal of Experimental Psychology: General 131(3): 412–423, 2002.

Hopkins, W. D.; Dahl, J. F.; Pilcher, D. Genetic influence on the expression of hand preferences in chimpanzees (*Pan troglodytes*): Evidence in support of the right-shift theory and developmental instability. Psychological Science 12(4): 299–303, 2001.

Hopkins, W. D.; Pilcher, D. L. Neuroanatomical localization of the motor hand area with magnetic resonance imaging: The left hemisphere is larger in great apes. Behavioral Neuroscience 115(5): 1159–1164, 2001.

Hopkins, W. D.; Rilling, J. K. A comparative MRI study of the relationship between neuroanatomical asymmetry and interhemispheric connectivity in primates: Implications for the evolution of functional asymmetries. Behavioral Neuroscience 114(4): 739–748, 2000.

Hopkins, W. D.; Russell, J. L. Further evidence of a right hand advantage in motor skill by chimpanzees (*Pan troglodytes*). Neuropsychologia 42(7): 990–996, 2004.

Hopkins, W. D.; Russell, J. L.; Cantalupo, C.; Freeman, H.; Schapiro, S. J. Factors influencing the prevalence and handedness for throwing in captive chimpanzees (*Pan troglodytes*). Journal of Comparative Psychology 119(4): 363–370, 2005a.

Hopkins, W. D.; Russell, J.; Freeman, H.; Buehler, N.; Reynolds, E.; Schapiro, S. J. The distribution and development of handedness for manual gestures in captive chimpanzees (*Pan troglodytes*). Psychological Science 16(6): 487–493, 2005b.

Hopkins, W. D.; Russell, J.; Hostetter, A.; Pilcher, D.; Dahl, J. Grip preference, dermatoglyphics and hand use in captive chimpanzees (*Pan troglodytes*). American Journal of Physical Anthrpology 128(1): 57–62, 2005c.

Hopkins, W. D.; Stoinski, T. S.; Lukas, K. E.; Ross, S. R.; Wesley, M. J. Comparative assessment of handedness for a coordinated bimanual task in chimpanzees (*Pan troglodytes*), gorillas (*Gorilla gorilla*) and orangutans (*Pongo pygmaeus*). Journal of Comparative Psychology 117(3): 302–308, 2003.

Hopkins, W. D.; Wesley, M. J. Gestural communication in chimpanzees (*Pan troglodytes*): The influence of experimenter position on gesture type and hand preference. Laterality 7(1): 19–30, 2002.

Hopkins, W. D.; Wesley, M. J.; Izard, K.; Hook, M.; Schapiro, S. J. Chimpanzees (*Pan troglodytes*) are predominately right-handed: Replication in three populations of apes. Behavioral Neuroscience 118(3): 659–663, 2004.

Jaffe, B.; Evans, T.; Howell, S.; Westergaard, G.; Snoy, P.; Higley, J. D. Left vs. right nipple preference in free-ranging infant rhesus macaques (*Macaca mulatta*). Developmental Psychobiology 48(3): 266–272, 2006.

Kalin, N. H.; Larson, C.; Shelton, S. E.; Davidson, R. J. Asymmetric frontal brain activity, cortisol, and behavior associated with fearful temperament in rhesus monkeys. Behavioral Neuroscience 112: 286–292, 1998.

Klar, A. J. S. Genetic models for handedness, brain lateralization, schizophrenia, and manic-depression. Schizophrenia Research 39: 207–218, 1999.

Laland, K. N.; Kumm, J.; Van Horn, J. D.; Feldman, M. W. A gene-culture model of human handedness. Behavioral Genetics 25(5): 433–455, 1995.

Lesch, O. M.; Kefer, J.; Lentner, S.; Madler, R. Diagnosis of chronic alcoholism-classificatory problems. Psychopharmacology 23: 88–96, 1990.

Lesch, O. M.; Walter, H. Subtypes of alcoholism and their role in therapy. Alcohol and Alcoholism 31(Suppl. 1): 63–67, 1996.

Levander, M.; Schalling, D.; Levander, S. E. Birth stress, handedness and cognitive performance. Cortex 25(4): 673–681, 1989.

Lewis, S. W. Obstetric complications, neurodevelopmental deviance and risk of schizophrenia. Journal of Psychiatric Research 21: 413–421, 1987.

London, W. P. Cerebral laterality and the study of alcoholism. Alcohol 4: 207–208, 1987.

Lonsdorf, E. V.; Hopkins, W. D. Wild chimpanzees show population-level handedness for tool use. Proceedings of the National Academy of Sciences 102(35): 12634–12638, 2005.

Loubinoux, I.; Boulanouar, K.; Ranjeva, J. P.; Carel, C.; Berry, I.; Rascol-Celsis, P.; Chollet, F. Cerebral functional magnetic resonance imaging activation modulated by a single dose of the monoamine neurotransmission enhancers fluoxetine and fenozolone during hand sensorimotor tasks. Journal of Cerebral Flow and Metabolism 19: 1365–1375, 1999.

Lubach, G. R.; Coe, L.; Karaszewski, J. W.; Ershler, W. B. Effector and target cells in the assessment of natural cytotoxic activity of rhesus monkeys. American Journal of Primatology 39: 275–287, 1996.

MacNeilage, P. F.; Studdert-Kennedy, M. G.; Lindblom, B. Primate handedness reconsidered. Behavioral and Brain Sciences 10: 247–303, 1987.

McManus, I. C. Handedness and birth stress. Psychological Medicine 11(3): 485–496, 1981.

Mednick, S. A.; Machon, R. A. Adult schizophrenia following prenatal exposure to an influenza epidemic. Archives of General Psychiatry 45: 189–192, 1988.

Mehlman, P.; Higley, J. D.; Faucher, I.; Lilly, A. A.; Taub, D. M.; Vickers, J. H.; Suomi, S.; Linnoila, M. Low CSF 5-HIAA concentrations and severe aggression and impaired impulse control in nonhuman primates. American Journal of Psychiatry 151: 1485–1491, 1994.

Mehlman, P.; Higley, J. D.; Faucher, I.; Lilly, A. A.; Taub, D. M.; Vickers, J. H.; Suomi, S.; Linnoila, M. Correlation of CSF 5-HIAA concentration with sociality and the timing of emigration in free-ranging primates. American Journal of Psychiatry 152: 907–913, 1995.

Mehlman, P. T.; Higley, J. D.; Fernald, B. J.; Sallee, F. R.; Suomi, S. J.; Linnoila, M. CSF 5-HIAA, testosterone, and sociosexual behaviors in free-ranging male rhesus macaques in the mating season. Psychiatry Research 72: 89–102, 1997.

Moffat, S. D.; Hampson, E. Salivary testosterone levels in left- and right-handed adults. Neuropsychologia 34: 225–233, 1996.

Orr, K. D. G.; Cannon, M.; Gilvarry, C. M.; Jones, P. B.; Murray, R. M. Schizophrenic patients and their first-degree relatives show an excess of mixed-handedness. Schizophrenia Research 39: 167–176, 1999.

Pilcher, D. K.; Hammock, E. A.; Hopkins, W. D. Cerebral volumetric asymmetries in nonhuman primates: A magnetic resonance imaging study. Laterality 6(2): 165–179, 2001.

Porac, C. Are age trends in adult hand preference best explained by developmental shifts or generational differences? Canadian Journal of Experimental Psychology 47: 697–713, 1993.

Porac, C.; Coren, S. Lateral Preferences and Human Behavior. New York, Springer, 1981.

Pujol, J.; Deus, J.; Losilla, J. M.; Capdevila, A. Cerebral lateralization of language in normal left-handed people studied by functional MRI. Neurology 52: 1038, 1999.

Satz, P.; Green, M. F. Atypical handedness in schizophrenia: Some methodological and theoretical issues. Schizophrenia Bulletin 25: 63–78, 1999.

Schneider, M. L.; Suomi, S. J. Neurobehavioral assessment in rhesus monkey neonates (*Macaca mulatta*): Developmental changes, behavioral stability, and early experience. Infant Behavior and Development 15: 155–177, 1992.

Schwartz, M. Handedness, prenatal stress and pregnancy complications. Neuropsychologia 26(6): 925–929, 1988.

Smith, E. L.; Coplan, J. D.; Trost, R. C.; Scharf, B. A.; Rosenblum, L. A. Neurobiological alterations in adult nonhuman primates exposed to unpredictable early rearing: Relevance to posttraumatic stress disorder. Annals of the New York Academy of Sciences 821: 545–548, 1997.

Sperling, W.; Frank, H.; Martus, P.; Mader, R.; Barocka, A.; Walter, H.; Lesch, O. M. The concept of abnormal hemispheric organization in addiction research. Alcohol and Alcoholism 35: 394–399, 2000.

Tan, L. E.; Nettleton, N. C. Left handedness, birth order and birth stress. Cortex 16(3): 363–373, 1980.

Tang, A. C.; Verstynen, T. Early life environment modulates 'handedness' in rats. Behavioral Brain Research 131(1–2): 1–7, 2000.

Tiefenbacher, S.; Novak, M. A.; Jorgensen, M. J.; Meyer, J. S. Physiological correlates of self-injurious behavior in captive, socially-reared rhesus monkeys. Psychoneuroendocrinology 25: 799–817, 2000.

van Praag, H. M. Faulty cortisol/serotonin interplay. Psychopathological and biological characterization of a new, hypothetical depression subtype (SeCA depression). Psychiatry Research 65(3): 143–157, 1996.

van Riel, E. Dysregulation of the HPA-Axis: Implication for Serotonin Responses in the Hippocampus. Doctoral Dissertation, Graduate School of Neurosciences, Amsterdam, Swammerdam Institute for Life Sciences, 2003.

van Strien, J. W.; Bouma, A.; Bakker, D. J. Birth stress, autoimmune diseases, and handedness. Journal of Clinical and Experimental Neuropsychology 9(6): 775–780, 1987.

Vauclair, J.; Meguerditchian, A.; Hopkins, W. D. Hand preferences for unimanual and coordinated bimanual tasks in baboons (*Papio anubis*). Brain Research: Cognitive Brain Research 25(1): 210–216, 2005.

Warren, J. M. Handedness and laterality in humans and other animals. Physiological Psychology 8: 351–359, 1980.

Westergaard, G. C. Hand preference in the use and manufacture of tools by tufted capuchin (*Cebus apella*) and lion-tailed macaque (*Macaca silenus*) monkeys. Journal of Comparative Psychology 105: 172–176, 1991.

Westergaard, G. C. Hand preference in the use of tools by infant baboons (*Papio cynocephalus anubis*). Perceptual and Motor Skills 76(2): 447–450, 1993.

Westergaard, G. C.; Champoux, M.; Suomi, S. J. Plasma cortisol is associated with handedness in infant rhesus macaques (*Macaca mulatta*). Developmental Psychobiology 38: 116–122, 2001.

Westergaard, G. C.; Chavanne, T. J.; Houser, L.; Cleveland, A.; Snoy, P. J.; Suomi, S. J.; higley, J. D. Biobehavioral correlates of hand preference in free-ranging female primates. Laterality 9(3): 267–285, 2004.

Westergaard, G. C.; Chavanne, T. J.; Lussier, I. D.; Suomi, S. J.; Higley, J. D. Hormonal correlates of hand preference in free-ranging primates. Neuropsychopharmacology 23: 502–507, 2000.

Westergaard, G. C.; Kuhn, H. E.; Suomi, S. J. Bipedal posture and hand preference in humans and other primates. Journal of Comparative Psychology 112(1): 55–64, 1998a.

Westergaard, G. C.; Kuhn, H. E.; Suomi, S. J. Effects of upright posture on hand preference for reaching vs. the use of probing tools by tufted capuchins (*Cebus apella*). American Journal of Primatology 44(2): 147–153, 1998b.

Westergaard, G. C.; Lussier, I. Left-handedness and longevity in primates. International Journal of Neuroscience 99: 79–87, 1999.

Westergaard, G. C.; Mehlman, P. T.; Suomi, S. J.; Higley, J. D. CSF 5-HIAA and aggression in female macaque monkeys: Species and interindividual differences. Psychopharmacology 146: 440–446, 1999.

Westergaard, G. C.; Suomi, S. J. Hand preference in capuchin monkeys varies with age. Primates 34: 295–299, 1993.

Westergaard, G. C.; Suomi, S. J. Asymmetrical manipulation in the use of tools by tufted capuchin monkeys (*Cebus apella*). Folia Primatologica 63: 96–98, 1994.

Westergaard, G. C.; Suomi, S. J. Hand preference for a bimanual task in tufted capuchins (*Cebus apella*) and rhesus macaques (*Macaca mulatta*). Journal of Comparative Psychology 110: 406–411, 1996a.

Westergaard, G. C.; Suomi, S. J. Lateral bias for rotational behavior in tufted capuchin monkeys (*Cebus apella*). Journal of Comparative Psychology 110: 199–202, 1996b.

Wittling, W.; Pfluger, M. Neuroendocrine hemisphere asymmetries: Salivary cortisol secretion during lateralized viewing of emotion-related and neutral films. Brain and Cognition 14: 243–265, 1990.

Author Index

C

D

E

F

I

J

K

M

N

O

P

Q

R

S

T

U

V

W

X

Y

Z

Subject Index